The Mystery of Life

What's It All About?

Discovering the Truth in a Skeptical World

DAVID E. PEEPLES

The Mystery of Life: What's It All About?
Discovering the Truth in a Skeptical World

The book's interior earth photo is a "Blue Marble" image taken by Apollo 17 astronauts traveling toward the moon in November 2007. It features the eastern coast of Africa, Madagascar, and the Arabian peninsula.

Credit: NASA used by permission.

The book's front cover background image was produced by Hubble and features a stunning image of the Tarantula Nebula. This region of space is home to many extreme conditions including supernova remnants and the heaviest star ever found. The Tarantula Nebula is the most luminous nebula of its type in the local Universe.

Credit: NASA, ESA used by permission

The book cover Earth image in the public domain was created solely by NASA.

file: NASA blue marble.jpg

The colorful image displays a composite photo of planet Earth featuring the Western hemisphere in great detail. This "blue marble" 2000 photo was downloaded in 2013. It is one of the most detailed Earth images ever made. The photo is an update of the original "blue marble" images taken by Apollo astronauts. This image is the result of a combination of data from three Earth-observing satellites. Hurricane Linda, a huge spiral storm system, appears at the center left of this remarkable photo image.

Credit: Earth Image Photo composite "blue marble" 2000: "Courtesy of NASA/JPL - Caltech" used by permission.

The back cover layout features the Hubble image of one of the most dynamic star-forming regions in space, located 210,000 light-years away in the small Magellanic Cloud, a satellite galaxy of our Milky Way. At the center is a brilliant star cluster NGC 346, which creates a fantasy of dust and gas.

Credits: NASA, ESA and A. Nota (ESA/STScl, STScI/AURA)

ISBN

Paperback	978-1-68547-255-9
Hardcover	978-1-68547-256-6
eBook	978-1-68547-257-3

Library of Congress Control Number: 2023914855

Printed in the United States of America
First Edition printed 2018, Second Edition published 2023

wordHOUSE

101 Foundry Dr,
West Lafayette, IN, 47906, USA
www.wordhousebp.com
+1-800-646-8124

DAVID E. PEEPLES

THE MYSTERY OF LIFE

What's It All About?

Discovering the Truth in a Skeptical World

* * *

2nd Edition

TABLE OF CONTENTS

ABOUT THE AUTHOR

Born in 1950, David Peeples grew up in a small Western Washington logging town near the coast of the Pacific Northwest. As a young boy he preferred to be outdoors whenever possible as this allowed him the opportunity to satisfy his growing curiosity for the natural world first hand. After his graduation from high school, he attended Lower Columbia College in Longview, Washington, where the focus of his academic study in the sciences would become the natural progression of his keen interest in a complex physical world and its many wonders. This led him to the discovery and understanding of many fascinating details that define the disciplines of physics, chemistry, biology, and the natural sciences. David soon transferred to the University of Washington in Seattle where he studied and received an academic degree in pharmaceutical science in 1973. Years later, after combining his science background with a personal interest in history and theology, he turned to writing as his way of sharing the knowledge gained through many years of study, research, observation, and personal experience. David worked for many years as a healthcare professional while raising a family in Olympia, Washington. He always felt inclined to share the knowledge gained over years of personal study and recognized that writing and teaching were the most effective ways to do that. In 2004 David and his wife Linda moved to Arizona where he continued his pharmacy career while investing additional hours in study, writing, and research. After many years of following his interest in writing, David began editing what would eventually become the completed book narrative he dreamed of sharing throughout much of his adult life. *The Mystery of Life* book project is not a casual stab in the dark at a few puzzling questions inherent in a complex physical world, but contains documentation of relevant and reliable scientific and historical facts that has effectively driven home the message of the book. It is unfortunate that many of these important and life-changing scientific facts have been, all too often, conveniently overlooked and even carelessly disregarded by a skeptical world.

becominglikehim50@gmail.com

To offer reader comments or your impression of this material you may use the email contact listed here or visit the book's official website at (istruthrelative.com) where you will find further information on this material.

Dedication and Acknowledgments

This book is dedicated to my dear wife Linda, my best friend and lifelong companion without whose love, patience, and encouragement this book would never have become a reality. And to my four wonderful children—Heidi, Aaron, Heather, and Travis with whom my wife and I have shared many joys, and the challenges of life that have shaped us into the people we are today.

I would also like to acknowledge and offer a special thank you to several individuals who have helped me with various aspects of this lengthy book project including editorial review, technical assistance, and project management. A sincere thank you to Linda Peeples, Paul Mcintire, Heidi Codorniz, Mark Ruttan, and to the book publishing staff and marketing team who have worked tirelessly to make this project a success.

PREFACE

Have you ever tried to make sense of the routine activities we engage in day after day? The sun comes up. We go to work or school. We eat, we play, we watch television, interact with family and friends, and we sleep. We do this day after day, and while we are doing it, we grow old, and eventually we die. Sounds rather mundane does it not; but from the cradle to the grave, many are going through the repetitious routines of life without giving pause to seriously consider the most important questions presented to our minds through the struggles and triumphs people experience on life's journey. We have come a long way in terms of human progress, but some very basic questions that have plagued mankind throughout history remain with us today. Many people are privileged to live in prosperity, and yet poverty is seen everywhere we look. We, who grew up in America, have had the opportunity for a great education coupled with unprecedented prosperity; and yet laziness, ignorance, hatred, and racial prejudice are far too pervasive in our society. The marvels of modern invention have made our lives easier in many ways, but what good is new technology when our relationships are falling apart? Many people have looked to America for diplomacy during times of human conflict, and yet war, hatred, greed, racism, lawlessness, and widespread corruption continue to dominate the headlines worldwide.

The twenty-first century is characterized by a new threat to civilized culture, which could never have been imagined by the parents and grandparents of generations past. The news media is dominated by civil unrest and crimes against humanity as radicalized religious zealots as well as domestic terrorists spread hatred, murder, and chaos in the hope of subjugating those who view life differently than they. In committing acts of hatred without conscience, they show no mercy to the young and old alike. We live in a dangerous world of extreme political, religious, and moral ideology inciting hateful violence of a type that is unprecedented in recent memory. Attacks of deadly brutality are waged against western targets and urban centers because of bitter hatred and a misguided ambition to dictate radical views and to cancel those who do not comply. We should have provided a safer world for our children and grandchildren to grow up in; but the truth is we are living in a perilous time. Through self-indulgence and prideful living, many have pursued pleasure and wealth only to discover these objectives have left them dreadfully empty of the fulfillment they desire. What is wrong with today's modern world? Why does life seem terribly void of purpose to so many, even in a culture as affluent as ours? I invite you to consider the content of these chapters. Perhaps you will discover a worldview that can transform your life and bring hope, purpose, and meaning to all your tomorrows.

Life is a mystery, but it does not have to remain so. The word means: "something not fully understood; a religious truth comprehensible only by divine revelation;" or "something beyond human understanding, that remains obscure as to excite curiosity;" or "that which is incapable of understanding by mere reason, but must be revealed." *The Mystery of Life* has been revealed! You are about to discover the systematic unveiling of a philosophical enigma that has plagued our world from the beginning of time and remains heavily upon our hearts and minds yet today.

INTRODUCTION

In the technologically advanced digital world of the twenty-first century, many people find themselves preoccupied with busy schedules and time-consuming activities causing them to disregard the most important questions of our existence. We humans are naturally drawn to the question of life's meaning, although humanity's deep-seated longing for truth and purpose is easily suppressed by the hectic pace of today's modern lifestyles. We busy ourselves with many time-consuming pursuits, which may reflect an inward desire to avoid or perhaps postpone the vital question of our own mortality. People are allowing the years to slip away without giving serious consideration to life's most pressing questions. They experience overstimulated lives and hurried schedules that often force the search for meaning and the divine to the back of their minds, where it is typically reserved for future consideration. What is this life all about? What on earth am I here for? What is my purpose for living? Why do we live in such a broken world?

Are these questions too difficult to consider, or can we lay hold of answers that will satisfy our deepest longing for truth? Perhaps many of us are simply going through the motions without ever seeking the purpose and significance suggested by struggles in which we engage. Until people can recognize the real meaning intended for their lives, many will continue to struggle. We press onward in a broken world that is reeling from chaos and heartache inherently caused by death and disease. If you are willing to consider life's all-important questions, and do so with an open, honest mind, then you are about to discover answers that could decisively reveal your purpose, alter your view of life, and forever shape your destiny. For you and me to experience the genuine purpose for our lives, it would be helpful to discover the truth about some very basic questions. Before we begin to consider these all-important questions, however, we should briefly examine the condition in which we find our world today.

Many, who live in America, often considered the most affluent, and advanced culture in the world, are still desperately searching for that elusive thing called happiness. Even though we have seen tremendous progress in modern society, which has greatly impacted the quality and comfort of daily living, we are still forced to admit that our world today is more corrupt, unsafe, and in greater peril than most of us can remember. Tragic events on American shores have confirmed to the whole world that the stability and peace we humans desire is still eluding us.

Extraordinary advances in modern science, medicine, travel, and communication have greatly impacted all our lives, but this has not provided a better and safer world in which to raise our families. The emptiness and despair all around us is revealed by the high incidence of anxiety, depression, insomnia, and other mental health issues people are facing today.

Why, in a nation of unprecedented prosperity, are so many people in great despair? It appears many Americans find themselves unhappy and dissatisfied with their lives, even though they enjoy many material blessings and comforts. Why does today's divorce rate exceed one-half of all marriages in America? Why do we find ourselves still at war in a world with brutal tyrants who impose their selfish will upon others? Why is the insidious evil of radical terrorism a threat to the national security of America and many other freedom-loving nations throughout the world? Why so much corruption, shameful immorality, and abuse of power in government and business? And why is there so much sexual perversion, uncontrolled violence, hate crimes, drug abuse, poverty, homelessness, and debt? These and other harsh realities are only symptoms of the most basic problem still facing humanity today.

Man's dilemma is not in the mind to be analyzed! We surely do not need another nifty invention or bigger government. The problem we have is in the heart of man, and it needs to be healed! This struggle we are facing stems from a lack of understanding of what life is really all about and what is necessary to fulfill the basic needs that we all have. How much longer can we continue this downward spiral of moral corruption we have experienced in Western culture over recent decades? I sense, more than ever, a desperate need in America and throughout the world for individuals to take a needed pause from the hectic pace of modern living. Overly driven by time-consuming technology, we must stop to consider the most important questions we will face before our days on earth are over. These simple, yet profound, questions must be adequately addressed if we expect to know the lasting fulfillment and happiness, we all seek. Who am I? Where did I come from? Why am I here? Where am I going? Can you honestly say that you have found reliable answers to these puzzling questions about human existence?

Millions of people are wandering aimlessly through life without a true sense of direction while desperately lacking the contentment they were meant to enjoy. Many people are seeking after immediate gratification as they live for the moment, driven by self-serving desires, feelings, and emotions. These troubled souls are either misinformed or unaware of the actual purpose for their existence. When considering the deeper questions of life, one such issue that we must all face is death. No matter what you believe, I think we can all agree that life, as we know it, will end for all of us one day. Some people are taken in death quite suddenly and without warning. Countless men and women working at the Twin Towers in New York City on September 11, 2001, had no idea it would be their last day alive. Sudden death was waiting for and stalking so many innocent lives on that infamous and ominous day in American history.

We prepare for many things in life, such as career, marriage, and retirement. But I wonder how many people have prepared themselves for the final call when they will face leaving this world behind forever? Having considered, studied, and researched these questions for many years, I would like to share with you answers to the troubling questions that we must all face at some time during our lives— questions such as, how did life originate? What is the true nature of man? Is there really a supreme, transcendent God? As we examine these all-important matters, there remains a guiding principle we must apply to our diligent search for understanding, if we expect to discover the actual truth about life. We must follow the evidence wherever it leads! Why? Because the real answers we seek are found in historical facts and observed data not in wild speculation that cannot be verified. I have spent a good portion of the last ten years gathering and organizing an abundance of objective details that were collected and recorded over many years of following the evidence where it was leading me. I have written this material so that others, searching for understanding, could benefit from this knowledge. As you read through a carefully researched book, I only ask that you do so with an open mind. This means that you must be open to the possibility that many things you were taught about life, via secondary and higher education in a culture saturated with deception, were and are untrue.

This reality will become evident as we undertake a thorough evaluation of the basic questions I am calling to your attention here. You will discover in the following pages that things are not always as they seem. In a modern world that is mostly dominated by the wisdom of man (or lack thereof), this will become obvious when interpreting and documenting the truth about life's most pressing questions. In a highly secularized world, there have been far too many scientific and historical facts that are deliberately hidden from public view for the purpose of shaping within the populace a worldview controlled by modern secular institutions. When we consider the most critical questions of life, you and I cannot afford to be hoodwinked by the elite establishment in concert with mass media bias, because it is our personal fulfillment and future that is at stake here. We must discover for ourselves the true meaning of life, which has been cleverly veiled behind a smokescreen of misinformation obscuring the view of many for far too long.

In this publication we will follow the evidence to see where it leads. And, as we do so, it will become apparent that a cloud of deception has been hanging heavily over Western civilization for so long that even many very well-respected scientists, historians, and educators remain in denial when authentic facts are finally laid out before them. For those who read this material with an open mind and a searching heart, the predominant public mind-set that has drastically obscured an accurate view of life is about to change. Let us begin with a thorough examination of the evidence associated with these and other important and related questions. It remains my sincere hope and prayer that this material will somehow assist those who truly desire to unravel *The Mystery of Life*.

Chapter One

* * *

The Origin of Life

Setting the Stage

Somehow life appeared on planet earth in the distant past. Without any doubt, the great mystery of our existence starts with the question: how did it all begin? At some point in our lives, all of us will have to consider the crucial question of origins.

How did we get here? Everyone has an opinion about the origin of life, but I wonder how many people can point to valid evidence that will support his or her personal conviction. It seems few people really care enough to examine the evidence on display in the physical world—evidence that will likely confirm or deny what they believe. I suppose that people are too preoccupied with the rapid pace of modern living, and the hearts of many in affluent cultures are overly stimulated by the time-consuming affairs of life, preventing pause to consider critical questions seriously. But why is this subject of such great importance? Some might even say, what difference does it matter where we came from, or how we got here? The obvious fact is we are all here, so we should just enjoy it while we can! This cavalier approach to life will leave a serious void in the human heart that can never be fully satisfied or at peace. I would submit that not knowing where we came from will leave us with little understanding about who we are, or why we are living on this earth. Without this fundamental knowledge, people are left with a lack of appreciation for the real purpose and meaning that was intended for their lives.

In this opening chapter, we will begin to investigate important considerations regarding the origin of life, and making a thorough examination of this topic will lead us toward a greater understanding of exactly why we are alive. We should attempt to comprehend the truth to the extent that it may be embraced by our feeble minds. I want to know whether we decide our own future, or if the consummation of our earthly days is simply an abrupt dead end to a difficult and exhausting journey. Anyone wanting to live a satisfying life with purpose-filled tomorrows must be willing to examine the truth about the all-important question of origins.

The two commonly held views about life's origin, which have dominated the thinking of most people in Western culture, are naturalism and creationism. Which of these ideas represent the correct worldview? Why do these thoughts present such conflicting concepts? Let us examine the critical topic of origins, and specifically the vital question: how did life begin? Implications inferred by the truth about our past will, ultimately, become far reaching and even life changing regarding our future. Whatever we come to accept about our beginning will have profound influence in determining our entire worldview, and as a result, will determine the way in which we live our lives. Are we the accidental outcome of unguided, random chance events in an impersonal universe? Are we are simply set adrift to fend for ourselves in a hostile, unforgiving world? Is a life plagued with pain, suffering, sorrow, and death all there is? Do we have anything to look forward to as we all grow old and eventually lose our will to live? Are we the unplanned product of blind chance, or do our lives have purpose and meaning, allowing us to make sense of a troubled existence into which we were born?

Western civilization has become saturated with the strange notion that the evolution model is the only possible explanation for life as we know it; but does the evidence confirm this point of view, or have we become indoctrinated by the voices of those who reject the idea of supreme intelligence? This intelligence, many believe, speaks to us of a loving, personal God who has the power to create all things with purpose. As you consider the material presented here, I hope it will assist you in your own personal quest for truth and understanding.

Creation and Evolution Defined

To refresh our thinking, let us define the concept of biological evolution that has saturated Western culture for more than a century and is now taught exclusively in public schools and universities throughout America and many nations of the world.

The concept of evolution, with regard to organic life-forms, has been defined and redefined in a variety of ways over the last century. Sometimes it refers to basic thoughts like change over time, species variation, and even artificial selection. Therefore, it is important for us to understand the basic concepts I am referencing when I mention evolution in this publication. The wide variation within established genetic boundaries is sometimes referred to by the term "microevolution." I do not know of anyone who would deny that genetic variation within prescribed limits has been written into the DNA code of all living organisms. In the words of MIT grad Dr. Lee Spetner, who once served as the chief professor of information and communication theory at Johns Hopkins University: "I have studied the implications of random variations on evolution for many years, and have found contradiction after contradiction, difficulty after difficulty, with neo-Darwinian theory...A large body of evidence points to the suggestion that organisms come equipped with the ability to make heritable changes in their organs and functions in response to environmental cues."[1]

"Macroevolution," on the other hand, is a clear reference to what Charles Darwin was initially proposing. His original hypothesis has conjured up the strange notion of large-scale physical transformation giving rise to many new improved species over time, driven supposedly by a mechanism incorporating mutation and natural selection. The perceived appearance of new organs, body structures, and innovative survival features, he believed, were due to random mutations leading to new genetic information. Finally, there is the concept of "molecular evolution." This terminology refers to the initial emergence of a living cell from nonliving chemical elements in a primordial state. This is also called "spontaneous generation."

The hypotheses of macroevolution and molecular evolution (spontaneous generation) are the popular ideas that I would earnestly question. I believe these popular concepts are dangerous and deceiving notions that have led toward an unparalleled decline in Western civilization by virtue of a steady descent toward atheism. These ideas are highly offensive to a great many Americans who have based their lifestyle choices and ethical ideals upon tradition grounded in our nation's Judeo-Christian heritage. The most troubling thing about these bizarre notions, which have become firmly embedded in the American mindset, is that no reliable scientific evidence is found to support them, as you will soon discover. I would even suggest that anyone who firmly aligns themselves with the concepts of macroevolution and spontaneous generation is walking on very thin ice scientifically, and must rely on unfounded speculation to generate the faith to maintain these intellectual commitments. The following general ideas are commonly included in the basic tenets of the evolution model as I have come to understand it.

1. Perhaps ten or twenty billion years ago, all matter was concentrated in one place. While spinning rapidly, it somehow exploded in what has come to be known as "the big bang." The cooling gases and debris from this spontaneous explosion of light, matter, and energy resulted in the formation of a vast cosmos populated by billions of star-studded galaxies, including the spectacular Milky Way, serving as host to a fascinating solar system as we know it today. The third planet

from the sun, in this group of celestial bodies, is where the story of life begins.

2. As the young universe expanded and the earth cooled, water mingled with the elements, provided a primordial soup teeming with raw chemicals. By some accidental process, these simple inorganic compounds began to randomly fuse together to form complex molecular structures. Spontaneously generated molecular aggregates somehow became highly energized and subsequently able to miraculously organize under the most improbable conditions. Some scientists believe that this random series of "lucky breaks" was necessary to produce the very first self-replicating cellular life-form.

3. Over many millions of years, this spontaneously generated, primitive cellular organism evolved to became a multifaceted life-form by random transmutation. After accumulating an exhaustive series of mutations, this primitive life-form gradually become more complex in its overall anatomy, and better suited to survive in a hostile world.

4. After eons of time, through endless gradual change by natural processes alone, primitive organisms began to randomly evolve into life-forms with greater complexity becoming more capable of environmental adaptation and reproductive fitness.

5. Through the painfully slow process of mutation and natural selection, by random chance, these early life-forms were

gaining advanced structures allowing one species to change itself into another in a continuous transformation from simple to complex. In time, these emerging life-forms would gain the fitness required to thrive amid the harsh and unforgiving climates of earth.

6. Over several billion years of relentless physiological change, much more robust organisms appeared. This drive toward transmutation led to many new advanced species that were more able to adapt and reproduce over time. Vast ages produced increasingly unique and amazingly complex living things until finally, ultimately, human life had evolved.

Evolution is based on the concept of naturalism, which teaches that all organic change occurs naturally as a result of random chance events acting totally independent of any divine force or intelligent agency. The evolution model requires that mutational change, occurring randomly in living organisms, must provide physical advantages that will, over time, result in significant improvements in the structure and function of all living things. These perceived physiological improvements, which many believe will become concentrated in the population, would result in progeny exhibiting greater fitness, and a higher probability of adaptive survival in the highly competitive and ever-changing environments of a hostile planet.

The other well accepted view of the origin of life is that of "special creation," which brings to mind the concept of "intelligent design." Despite the widespread dissemination of evolutionary ideas in American, national surveys suggest that over 60 percent of Americans still believe in this traditional view of origins. The creation model implies that an all-powerful Supreme Being is solely responsible for intelligently planning and purposefully designing the entire universe, including planet earth, and the complex life-forms it supports. Special creation teaches that life on earth is not accidental and unplanned but the ingenious design of a purposeful God. A written and very detailed record of the creation account is precisely summarized for our examination in the first two chapters of the biblical book of Genesis.

When considering the question of origins, the concept of evolution is only a recent development. At the onset of the scientific revolution, Christianity and science were totally compatible. For the first 300 years following the rise of the scientific age, Christianity offered the most predominant worldview, and the majority of early scientists were devoted believers. It has only been within the last 150 years or less that an evolutionary worldview has taken a strong grip upon Western civilization. This popular view of origins now represents the official position of public policy in the most advanced cultures of the world.

In the first-generation following World War I, a survey of more than five hundred National Academy of Science members revealed that only 7 percent of them believed in the concept of a personal creator God. Nevertheless, despite over 150 years of public debate, it was men who held a strong belief in a Creator that founded the primary branches of science, as we know them today. The long list of creation-minded scientific pioneers is quite impressive when you consider names like Sir Isaac Newton,

Michael Faraday, Louis Pasteur, William Kelvin, Robert Boyle, Gregor Mendel, Nicolaus Copernicus, Galileo Galilei, Johannes Kepler, and Blaise Pascal, just to name a few of the most impressive scientific figures in world history. Even classical Greek philosophers like Plato and Aristotle, whose influential ideas have given rise to a materialistic view of the world, recognized the obvious fact that the universe is far from self-explanatory, requiring an explanation well beyond itself.

In my exhaustive search for truth and knowledge, I discovered that many men and women of contemporary science are continuing in the time-honored tradition of these early pioneers who laid a foundation for the scientific method. Because of their insatiable appetite for discovery and understanding, distinguished scientists the world over have given us a remarkable glimpse of the unknown with some surprisingly spectacular insights into the characteristics of the universe we inhabit. On a grand scale of the incomprehensible vastness of remote outer space, the technologically advanced Hubble Telescope continually transmits to earth stunning images of the starry heavens above. At the opposite end of the physical spectrum, we discover another scientific discipline of modern biochemistry that utilizes electron microscope technology to map the complexity of minute cellular physiology. This microscopic world of macromolecules is constantly at work generating microminiaturized molecular machinery within every living cell. These cellular components are specifically calibrated to manufacture a wide array of essential proteins with unbelievable precision and efficiency. The complex chemical factory that science has discovered within the living cell makes the most advanced engineering designs of human technology appear crude by comparison.

In recent decades, evolutionary thought has taken Western civilization by storm, but does that necessarily make it true? When we consider what is suggested by the big bang theory, it is not reasonable for rational man to assume that a random explosion, absent of intelligent oversight, could produce complex order and design. This is especially true when one considers the precision and mathematical order in the physical laws that are governing a vast array of heavenly bodies, such as planets, stars, and galaxies. Random explosions, lacking intelligent supervision, will always produce chaos and disorder. They will never produce organization, beauty, and useful complexity. That, they say, is how the universe began, and from that point on it was merely one impossible accident after another that supposedly produced the detail and undeniable beauty of the material world in which we live.

The proponents of evolution theory would have us believe that several billion years ago random chemicals in some imagined primordial soup were spontaneously arranged, fully integrated and energized to become a living organism. This random series of colossal accidents, they believe, resulted in the initial 'spontaneous generation' of the first primitive life-form. They will not tell you that the chance generation of the first living cell would have required thousands of impossible biological accidents occurring randomly, all at the same time. By simply evoking materialism as the fundamental basis of their beloved theory, evolutionists infer that all things became as they are without any intelligent oversight and planning. They aggressively and dogmatically promote the idea that organization and order in the known universe has required no intelligent agency and, therefore, the vast material world we know and experience has had no specific reason for existing. On the other hand, intelligent design suggests that a transcendent God had a definite plan and a specific purpose in mind when he chose to uniquely order and intentionally fashion a vast and remarkable universe. This exceptional creation included an extraordinary planet like earth teeming with abundant life—a planet we are privileged to call our home.

There is a considerable degree of contradiction between the concepts of creation and evolution. How can one honestly believe in a literal interpretation of the biblical record of creation and, at the same time, accommodate evolution theory. In order to believe in creation, one must believe in a Creator as well—perhaps the one who left a detailed record of his impressive handiwork. What may surprise a lot of people is that one cannot logically believe in the known facts of empirical science and accept evolutionary theory as well. The reason for this is revealed in that the facts of science do not adequately support the evolutionary model, as we will soon discover. Unfortunately, however, many people frame their position on origins based on the conclusions of others, instead of carefully examining the evidence for themselves.

Before we begin to explore this detailed material, let me state here, that the conflict between these two worldviews is so great in scope that any proposed compromise, embracing aspects of both views in an attempt to satisfy everyone, is totally out of the question. If one of these views is found to be truthful, based on a consensus of scientific data and historical evidence, the other view must remain utterly false and should be completely discarded.

Doubting Darwin

For thousands of years philosophers have expounded a materialistic view of life. However, it was Mr. Charles Darwin who is credited with popularizing this view of origins. In his two major works, *On the Origin of Species* and *The Descent of Man*, Darwin laid out the basic ideas that have shaped the evolutionary model widely accepted today. Darwin strongly believed that the mechanism of natural selection could account for the development of all the species over vast ages of time. This is evident from some of the statements recorded in his original written works. Darwin stated that, "Natural selection is daily and hourly scrutinizing ... every variation, even the slightest; rejecting that which is bad, preserving and adding up all that is good; silently and insensibly working at the improvement of each organic being."[2]

Even though Darwin offered his initial conclusions as only a possible explanation for the origin of living things, his early works have received such wide acclaim that he became, to some degree, misrepresented over the years. Many believe Darwin was dogmatic about his proposed ideas and offered sufficient evidence to support them. However, after considering his personal statements, I think you will agree that he was somewhat uncertain and careful not to overstate his intentions. From the very outset of his major work, *On the Origin of Species*, it is obvious that Darwin himself harbored serious doubts about the validity of his claims, thus setting a tone of uncertainty about his conclusions.

In the introductory material of his initial publication, Darwin makes this prudent statement: "A fair result can be obtained only by fully stating and balancing the facts and arguments on both sides of each question; and this is here impossible."[3] Darwin knew that his proposals could not be tested for accuracy, and therefore, he was careful to convey his conclusions as supposition only. His cautious conduct, compared to the dogmatic approach of modern-day naturalists, leads us to the conclusion that contemporary evolutionists show unwarranted bias in their public approach to origins. The suggestion that so-called "beneficial mutations" will lead to improved species over time has become linked with natural selection. In adopting the overt mechanism of fitness through mutation, evolution teaches that nature will preserve the more robust species, while those, not favored by natural selection, will die off and become extinct over time.

From his observations, Darwin presumed that when a mutant variation occurred, in a given species, which could possibly improve its chance of survival, the resulting trait would take precedence over less desirable physical traits and in time would become concentrated in the population. By this process, known as the "variation of species," Darwin concluded that over millions of years, an accumulation of minor changes could, perhaps, allow for one species to transform into another. However, his initial doubts about the accuracy of this prediction are apparent from these remarks: "Why, if species have descended from other species by fine gradations, do we not everywhere see innumerable transitional forms? Why is not all nature in confusion, instead of species being, as we see them, well-defined?"[4] He continues with this perplexing observation: "But, as by this theory, innumerable transitional forms must have existed, why do we not find them embedded in countless numbers in the crust of the earth?"[5] The absence of so-called transitional forms (missing links) has been a constant source of confusion for today's evolutionists. Many scientists simply assume these intermediate forms have existed, despite the overwhelming lack of evidence in the fossil record. Again, Darwin expresses serious misgivings after having observed various species in the normal process of sexual reproduction. In response to these observations, he posed another intuitive question: "How can we account for species, when crossed, being sterile and producing sterile offspring, whereas, when varieties are crossed, their fertility is unimpaired?"[6]

He was questioning how his theory could hold up to scrutiny in light of obvious fixed boundaries between various animal kinds. DNA chemistry in living organisms was designed to ensure that each kind of plant and animal would retain its specific identity, while expressing a wide variation in certain physical traits, making each variety adaptive to environmental change. Before his death, Darwin already knew his theory was in serious trouble and could not account for certain unpredictable anomalies. He was dismayed by a fossil record that failed to demonstrate any tangible evidence for transitional intermediates. This prompted regretful remarks, and added to speculation that Darwin had some serious doubts about his own conclusions: "The number of intermediate varieties which have formerly existed on the Earth must be truly enormous. Why then is not every geological formation and every stratum full of such intermediate links? Geology assuredly does not reveal any such finely graduated organic chain; and this is perhaps the most obvious and gravest objection that can be urged against my theory."[7]

Many people today choose to overlook the disturbing lack of intermediate transitions in the fossil record and dogmatically convey evolution theory, publicly, as a scientific fact. This unwarranted over-confidence is offered consistently in the public domain despite the serious doubts Darwin expressed in his own conclusions. When Darwin initially advanced his theory in 1859, modern science was still in its formative years. Today, however, with amazing advances in technology, opportunities to discover the features of our world have accelerated at a blistering pace. With the aid of the electron microscope and digital computer technology, we are now able to map the remarkable complexity of a single cell. The fascinating physiology and complex design man has discovered in this tiny speck of life remains light-years beyond the imagination of the greatest scientists of Darwin's day, who considered a single cell to be a simple, unremarkable blob of protoplasm. Captivating recent discoveries in the technical world of empirical science have led mankind toward numerous innovative frontiers of research, including molecular microbiology, astrophysics, quantum electrodynamics, and probability analysis. Advanced research into these, and other scientific disciplines, is now proving to be disastrous for evolutionary theory.

Many highly qualified and well-equipped researchers in the ongoing study of origins, unlike the disadvantaged and ill-equipped investigators of Darwin's day, are finding the popular concept of evolution to be quite impractical and are no longer taking it seriously. We are living in a modern society in which evolutionary theory is assumed by multitudes to be factual. This commonly held position exists because a materialistic view of origins is continually and consistently embellished by mass-media bias, while forced upon American citizens through a massive public-education campaign. This dogma has been presented to us repeatedly as objective truth and scientific fact. Now, however, we are discovering that modern science has exposed major flaws in "the theory," having uncovered a serious lack of convincing evidence to support the evolution model. Even so, American families are finding their children repeatedly subjected to a blatant form of indoctrination funded by the government at taxpayers' expense. As it turns out historically, those who chose to dismiss God from their life were standing by, ready to believe that Darwin's theory had eliminated the need for an "intelligent designer," by demonstrating a process through which naturalistic events could account for the remarkable complexity and diversity of living things.

We now understand that Darwin's theory was seriously flawed. His presumptuous speculation and premature guesswork about the origin of life is far from satisfactory when we consider the compelling evidence uncovered by mankind's search for knowledge over the last 150 years of scientific investigation. By mapping cellular physiology, we are recognizing scientific realities Darwin never dreamed of. With fascinating advances in new fields of research, scientists have unveiled remarkable complexity in cellular design that Darwin's outdated theory utterly failed to account for. The living cell's delicately detailed interlocking machinery of sensors, pumps, gates, and delivery systems revealed by microscopic cellular technology requires that all essential components must be in place simultaneously and integrated together for active cellular function to occur. Darwin's fatal error was twofold regarding the mechanisms of living cellular design. First, he wrongly assumed that single-celled organisms were very primitive in design. Second, Darwin believed that vital structures and interlocking components could be slowly added, over millions of years, by the piecemeal process of mutation and natural selection.

Because the technology of Darwin's day was so limited, he was able to convince the scientific world of the nineteenth century that simple design in cellular physiology would allow for gradual change and increasing complexity over vast ages. Darwin had referred to this process as the "transmutation of species." and believed that these continuous random changes would slowly move evolutionary biology toward increasingly complex living organisms.

Outspoken Evolutionists

In Western culture, prior to Darwin, biblical Christianity was the foundation of formal education in all areas of study. After Darwin introduced his theory, however, the world of modern academia was ready and eager to cast aside the Bible in favor of a worldview that denied God. Since that time, with passionate zeal, progressive reformers have gradually infiltrated all areas of public influence in American culture. These power centers include public education, government, entertainment, science, journalism, and to some degree, even the church.

The dominant public institutions within modern culture have gradually drawn the unsuspecting public away from biblical truth in favor of a secular, materialistic view of the world. Evolutionists have generated this bias with overzealous reports, and, in some cases, downright deception in representing the true nature of their position. Years ago, *Reader's Digest* magazine printed a review of the book entitled *The Naked Ape*. In the very first paragraph of this article, zoologist Dr. Desmond Morris bluntly states that it is a known fact that man evolved from ancestors of apelike intelligence.[8] Enormously popular *Time magazine* has printed several features on the origin of life over the years, but always with an evolutionary bias. One cover story was published in 1977 with the title "*How Man Became Man*."[9] Typical of news magazine coverage on this subject, the Time writers make it perfectly clear that evolution is the key to understanding the origin of the human species. However, throughout the article, a serious lack of convincing evidence forces anthropologists to use words of conjecture to describe their work, such as 'we believe,' 'we theorize,' and 'we speculate.' These phrases tell us that a great deal of educated guesswork has gone into today's "expert" scientific thought on origin-of-man research. The article ends with the following major concessions. "Still, doubts about the sequence of man's emergence remain. Scientists concede that even their most cherished theories are based on embarrassingly few fossil fragments, and that huge gaps exist in the fossil record."[10]

Even the world's so-called greatest experts are baffled by the lack of compelling evidence from scientific research, and investigation over recent generations. After more than 150 years from the time of Darwin's classic publication, intense study and painstaking research has left evolutionary theory in the poorhouse. Nevertheless, faithful proponents continue to make profound yet unverifiable claims about their cherished ideology. The following statements present dogmatic assertions evolutionists have made to the world over the public airwaves in recent years.

1. "The first point to make about Darwin's theory is that it is no longer a theory but a fact ... Darwinism has come of age, so to speak. We are no longer having to bother about establishing the fact of evolution."[11] (Julian Huxley, biologist)

2. "Modern science must rule out special creation or divine guidance."[12] (Julian Huxley)

3. "Evolution is a fact, not a theory."[13] (Carl Sagan, astronomer)

4. "Natural selection is a successful theory devised to explain the fact of evolution."[14] (Carl Sagan)

5. "To those who were trained in science, creationism seems like a bad dream, a sudden reliving of a nightmare, a renewed march of an army of the night risen to challenge free thought and enlightenment."[15] (Isaac Asimov, biochemist)

6. "The process of natural selection and survival of the fittest, with the many mutations that occur over hundreds of millions

of years, adequately account for the origin and development of the species."[16] (Corliss Lamont, philosopher)

7. "Man is the result of a purposeless and natural process that did not have him in mind. He was not planned. He is a state of matter, a form of life, a sort of animal and a species of the order Primates, akin nearly or remotely to all life and indeed to all that is material."[17] (George Gaylord Simpson, paleontologist)

8. "Most enlightened persons now accept as a fact that everything in the cosmos from heavenly bodies to human beings, has developed and continues to develop through evolutionary processes."[18] (René Dubos, microbiologist)

9. "Evolution ... is the most powerful and the most comprehensive idea that has ever arisen on earth."[19] (Julian Huxley)

10. "Humanism believes that man is a part of nature, and that he has emerged as the result of a continuous process."[20] (Humanist Manifesto I)

11. "Science affirms that the human species is an emergence from natural evolutionary forces."[21] (Humanist Manifesto Il)

These unconfirmed statements and dogmatic assertions by Huxley, Sagan, and others have directly challenged the intelligence of many people who remain unconvinced that materialism holds the key to explaining the origin, complexity, and diversity of living things. Publicly making unscientific statements speaks volumes about the dogmatism and overconfidence of those who belittle, demean, and harass anyone who does not concur with their atheistic worldview.

In his 1980 television series Cosmos, atheist astronomer Carl Sagan boldly declared, "The cosmos is all there is, or was, or ever shall be."[22] Naturalism proposes that there is nothing beyond nature. Atheists insist there is no Agency that has altered the natural order by way of transcendent supernatural power. Sagan's assertion is not birthed in science but is merely a philosophical view. Therefore, the question we must answer is: which worldview does empirical science support—naturalism or theism? To modern-day evolutionists, the unnatural idea of an all-powerful creator God, to whom they are accountable, is unthinkable. Therefore, because evolution theory is the only explanation of origins that excludes supernatural agency, those who reject God embrace it with great tenacity. So, while most naturalists persist in their ceaseless effort to explain life without God, we will investigate the revealed facts, expert testimony, and empirical scientific data to see if this information will support their bold and provocative claims.

Design Demands a Designer

We live in a world filled with amazing design and mind-blowing complexity. Just try to explain the explosive defense mechanism of the tiny bombardier beetle, the visual accommodation apparatus of the archerfish, or the homing instinct of a carrier pigeon. Australian molecular biochemist Dr. Michael Denton has concluded that: “It is the sheer universality of perfection, the fact that everywhere we look, to whatever depth we look, we find an elegance and integrity of an absolutely transcending quality, which so mitigates against the idea of chance.”[23]

One of the world's leading evolution advocates, Dr. Richard Dawkins of Oxford University, said this in his book *The Blind Watchmaker*. “We have seen that living things are too improbable and too beautifully 'designed' to have come into existence by chance.”[24] Although we understand that Dawkins is not advocating for intelligent design, this acknowledgement, from a world-class atheist, offers at least the vague notion that nature is far too remarkable to be explained by a long series of random accidents. In his book entitled *Darwin's Black Box*, Dr. Michael Behe, a molecular biochemistry researcher from Lehigh University, has introduced the term, “irreducible complexity,” which refers to a strict requirement in nature that all major subcomponents of any vital organ system must be found simultaneously present, fully functioning, and all networked together for any complex integrated system to work. A simple design that will illustrate this point can be found in the common ‘mousetrap’. This device will only function for its intended purpose if all five parts are present simultaneously. It requires a platform, spring, catch, hammer, and hold-down bar all working together at the same time. If any part is missing or broken, the device is useless for mouse removal. In a more complex illustration, all the essential components of a Boeing 747 must be present, networked together and fully integrated with extreme precision before the plane can fly successfully. In the same sense, all the parts of any biological organ system, such as digestion or vision, must be simultaneously present and working in unison before active, useful function can occur. This suggests that the vital parts of a living biological system could not possibly have resulted from a piecemeal process like natural selection!

Consider Dr. Behe's comments on cellular design. "Now it's the turn of the fundamental science of life, modern biochemistry, to disturb ... The simplicity that was once expected to be the foundation of life has proven to be a phantom; instead, systems of horrendous, irreducible complexity inhabit the living cell ... The resulting realization that life was designed by an intelligence is a shock to us in the 20th century who have gotten used to thinking of life as the result of simple, natural laws. But other centuries have had their shocks, and there is no reason to suppose that we should escape them."[25]

The design principle of irreducible complexity must have caused great anxiety in Darwin, as he considered the intricacy, unique design, and vast complexity of living systems, and how his theory utterly failed to account for them. He once wrote, "If it could be demonstrated that any complex organ existed, which could not possibly have been formed by numerous, successive, slight modifications, my theory would absolutely break down."[26]

In the case of cellular design, science has identified a minimum of seventeen complex systems that must be simultaneously integrated to produce viable cellular function. Otherwise, the cell would be unable to generate the minimum essential proteins required for life.

Anatomically speaking, the visual system represents a pronounced example of "irreducible complexity" in living organisms. Complex organ systems like vision, present major problems for evolution. Sight allows electromagnetic radiation (visible light energy) to be detected by a specialized organ system and converted into chemically mediated nerve impulses that are interpreted by the brain as visual images. This occurs as visible light is reflected off an object, enters the cornea of the eye, where it passes through the lens, and is focused on the retina. The refraction of light through the lens causes the image to be displayed on the retina upside down. This energy from visible light is then absorbed by highly sensitive pigment cells on the surface of the rods and cones lining the retina. The inner portion of the retina converts this light energy into chemically mediated nerve impulses (electronic signals), which are transmitted by the optic nerve to the brain, where the visual images are received at the geniculate body. The iris muscle acts like a small mechanical diaphragm as it alters the size of the pupil and controls the amount of light entering the cornea. As the optic nerve transmits these neural impulses to the brain, an electronic message is received. From there, electronic impulses are sent to the primary and secondary visual cortex (occipital cortex) at the back of the brain, where the image is displayed right-side up. Each image is received at the frontal lobes of the brain where visual acuity and pattern recognition occurs and the object is identified. Eyelids, nerves cells, blood vessels, tear ducts, aqueous humor, and vitreous humor are other features that play an important role in this complex system that is vital to a wide variety of living organisms. If any primary component of the visual apparatus is missing or defective, this could negate the proper function of the entire system. It is certain that the piecemeal process of evolution cannot possibly account for this complex organic system in which all major components must appear in unison and be fully integrated to allow for visual recognition. Mr. Charles Darwin knew that irreducible complexity presented a major obstacle for his theory. He even referenced this specialized design principle. "To suppose that the eye (with so many parts, all working together) could have been formed by natural selection, seems, I freely confess, absurd to the highest degree."[27] Without question, irreducible complexity reveals an undeniably complex feature created within living organisms. This speaks volumes about the necessity for a complex-minded, ingenious design engineer. After thoroughly examining evolutionary claims biochemist Michael Behe, offers this compelling statement: "I conclude that another possibility is more likely: the elegant, coherent, functional systems upon which life depends are the result of deliberate, intelligent design."[28]

Have you ever stopped to consider our remarkable solar system? It is a grand geometric design that works by mathematical order based upon the fundamental laws of physics. The movement of the heavenly bodies is so precise that we can determine the exact location of each planet at any given moment past, present, or future. We can predict exactly when a lunar or solar eclipse will occur and just how long it will last. Halley's Comet, which speeds around the solar system on a very predictable path, was last viewed in 1985. We can now calculate exactly when it will appear again, in the year 2060, because its passage across the solar system brings it near our planet on a reliable seventy-five-year cycle. A rare astronomical event occurred on June 5, 2012 that allowed many to observe the *Transit of Venus* as it passed directly between the earth and sun. Astronomers predicted exactly when this event would begin and exactly when it would end. You and I will be long gone the next time it occurs on December 10 of the year 2117. We can predict astronomical events with precise accuracy on any given day of the year at any location on the planet, past, present, or future. This is because the universe obeys physical laws defined by mathematical order. Does it make sense to think of a delicately designed watch in terms of a freak accident? Our fine-tuned and mathematically arranged solar system speaks no less of a creator. Galileo and Kepler investigated this mathematical order, and concluded that the solar system, operating with remarkable precision, was clearly the work of 'intelligent design'.

Life-giving water and breathable oxygen, noticeably absent elsewhere in the solar system, appear to be purposefully designed to sustain life on earth. Do you think this was an accident? While oxygen is so critical for sustaining life on the earth's surface, there is another important need for oxygen to sustain life on earth. The intense ultraviolet light from solar radiation would be deadly if it were not for an oxygen blanket covering earth's upper atmosphere. Unlike surface oxygen, the ozone layer high above us is also composed of pure oxygen, but its molecular structure consists of three oxygen atoms resulting in a physical feature that was designed for another very essential purpose. The earth's ozone blocks harmful ultraviolet light. Without this special filtering mechanism, high-energy UV waves would become lethal to life on the planet's surface. On the other hand, the sun's visible light waves, essential for the energy and growth of plants, along with supporting all life-forms, pass on through the ozone layer. I am quite certain that this unique, life-saving feature of planet earth is not just another random accident!

Whose idea was photosynthesis? This unique design feature allows plants to capture light energy from the sun and convert it into useful chemical energy. Webster's Dictionary defines photosynthesis as "the process by which chlorophyll-containing cells in green plants are able to convert the sun's visible light energy into chemical energy and synthesize organic compounds from inorganic compounds." Thanks to this complex photo-biochemical process, plants can efficiently produce vital organic compounds by utilizing minerals from the soil. Carbohydrates are produced from carbon dioxide and water, accompanied by the simultaneous release of oxygen. Photosynthesis is the only naturally occurring process utilizing large amounts of solar energy to sustain plant and animal life on earth. With a bounty of stored chemical energy plants manufacture essential fatty acids, sugars, proteins, carbohydrates, amino acids, hormones, and a great variety of complex phytochemicals. This critical manufacturing system had to be in place from the very beginning, when plants were first created, because without this genius, irreducibly complex design mechanism, living things could not survive. It was probably not an accident!

The overwhelming impression of purposeful design in nature is causing many scientists to abandon their initial allegiance to impersonal, unguided materialism. Regarding the design argument, famous atheist Bertrand Russell was quite impressed with the logical implications of the specialized designs he observed. Consequently, he offered the following response: "This argument contends that, on a survey of the known world, we find things which cannot plausibly be explained as the product of blind natural forces, but are much more reasonably to be regarded as evidences of a beneficent purpose. This argument has no formal logical defect."[29]

What is True Science?

Evolution is a theory in crisis! Consider this comment from Dr. Colin Patterson, a former senior paleontologist of the British Museum of Natural History. "A theory that simply accepts the diversity of life, without offering any mechanism to explain how that diversity came into being, cannot be considered a scientific theory at all."[30] Leaders of evolutionary thought object to the idea of a Judeo-Christian God, as revealed in the Holy Scriptures. Their refusal to seriously consider the possibility of intelligent design reveals that their interpretation of the evidence is misleading and intellectually dishonest. Many people have confused objective knowledge with theoretical science, and will blindly trust the misguided philosophy of the world's so-called experts. These individuals are often instructing from textbooks, not from empirical data that was gathered through firsthand experience in disciplined areas of serious research. Theoretical science deals with speculation about what may have happened based upon a line of reasoning the investigator is accustomed to and comfortable with. True empirical science is defined as knowledge derived from observation. Even though evolution is often debated in scientific circles, the theory is, in fact, not scientific in the true sense, because it has never been observed.

An American biochemist, the late Dr. Duane T. Gish, was a teaching associate with the *Institute for Creation Research.* As an expert authority on the fossil record, He said: "It is obvious, for example, that no one has observed the origin of the universe, the origin of life, the conversion of a fish into an amphibian or an ape into a man. No one, as a matter of fact, has even observed the origin of a species by naturally occurring processes. Evolution has been postulated, but it has never been observed."[31] Evolution is speculation about the past that crowns man supreme and relegates God and His Word to the realm of religious fantasy. Creation by intelligent design is a reasonable faith that fits the known facts—verifiable facts observed in our world today and painstakingly uncovered from the historical world of the past. To call the evolution model a valid theory is a clear contradiction. A credible scientific hypothesis must be put to the test of impartial scientific investigation. The data gathered from these experiments should either confirm or deny the validity of any such claim. In the case of evolution theory, as well as creation, no observable scientific experiments have ever been devised that could validate either popular claim in a truly empirical sense; and yet those who stubbornly advocate for evolution as the correct mechanism for explaining life's origin have consistently insisted on calling it factual.

So, what is science? The word means "knowledge," inferring testable knowledge that can be logically gathered through an ongoing scientific process, and theoretical knowledge that must be carefully weighed in the balance to determine if the evidence supports a given hypothesis or not. It is an absolute misuse of scientific protocol to insist on promoting one's preferred ideology as factual, when the scientific community is utterly incapable of producing sufficient evidence to authenticate and legitimize such a claim. Science affords us a reliable way to explore and to analyze features of the natural world, but true science is powerless to embrace questions like, "Why does the universe exist?" "Why is the physical world governed by the universal laws?" "What is the meaning and purpose for human life?" "Why is death a reality in the natural world, and what lies beyond the grave?" There are many ethical questions that we wrestle with throughout our lives. These philosophical issues remain outside the scope of science.

Consider the following clear statement from a well-known Christian scientist: "By far the most influential argument against the Bible is the widespread belief that science has proven evolution, and therefore disproves the Biblical account of creation."[32] Dr. Henry M. Morris, PhD, is a scientist, educator, and author of more than twenty fascinating books on creation, design, and evidence. The late Dr. Morris, who served on the faculty of five major universities, was one of the most accomplished men of science over the last century. He offered the following reaction to the bold and demanding claim of evolution theorists, who insist on circumventing standard scientific protocol in their relentless effort to promote a baseless and controversial theory. Dr. Morris explains that "one of the strangest paradoxes in the history of science is that a theory so barren of scientific proof as evolution could be so universally accepted as scientific fact. Science is knowledge, and the essence of the scientific method is experimentation and observation."[33] Empirical science, by definition, only deals with observable processes in the present. It does not allow for speculation about unrepeatable events in the past. Given this authentic definition for science, evolution can only be described as nonscientific dogma that has managed to capture the minds of vulnerable people the world over, including many Americans.

Those who blindly follow this ideology have become thoroughly indoctrinated, thinking that what they believe amounts to real science. However, early scientific pioneers like Copernicus, Kepler, Newton, and Galileo, all believed that their monumental discoveries, which sparked a scientific revolution, were revealing the magnificent design of a purposeful God. How sad they would surely be, if they knew that scientific authorities in today's world, who have fallen away from belief in a personal God, are now crediting their epic insights to blind-chance events and nothing more. This unfortunate reality occurred long after these brilliant men of early science revealed the amazing outcomes of their monolithic search for knowledge—a deep search that ultimately delivered such accurate and astonishing information to the modern world about the fundamental laws that are consistently governing the vast physical dimension.

At the core of science lies a firm conviction that the universe was fashioned in an orderly manner based on the universal laws of physics. The ancient Hebrew prophets were the first to articulate that the universe is governed by an all-powerful God. It was this original monotheistic worldview that laid the historical foundation upon which modern science sprang forth. Sir Francis Bacon, regarded by many as the founding father of the scientific method, taught that having a proper education required a thorough knowledge of both fascinating books the one true God had provided—the book of nature and the Bible. When the modern scientific revolution commenced in sixteenth-century Europe, it was God-fearing men professing profound Christian faith whose epic scientific discoveries and mathematical calculations had forged a successful paradigm shift in mankind's view the solar system. The movement of the planets could finally be understood by discovering the earth's rotation around the sun. For centuries tradition held that the sun rotated around the earth through the influence of the classical philosopher Aristotle.

As an early investigator, Galileo was captivated by scientific study and argued that endeavors of research were not only acceptable but constituted the noblest course of action for a believer. In a famous remark that inspired many Christians to pursue a life of science, he once said, "I do not feel obliged to believe that the same God who has endowed us with sense, reason, and intellect has intended for us to forgo their use."[34] In a gallant effort to enhance his scientific knowledge, Galileo chose to develop theories about the universe that were based squarely on physical evidence. He criticized the prevailing worldview held by the secular philosophers of his day whose ideas were based on outdated traditions. As a student of Plato, Aristotle taught that the earth was the center of the universe and everything else revolved around it. He believed that ether was the substance that filled the regions of space in which the celestial bodies floated. This worldview denied a personal creator and led the masses toward a materialistic view of the cosmos, in which man is supreme. Galileo's conviction that scientific observation should always take precedence over philosophical tradition was not well received by the thinkers of his day or by the Roman Catholic

Church, and he experienced persecution because of it. Even in our time, criticism of the prevailing scientific paradigm is an open invitation for censorship, persecution, and isolation. Tensions seem to run high whenever empirically derived evidence is found in conflict with the established and accepted scientific framework. It takes a long time before the collective sum of evidence favoring a new paradigm can supersede an existing one. This is especially true in the realm of science. The ideological battleground within our culture is very real, but has often been mistakenly characterized as a war between science and religion. This monumental conflict is the inevitable collision between two powerful, diametrically opposing worldviews that are best described as *theism* and *naturalism*.

The Laws of Thermodynamics

The first law of thermodynamics (the law of conservation of mass and energy) has stated that neither matter nor energy is being created or destroyed. The book of Genesis says that God ended his work of creation (Genesis 2:2). This means all matter and energy resulted from an initial "first cause", which implies a creator! There is nothing we understand in natural law that would support the idea that matter either created itself or has always existed. Therefore, a "first cause" must be considered. Judging by the quantity of mass and energy in the universe, this ingenious "causing agent" should be esteemed worthy of the devoted worship of his creation. The second universal law of thermodynamics teaches that all matter is moving from a state of order and useful purpose toward a state of disorder, uselessness, or death. The world you and I experience is one in which entropy (the degree of disorder) is always increasing. This reality is demonstrated everywhere as a fundamental law of physics. All things, including organic life-forms are subject to a constant shift toward physical decline, decay, death, or the end of useful purpose. Evolution, however, calls for a universe that is becoming progressively more orderly, advanced, and more resistant to disease (advancing from a state of chaos toward perfection by random chance). The 2nd law of thermodynamics would flatly deny this erroneous claim!

The effect of the second law is seen everywhere in our world. These laws teach us that the universe had a beginning and what was created new is now growing old, falling apart, dying, or becoming less useful over time. Everywhere we look we see a constant shift toward disorder, life toward death, useful toward useless. There is no means of escape from its relentless power over us on this side of eternity. We observe this insidious curse when we visit a sick and dying friend or when we simply throw away a worn-out appliance. Yet, the incredible claims of the evolution model would imply a very different reality. They imagine a world where integrated life-forms are improving over time, developing greater complexity, becoming ever more resistant to disease, and better fit to survive while gradually moving toward an ideal state. Without constant intervention, it is obvious that everything left to itself is in decline. The evidence suggests that evolution's unproven dogma is simply wishful thinking. It is really a wasted dream about some imagined utopia (an escape from reality) denied by observation everywhere on this planet.

What young people really need today is honest science

that makes no attempt to manipulate the truth and does not withhold the evidence of recent discoveries in order to protect a strictly materialistic view of life. Evolutionists refuse to consider any evidence deemed unsupportive of their cherished theory. Consider the following statement from Dr. James Tour, a nano-scientist and professor at Rice University's Department of Chemistry and Center for Nanoscale Science and Technology. Dr. Tour, who makes molecules for a living, made this remarkable statement after many years of devoted research at the molecular level: "I stand in awe of God, because of what he has done through his creation. Only a rookie, who knows nothing about science, would say science takes away from faith. If you really study science, it will bring you closer to God."[35]

Chapter Two

* * *

Evolution: A Theory in Crisis

What about Mutations?

For evolution to be true, there must be some mechanism that drives the process forward resulting in new improved species over time. Evolutionists have claimed this apparatus is mutation and natural selection. Dr. Ernst Mayr, a former professor of zoology at Harvard and considered a leading authority on evolution during the 1970s and 80s, stresses the absolute necessity of mutations as the driving force behind evolutionary change. He emphasized this claim in the following statement: "It must not be forgotten that mutation is the ultimate source of all genetic variation found in natural populations, and the only new material available for natural selection to work on."[1] As a leading authority, he is telling us that, for evolution theory to work, major variation must occur randomly over time contributing new information through mutational change. Theoretically speaking, so-called beneficial mutations could be passed to progeny, but only if they appear in germ cell DNA. Of the trillions of cells that make up complex biological systems, only reproductive cells can pass the information from a mutated gene to offspring. Also, keep in mind that the chance of a mutation occurring within germ cell DNA is very low. How then can mutation be the machinery that drives evolutionary change? By adding billions or even millions of years to the final equation of life, evolutionists believe that random mutations could eventually add up to significant change required by the theory. Eons of time have made evolutionary thought seem very obscure and far beyond the comprehension of the average person. Most people simply rely on the dogmatic views of perceived experts, who, by the way, were not there to observe naturalism occurring when life began. Again, I refer to the assertive statement of philosopher Corliss Lamont, which is representative of the unproven and overconfident remarks declared publicly by contemporary evolutionists: "The process of natural selection and survival of the fittest, with the many mutations that occur over hundreds of millions of years, adequately account for the origin and development of the species."[2]

In this chapter, we examine the topic of mutations in an honest attempt to determine if they really could be the mechanism advancing evolution as so many proponents have claimed. Evolution requires that mutations must lead to improved species over time that are better able to adapt, survive, and reproduce in the most hostile environments. Virtually all biology and science textbooks found in public education will claim that mutation/

selection is the driving force behind evolutionary change. What we know is that without these so-called beneficial mutations, there can be no evolution. Many people believe and will explain that occasional copying errors in the DNA genetic code are a probable source of mutations. However, these mistakes are very rare primarily because biological DNA comes equipped with its own error-correcting system that specifically detects and corrects the occasional occurrence of genetic misinformation resulting from normal cellular replication.

We should think of DNA like Morse code, only instead of dots and dashes genes use a four-letter chemical alphabet ATCG to communicate. If a letter is altered or lost in the replication process, the resulting genetic signal has changed. Virtually all mutations will generate a net loss of information. However, on some occasions this net loss could possibly generate a perceived beneficial outcome. For example, you and I may prefer to purchase seedless grapes, oranges, or watermelons. We are benefiting from chemically mutated plant hybrids, but these plants cannot reproduce, so is the plant better off? No! These plants have had a net loss of genetic information. Over eons of time, evolving from an amoeba to a man would require billions of beneficial mutations; but remember that nearly all mutations produce a net loss of information resulting in an overall negative outcome. Also, we must keep in mind that in nature, every potentially useful mutation would be accompanied by millions of harmful ones, so organic change is clearly moving in the wrong direction, generating an overall loss of fitness, thereby working against the far-reaching claims of evolution. The theory promises progressive change over time with positive transformation leading to new, improved species.

According to information specialist Dr. Werner Gitt, a former director at the German Federal Institute of Physics and Technology, the replication of DNA in an actively dividing human cell is unbelievably accurate. He once wrote: "This replication is so precise, that it can be compared to 280 clerks copying the entire Bible sequentially each one from the previous one, with, at most, one single letter being transposed erroneously in the entire copying process."[3] Why, then, do mutations happen? It's because cellular division and DNA replication occurs constantly in all organic populations, making mutations inevitable. A harmful mutation may result from the corruption of one or more letters within a genetic sequence. A problem with random mutations, as the hypothetical engine that propels evolutionary change, is they are generally unable to add new information to genetic code as required for the theory to work. Mutations can only scramble or corrupt existing information present in cellular DNA. This is why genetic mistakes will typically produce fatal, harmful, or useless outcomes. In other words, given the negative outcome of randomly mutated genes, the fantom concept of a "beneficial mutation" is an oxymoron.

This presents a serious dilemma for evolution theory, because mutations are totally relied upon as the only instrument that drives organic change. Also, we should keep in mind that naturally occurring mutations are very rare, and those that do randomly occur virtually always produce a negative outcome, creating major problems and significant disadvantages for future offspring. Thus, making them less fit to survive. Remember that mutations are just mistakes in the genetic code. It is hard to imagine how numerous mistakes spread out over thousands of generations could possibly add up to positive, beneficial change. We might simply think of mutations as something like typing errors. We must recognize that typing errors generally subtract information by corrupting a sequence of letters, leaving a message less intelligible than it was previously. Cellular DNA contains computer-code-like messages that spell out every last detail of life's biological function, and also directs the developing physical characteristics of each living organism. Evolution theory must rely upon the hope that mutations can somehow add new intelligible information to germ cell DNA in a parent organism. Paleontologist George Gaylord Simpson once wrote: "If evolution is to occur, living things must be capable of acquiring new information or alteration of the stored information."[4]

Consider these comments from Dr. Lee Spetner, a retired physicist and former professor of information and communication theory at Johns Hopkins University. In Spetner's thoroughly researched book, *Not by Chance*, he makes a startling assertion about mutations: "In this chapter I'll bring several examples of evolution, particularly mutations, and show that information is not increased ... In all the reading I've done in the life-sciences literature, I've never found a mutation that added information."[5] His continues with this: "All point mutations that have been studied on the molecular level turn out to reduce the genetic information and not to increase it."[6] Mutations can, in fact, produce changes in stored information, but they are not able to add new information to that which is already present in the DNA. Because of this condition, the outcome is generally injurious. Science has demonstrated repeatedly that mutations do not provide the new and beneficial structures required to develop advanced species. Without a regular addition of new information to the DNA of an organic genome, Darwin's dream has become a nightmare. Not only is this a failure to support the theory, but it actually amounts to strong evidence against the evolution model. Therefore, the loss of information from mutated DNA will likely lead to a flawed trait within the affected organism, resulting in diminished overall fitness.

Evolutionists are hard-pressed to explain how major evolutionary transition occurs, and the random mistakes that have resulted from occasional mutations certainly do not appear to be supporting their case. Because science has never clearly demonstrated just how mutations are able to break the limits of change required to advance new species, the proponents of evolution will have to look for some other mechanism to explain their theory. After studying mutations in microorganisms for many years, French zoologist Dr. Pierre-Paul Grasse offers the following assessment: "What is the use of their unceasing mutations, if they do not change? In summary, the mutations of bacteria and viruses are merely hereditary fluctuations around a median position; a swing to the right, a swing to the left, with no final evolutionary effect."[7] Dr. Grasse flatly rejects mutation-selection as the mechanism for evolutionary change. He further explains that mutations "are not complementary, nor are they cumulative."[8] In other words, mutations do not harmonize with or complement each other. They are simply random and independent– not able to work together, adding up to anything significant by way of synergism. "They modify what pre-exists," says Dr. Grasse– meaning you get nothing but variation within a kind. He adds that: "The opportune appearance of mutations permitting animals and plants to meet their needs seems hard to believe. Yet the Darwinian theory is even more demanding: a single plant, a single animal would require thousands and thousands of lucky, appropriate events. Thus, miracles would become the rule: events with an infinitesimal probability, could not fail to occur ... there is no law against daydreaming, but science must not indulge in it."[9] Dr. Grasse also reminds us that "some contemporary biologists, as soon as they observe a mutation talk about evolution. They are implicitly supporting the following syllogism: 'mutations are the only evolutionary variations, all living beings undergo mutations, and therefore all living beings evolve.' This logical scheme is, however, unacceptable: first, because its major premise is neither obvious nor general; and second, because its conclusion does not agree with the facts. No matter how numerous they may be, mutations do not produce any kind of evolution."[10]

Some have pointed to the occurrence of antibiotic resistance as genuine evidence of evolution working today. However, it is interesting to note that bacteria samples from organic specimens, which were placed in frozen storage prior to the development of antibiotics, were subsequently determined to be totally antibiotic resistant. This means that bacteria cells have been losing their natural resistance over time, resulting in a diminished ability to survive, and indicating a loss of fitness. Antibiotic resistance, again, is variation within a kind with a tendency to revert to the original genetic blueprint.[11] Even in the case of sickle cell anemia, when examined closely, the long-term outcome of the offending mutation produces a negative overall effect on the host, even while providing added protection against malaria infection. In this instance, the survival, fitness, and longevity of the host is significantly impaired in the individuals whose lives have been affected by this mutation. Genetic mistakes, often thought to be beneficial, have never demonstrated an ability to add the genetic information required to propel the evolution model. This reality has proven devastating to evolution theory because it must depend on the hopeful proposition of billions of beneficial mutations leading to positive evolutionary change.

Due to the harmful nature of genetic mistakes, placing one's hope in favorable mutations leading to positive outcomes is like believing an electronic virus will somehow improve the working performance of your personal computer. That said, it is hard to understand how mutations could be the driving force behind evolutionary change. Albert Szent-Gyorgyi, the Hungarian-born, Cambridge-educated biochemist who became the 1937 Nobel Prize winner, informs us that "the origin of complex traits by random mutation has the probability of zero."[12] Former biology professor, former atheist, and author of the book *Creation Facts of Life*, Dr. Gary Parker, sums up the evolution problem this way: "Evolution demands an increase in the quantity and quality of genetic information and mutation-selection, no matter how long you wait, cannot provide it."[13] Atheist, Thomas Nagel also appears to be impressed with the powerful argument for intelligent design. He commented that "evolutionary biologists regularly say they are very confident that random mutations are sufficient to account for the complex chemical systems we observe in living things, yet he feels that there is a great deal of pure rhetoric in their arguments, and judges that the evidence is not sufficient to rule out the input of intelligence."[14]

The Fruit Fly Fiasco

Because of their rapid reproduction, the fruit fly (Drosophila) has been one of researchers' favorite specimens to experiment on. Fruit flies are complex organisms that scientists love to study because of their rate of growth to maturity. A fruit fly generation, from egg to adult, takes only nine days. The manipulation of these flies by laboratory experimentation has produced a wide variation of body features. By bombarding them with high doses of radiation, scientists have attempted to demonstrate evidence for evolution through the driving force of mutational change. The outcome of these experiments has resulted in a variation of eye color, wing size, body length, and so on. However, the physical changes observed in these small insects never adds up to any significant change, as Darwin's theory requires. This is because mutant flies have never had new information added to their DNA, and because of this, they always remain fruit flies. After experimenting with these flies for nearly half a century, famous Cal-Berkley geneticist, Dr. Richard Goldschmidt, virtually admitted a complete failure to demonstrate the evolutionary process when he finally threw his hands up and exclaimed: "Even if you could accumulate a thousand mutations in a single fruit fly, it would still be nothing but an extremely odd fruit fly."[15] He expressed that "it is good to keep in mind that nobody has ever succeeded in producing even one new species by the accumulation of micro-mutations. Darwin's theory of natural selection has never had any proof, yet it has been universally accepted."[16] After the laborious study of many thousands of radiated insect generations, consider the conclusion of environmental activist and author Jeremy Rifkin: "All in all, scientists have been able to catalyze the fruit fly evolutionary process such that what has been seen to occur in 'Drosophila' (fruit fly) is the equivalent of many millions of years of normal mutations and evolution ... Even with this tremendous speedup of mutations, scientists have never been able to come up with anything other than another fruit fly."[17]

After studying countless generations of bacteria and flies, the common conclusion that evolution is not happening today strongly suggests that it has never happened in the past. According to well-known microbiologist, Dr. Gordon Taylor, "In all the thousands of fly-breeding experiments carried out all over the world for more than fifty years, a distinct new species has never been seen to emerge ... or even a new enzyme."[18] If no evolutionary change has ever appeared in *Drosophila*, after so many years of forced manipulation, we are left with powerful evidence that evolutionary change in organic species has never occurred anywhere at all. Other research experimentation involving E coli bacteria has produced the same result. After observing thirty thousand generations of E coli reproduction in the laboratory (equivalent to one million human years in evolutionary terms), these experiments have also produced no significant innovative change that could offer any support for the evolutionary hypothesis.[19]

If sporadic mutations drive evolutionary change, they do not appear to be taking this fashionable notion anywhere to speak of, and, in fact, are clearly working against the theory. Harvard's evolutionary biologist Dr. Stephen J. Gould explains: "The essence of Darwinism lies in a single phrase: natural selection is the creative force of evolutionary change. No one denies that natural selection will play a negative role in eliminating the unfit. Darwinian theories require that it create the fit as well."[20] Logically speaking, if natural selection favors stronger organisms, mutations cannot possibly work to advance positive evolutionary change. Mutated structures are typically not beneficial, and seldom, if ever, add value to the overall fitness of the organisms from which they originate. Consequently, natural selection will simply wipe out unhealthy organisms, and they will be lost from the population. Evolutionary geneticist N. Heribert-Nilsson explains the dilemma that scientists face when they depend on mutations to explain evolutionary transition: "If one allows the unquestionably largest experimenter to speak, namely nature, one gets a clear and incontrovertible answer to the question about the significance of mutations for the formation of species and evolution. They disappear under the competitive conditions of natural selection, as soap bubbles burst in the breeze."[21] Evolutionists are beginning to admit this poses a problem. Stephen J. Gould laments the dilemma: "What good is half a jaw or half a wing?" Gould has been critical of various artists employed by museums and textbook publishers who present publicly what amounts to hypothetical guesswork in their clever charts and fancy drawings creating a delusion of gradual change in organic species over time. These illustrations, highlighted with evolutionary comments, are steeped in imagination pointing toward biased speculation. Dr. Gould responds with this scathing rebuke: "These tales, in the 'just-so-stories' tradition of evolutionary natural history, do not prove anything; concepts salvaged only by facial speculation, do not appeal to me."[22] How disturbing that our children are consistently exposed to biased conjecture when it comes to their fragile understanding of how life began. And to make matters worse, this bizarre speculation is constantly presented to them as factual.

The Fossil Record Speaks

When considering the fossil record, it is important to understand exactly what we are dealing with. Fossils, by definition, include "the remains, or visible trace, of a plant or animal organism from the past that have been, to some degree, preserved in the crust of the earth." The fossilization of an organic specimen will occur only under ideal conditions, and apparently those conditions are scarcely seen on the earth today. Fossil formation requires rapid death and sudden burial through the action of water-laid sediments. This process occurs when ingredients of lime, sand, water, pressure, and fossilizing chemicals are present in sediment deposits. This would be the conditions resulting from severe flood devastation. "To become fossilized, a plant or animal must usually have hard parts, such as bone, shell or wood. It must be buried quickly to prevent decay, and must be undisturbed throughout a long process."[23] The entombing of plant and animal remains must occur rapidly—otherwise, agents of decay, such as rotting by scavenger, bacteria, and moisture will quickly destroy organic remnants. After being buried, however, each specimen must lie completely undisturbed for an extended period, allowing the fossilization and rock-forming action to occur. Fossil formation would typically require an inescapable cataclysmic event in order to rapidly bury and preserve organic remains. If the present is a window to the past, as naturalists claim, how can isolated local floods throughout history explain the massive worldwide deposition of fossil-bearing sediments we observe today.

Some unusual fossil deposits have been reported that contain the remains of jellyfish and other soft-tissue organisms. This unexpected phenomenon can only be explained by an extremely rapid process under catastrophic conditions. Evolutionists would have us believe that isolated local flooding in history could account for the vast museum of natural history we find preserved in the earth's crust. However, judging from our current data, local flooding cannot account for the massive fossil graveyards we have uncovered from rock formations the world over. If the fossil record supports the gradualist claim that the simplest life-forms were buried first, moving toward complex life forms near the surface, then we have a vital question to ask. Why then is the fossil record so far out of its predicted order, according to naturalist claims, almost without exception, throughout the world? Many today are hard-pressed to explain why millions of years of anticipated geologic column are completely missing, in many locations, the world over.

According to geology expert and science educator John Woodmorappe, "80 to 85 percent of the earth's land surface does not have even three geological periods appearing in 'correct consecutive order; ... it becomes an overall exercise of gargantuan special pleading and imagination for the evolutionary uniformitarian paradigm to maintain that there ever were geologic periods."[24] At one particular site in England, evolutionary geologists proudly pointed out the correct order of organic fossil deposition as predicted by the evolutionary forecast. However, workers subsequently discovered, during a rock quarrying operation, that a large petrified tree was embedded across this geologic deposit at a forty-degree angle. If it had taken a billion years to deposit the fossil-bearing strata at this location, as evolutionists have claimed, then how could an intact tree possibly be preserved while lying across the entire bed of fossil deposits without any sign of decay? This fascinating discovery, along with similar evidence involving coal deposits containing embedded trees, would suggest that the various geological formations of fossil-bearing sediments were, most likely, all deposited within a short period of time. In fact, the worldwide museum of fossil history likely resulted from a single cataclysmic event of epic proportion— an event like the universal flood of the well-known Genesis account.

Unfortunately for evolution, the highly regarded fossil record has proven to be a disappointment indeed. Consider this admission of Dr. David Raup, an evolutionary paleontologist from the University of Chicago. He conceded that: "Most people assume that fossils provide a very important part of the general argument made in favor of Darwinian interpretation of the history of life. Unfortunately, this is not strictly true."[25] In fact, historical data produced by fossil remains is far more consistent with the Genesis record of creation and a universal flood. The fossil record, which at one time was believed to be the most powerful evidence for Darwin's evolutionary transition, has failed to live up to the anticipated claim of gradualism. Consider the words of American biochemist, Dr. Duane T. Gish, who said: "It is our contention that the fossil record is much more in accord with the predictions based on creation rather than those based on the theory of evolution, and actually strongly contradicts evolution theory."[26] He continues: "The fossil record, rather than being a record of transformation, is a record of mass destruction, death, and burial by water and its contained sediments."[27] According to earth scientist Henry M. Morris, typical evolutionary claims are not supported by evidence from the fossil record. Dr. Morris describes the problem this presents for evolution theory. "The very existence of fossils in any significant size and number seems to require rapid processes of sedimentary deposition, burial, compaction, and lithification. Otherwise, normal decay processes would soon destroy and dissipate such organic remains. Furthermore, the fossil record does not show a continuous evolutionary progression at all as the theory requires. The same great gaps between the major kinds of plants and animals that exist in the present world are also found in the fossil world."[28]

The distinctive gaps that are found in the fossil record have also been acknowledged by those in the evolution camp. Consider this statement from evolutionary paleontologist Dr. David B. Kitts, former head curator of the Stoval Museum. "Despite the bright promise that paleontology provides a means of seeing' evolution, it has presented some nasty difficulties for evolutionists, the most notorious of which is the presence of 'gaps' in the fossil record. Evolution requires intermediate forms between species and paleontology does not provide them."[29]

According to Harvard's Dr. Stephen J. Gould, "The fossil record caused Darwin more grief than joy. Nothing distressed him more than the 'Cambrian explosion', the coincident appearance of almost all complex organic designs."[30] The fossil record confirms the wide deposition of a vast array of complex organic designs all buried together at about the same time in history. Scientific data from fossil remains has been cataloged over the last two hundred years. Dr. Gould reports that the earth history of most fossil species, uncovered to date, presents two basic characteristics that are readily conceded and are incompatible with the evolutionary hypothesis of gradualism.

1. "Stasis: Most species exhibit no directional change during their tenure on the earth. They appear in the fossil record looking pretty much the same as when they disappear; morphological change is usually limited and directionless."

2. "Sudden Appearance: In any local area, a species does not arise gradually by the steady transformation of its ancestors; it appears all at once and 'fully formed.'"[31]

These claims by the leading evolutionary voice of the 20th century are devastating to the dogmatic assertion of gradualism. The fossil record confirms that species have appeared suddenly on earth, are fully formed at their appearing, and have not arrived on the scene by some gradual transmutation of supposed ancestors. Evolutionary natural history teaches that mankind and apes descended from a common ancestor between thirty and seventy million years ago. To deliver this mandatory education, textbook publishers have included a variety of charts and suggestive drawings depicting imaginary transitional primates. A great deal of biased guesswork goes into the depictions of these primate figures commonly presented to unwary students in public school classrooms. These drawings are based on embarrassingly few fossil fragments, and represent pure conjecture influenced by evolutionary bias. These evolutionists are manipulating the occasional discovery of organic remains to support their preconceived notion about the origin of man. Fraudulent pictures and bogus claims appearing in middle-school and high-school textbooks are used to indoctrinate naive school children. It appears that unscrupulous investigators, along with thoroughly indoctrinated public school educators will present what they believe to be true whether the evidence confirms it or not.

The Coelacanth has long been touted by evolution science as a genuine transitional link; but this too has become a great disappointment. In the 20th century, this primitive looking fish was claimed by naturalists as a transitional link to land animals; they even reported that it became extinct before the dinosaurs appeared. This stout-looking fish supposedly first appeared in the Devonian period some 350 million years ago, before it evolved into some four-footed land animal. Imagine the total embarrassment of diehard evolutionists when this fossilized marine specimen, believed long extinct, was discovered alive off the coast of Africa in 1938, when it was netted by fishermen. In 1952, a second specimen was captured, and since then underwater researchers investigating the depths of the Western Indian Ocean have successfully filmed more of the same. I had the opportunity to observe this for myself via a Fox News broadcast that was aired on November 17, 2009, as they reported the latest discovery with live video footage of researchers filming the living Coelacanth off the coast of Indonesia. Imagine that—an extinct transitional form that supposedly had died out over two-hundred million years ago, being filmed alive with the exact same features that appear in the fossil record! These living Coelacanth specimens are a great example of a living fossil, but there are other examples (both plant and animal) that you will discover by simply searching under "living fossils."

Contrary to one-hundred-fifty years of evolutionary indoctrination, the basic kinds of plants and animals are distinctly represented in the fossil record. They are found to be completely formed and fully developed with distinct and undeniable gaps separating each kind from all the others. When we begin digging for reliable fossil evidence from the past, what we find, in the words of creation scientist Gary Parker, are "billions of dead things buried in rock layers laid down by water all over the earth."[32] This succinct phrase speaks very clearly to us of sudden death, destruction, and massive devastation on a worldwide scale; it certainly does not speak well for a popular theory that projects, as its core premise, a continuous and very gradual improvement in living organisms that are somehow randomly advancing toward greater fitness and complexity over vast, unconfirmed, and unbelievably prolonged periods of time.

Did you know that fossils are used to assign rocks to a geological age? Index fossils are typical, abundant, and representative of a certain geological period, as is laid out by the evolutionary model. If it can be shown that fossils are not nearly as old as they are claimed to be, the entire geologic-age timescale becomes useless speculation only necessary to prop up the rapidly declining philosophical view of "naturalism." Therefore, because fossils are used to assign geological ages, this dating system, which has become so deeply engrained in the public mind-set is known to be based entirely on the unproven assumption of evolution!

Creationists and evolutionists can both agree that the ideal conditions for the deposition of fossils in sedimentary rock layers are initiated by major flood and volcanic devastation. Such was the case in the explosive eruptions of Mount St. Helens beginning on May 18, 1980, in Washington State when this angry volcano blew off 1,300 feet of her upper north side, and left behind utter desolation and ruin over a wide remote area, while also claiming every life that was in her path. The initial explosion sent hot ash eighty thousand feet into the atmosphere. Some ash, we are told, circled the earth twice before coming to rest. Living with my young family in Western Washington at the time, I personally witnessed this devastating eruption from a vantage point some fifty miles away. A second major eruption occurred in June of 1982, triggering a massive mudflow, with glacier run-off, that spilled over the confines of Spirit Lake at the base of the mountain, carving out a 1/40-scale version of the Grand Canyon in just one afternoon. Imagine a geological formation, resulting from flood devastation, that would appear to have taken millions of years to form by standard geological interpretation; yet, this dramatic transformation was etched into the Western Washington landscape in only a few hours! If this rapid landscape transformation occurred in one afternoon, why should anyone believe that it took billions of years to form the Grand Canyon or any other remarkable geological feature?

How scientific is a theory whose main evidence is the assumption that it is true? Since there are no dependable ways to date geological formations, rocks are generally dated by the fossils they contain. This is why the evolution advocates have constructed the geological-period timescale. This earth-history dating system is based on a theory that is void of definitive and substantial physical evidence. The absence of proof leaves the geologic-age dating system without merit.

Evolution lives or dies by the fossil record, because of the belief that plant and animal remains became gradually entombed in rock-forming sediments over millions, if not billions of years. If true, fossils should demonstrate a gradual transformation of species from simple to complex without any gaps separating them. Paleontologist of the American Museum of Natural History, Niles Eldredge stated: "[We modern] ... paleontologists ever since Darwin have been searching ... (largely in vain) ... for the sequences of an insensibly graded series of fossils that would stand as examples of the sort of wholesale transformation of species that Darwin envisioned as the natural product of the evolutionary process. Few saw any reason to demur--though it is a startling fact that most species remain recognizably themselves, virtually unchanged throughout their occurrence in geological sentiments of various ages."[33] Dr. P. L. Forey added: "Many fossils have been collected since 1859, tons of them, yet the impact they have had on our understanding of the relationships between living organisms is barely perceptible ...In fact, I do not think it unfair to say that fossils, or at least the traditional interpretation of fossils, have clouded rather than clarified our attempts to reconstruct phylogeny."[34]

When we speak about fossils, people automatically think of preserved organic remains in the earth's crust from multiple millions of years ago. We have become conditioned, as the result of biased evolutionary indoctrination, to think of fossils, and especially dinosaur fossils, in terms of vast ages of earth history. It is important to note that dinosaurs have become a favorite topic of evolutionists because they generate fascination in the minds of young people. Evolutionary indoctrination convinces people that dinosaurs lived sixty to a hundred million years ago. Many young people have concluded that this disproves the Genesis account of earth history and, therefore, the Bible cannot be trusted. After all, how could the dinosaurs fit into Noah's ark?

Are dinosaurs really that old? Paleontologist, Dr. Mary Sweitzer, is currently on staff at the campus of North Carolina State University. She is known for leading a research team that discovered the remnants of blood cells and later identified other soft tissue in the fossil remains of a Tyrannosaurus rex, which was uncovered in the state of Montana. Sweitzer was the first researcher to identify and isolate what appeared to be dried red blood cells and other soft tissue remains in the fossilized bone of a dinosaur specimen. These startling tissue samples were isolated and classified from a T-rex leg bone dated at sixty-eight million years old. What is so remarkable about this discovery is the prior established consensus in science that DNA cannot remain intact much beyond ten thousand years. After publishing her discovery in 1993, a huge storm of controversy ensued as evolutionary biologists scrambled from all directions to dispute the claim and to protect a geological time scale and fossil-dating methodology that has come under greater attack than ever before. The ramifications of this discovery cannot be silenced, however, and the logical conclusion is to assert that these bones, and perhaps many others like them, are much younger than evolutionary scientists have claimed. Can you imagine a dinosaur believed to be this ancient having leg bones that are not even completely fossilized, containing red blood cells and other soft tissues? This discovery was particularly remarkable considering the sediment layer where the T-rex was found contained copious amounts of water that would have been gradually absorbed into the porous bone matrix of the dinosaur and, over time, would have thoroughly destroyed and mummified all existing soft tissue remains.

Researchers have uncovered interesting specimens at other locations that also cast suspicion on the accuracy of fossil-dating methods currently used. The Clarkia fossil beds of Idaho have produced impressive fossil leaves in exquisite condition. Some of the rock samples at this site have contained partially fresh leaves. In fact, some samples were not even fossilized at all and were found to retain their natural color. For example, a green magnolia leaf was discovered in these popular fossil-bearing formations that was dated at twenty million years old according to evolutionary assumption; this leaf, however, retained its green color, was fresh looking, and even pliable. Scientists were able to successfully extract DNA from this specimen, which has matched the DNA of modern magnolia trees. Scientists are found scratching their heads in amazement as to how this DNA could have survived intact for so long. Dr. Brian C. Sykes, an Oxford University genetics professor, reported in the journal *Nature* that the rate at which DNA breaks down in the laboratory would suggest that after ten thousand years, no intact DNA should be found.[35] This would suggest that a magnolia leaf specimen, found in fossil-bearing strata, could not possibly be dated at twenty million years old. Therefore, if the vast geological timescale dating method is found to be grossly inaccurate, the entire bizarre evolutionary scenario of naturalistic gradualism will ultimately collapse on itself like a house of cards.

Why do we have vast fossil deposits covering the entire earth, including shellfish deposits uncovered at high elevations on mountain ranges throughout the world? As a young boy of ten or twelve years, I can remember going on a field trip to the Willapa hills outside of my childhood hometown of Raymond, Washington. My friends and I spent the afternoon extracting clamshell fossils from the side of a mountain. How can we explain this in terms of local flood activity? Atheist, Carl Sagan had proclaimed that "Evolution is a fact, amply demonstrated by the fossil record."[36] Sagan's bold, although terribly inaccurate statement, is totally without credibility when held up to scrutiny against a well-researched fossil record that denies evolutionary gradualism.

Consider this remark from Dr. Steven M. Stanley, a leading paleontologist at Johns Hopkins in the 1970s. "The known fossil record fails to document a single example of phyletic evolution accomplishing a major morphological transition."[37] Professor David Woodruff from UC San Diego offers a similar conclusion. "But fossil species remain unchanged throughout most of their history and the record fails to contain a single example of significant transition."[38] Oxford zoologist Mark Ridley confirmed this admission. "In any case, no real evolutionists, whether gradualist or punctuationist, uses the fossil record as evidence in favor of the theory of evolution as opposed to special creation."[39] Since the fossil record does not provide evidence to support evolution, we can be certain this admission leaves the theory in the poor-house, as the voice of the fossil record speaks against Darwinism. Without this convincing support, the theory cannot even identify itself in the scientific method; therefore, evolution must be ultimately relegated to the realm of faith. The formidable evidence of vastly cataloged earth history has proven a great disappointment to mainstream naturalists, and, in fact, revealed data aligns consistently with the Genesis record of creation, and a worldwide flood occurring early on in human history.

Dr. Gary Parker, with expertise in biology and paleontology, began his career as a devoted atheist with great enthusiasm, believing he was ridding his students of outdated prescientific superstitions, such as Christianity. He was eventually challenged to examine the evidence, without preconceived bias, by a chemistry professor at the university. He courteously accepted the challenge to vindicate his worldview, and spent the next three years pouring over the evidence with an open-mind. As a result of this thorough investigation, he was compelled to renounce his deep-seated belief in evolution. He finally concluded, as have others in recent years, that the biblical framework offers the world a much more logical explanation for the existence of life, based on scientific discovery and observation. He has offered the following statements to summarize the many problems one encounters when attempting to use the current fossil record as evidence to defend evolution. "In most people's minds, fossils and evolution go hand in hand ... In reality, fossils are a great embarrassment to evolutionary theory and offer strong support for the concept of creation ... If evolution were true, we should find literally millions of fossils that show how one kind of life slowly and gradually changed into another kind of life ... But missing links are the trade secret, in a sense, of paleontology ... The point is, the links are still missing. What we really find are gaps that sharpen up the boundaries between kinds. It's those gaps, which provide us with the evidence of creation of separate kinds ... As a matter of fact; there are gaps between each of the major kinds of plants and animals. Transition forms are missing by the millions. What we do find are separate and complex kinds, pointing to creation."[40] This anomaly alone has proven to be a devastating blow to an old, worn-out theory that has now reached the very brink of extinction. Evolution? The fossils say no!

Chapter Three

* * *

Case of the Missing Links

Where Are the Transformers?

In recent years Hollywood producers have glorified the imagined story of mechanized vehicles that are able to magically transform themselves into robotic-looking, behemoth-like creatures, which commence the drama of fierce fighting in an epic battle of good versus evil. Survival of these emotionally driven mechanical creatures is linked to their ability to generate the defense mechanisms and survival features that will make them superior to their formidable rivals of like kind. Though this popular story line has made for some very interesting entertainment, the producers, of course, are dabbling in pure fantasy. When we search for change in the organic world leading to real-life transitional intermediates linking one species to another, we come up empty. Transforming links achieved over millions of years of mutation and natural selection simply are not to be found. Imagined transitions leading to the appearance of new and improved species over millions of years are as rare and fantastic as those found in the imagination of Hollywood writers and producers today. The current fossil record provides no clear examples of macroevolution, which may seem surprising to people who are accustomed to a strong public perception that the fossil record has provided the most powerful evidence for evolutionary change. In 1859, Darwin was unable to cite a single example of evolutionary transition in the fossil record; and today, more than 150 years later, with record activity in paleontology, the complete absence of bona fide 'missing links' still holds true. So where are the countless transformers of the organic world that are required to validate the long-standing tradition of evolutionary gradualism? As a coverup, contemporary paleontologists have proudly projected the public persona of a fossil record that supports Darwinism, while privately admitting that reliable evidence for a smooth transition is basically nonexistent. These people do a great disservice by disseminating publicly a view that their work does not support privately. Professor John Lennox of Oxford asks the obvious question: "What conceivable reason could there be for members of an academic community to suppress what they know to be the truth unless it were something which supported a worldview, which they had already decided was unacceptable?"[1]

Evolution theory teaches that microorganisms, over several billion years, evolved into complex invertebrate life-forms such as clams, sponges, and jellyfish. These invertebrates, in turn, evolved into fish, fish into amphibians, amphibians into reptiles, reptiles into birds, and so on until apes appeared, which eventually evolved into man. These claims certainly seem absurd considering the fossil record has not produced a single intermediate link connecting any of these uniquely different and incredibly complex life-forms. In fact, the evidence strongly supports the biblical view of distinct created plant and animal kinds with noticeably wide, insurmountable gaps between them. If the concept of evolution were true, what would we expect to observe in the plant and animal kingdoms today?

If all living organisms have appeared completely by random chance, through natural processes alone, we could predict a vast continuum of related organisms living together without any sharp distinguishing gaps separating them. With evolution theory, evidence for predictable transitional intermediates between various kinds of organisms should be obvious and abundant in both the fossil world and the world of living organisms today. If the present is a window to the past, what we observe today should be representative of organisms as they appeared in the ancient world. However, what we observe today is a total absence of 'transitional intermediates' of any kind. The missing links are still missing! What is undeniably true about planet earth both presently and in the past is the existence of distinct kinds of complex life-forms without any transitional links connecting them. Science journalist, Mr. Jerry Adler explains: "In the fossil record, missing links are the rule: the story of life is as disjointed as a silent newsreel, in which species succeed one another as abruptly as Balkan prime ministers. The more scientists have searched for the transitional forms that lie between species, the more they have been frustrated ... Evidence from fossils now points overwhelmingly away from the classical Darwinism, which most Americans learned in high school: [that new species evolve out of existing ones by the gradual accumulation of small changes... each of which helps the organism survive and compete in the environment]."[2] The undeniable and universal absence of transitional forms in the organic world is perhaps the most devastating indictment ever spoken against evolutionary thought. If we saw Darwinism as true science, doesn't it seem odd that 150 years of historical research has not produced even one transitional link between plant and animal kinds that can be fully verified? Paleontologists admit to this glaring void, and yet they persist in explaining their way around this disturbing flaw in their cherished theory. This complete absence of documented evolutionary transition remains the most monumental source of confusion and disappointment for modern evolutionists today. Contemporary naturalists simply assume these predicted intermediate links have existed, despite the overwhelming evidence to the contrary.

In November of 1980, Newsweek magazine reported on a scientific conference at the Chicago Field Museum of Natural History entitled "Macro Evolution." At this academic conference, many disillusioned paleontologists reported on their current findings. They reluctantly revealed that the fossil record does not and perhaps never will support the Darwinian scenario of a smooth and continuous progression of advancing life-forms from simple to complex. They were essentially forced to concede that cataloged fossil deposits have shown a pervasive pattern of undeniable gaps between plant and animal kinds with life-forms appearing suddenly, and no transitional links proceeding them. These creatures have simply appeared abruptly in the fossil record, followed by long periods of stability, during which they show very little or no change at all.[3]

The absence of animal transition is obvious, but consider the remarks of Chester A. Arnold of the University of Michigan. As a paleobotanist referencing the Darwinian progression of plant life, he records a profound statement: "It has long been hoped that extinct plants will ultimately reveal some of the stages through which existing groups have passed during the course of their development, but it must be freely admitted that this aspiration has been fulfilled to a very slight extent, even though paleo-botanical research has been in progress for more than a hundred years. As yet, we have not been able to trace the phylogenetic history of a single group of modern plants from its beginning to the present." [4] Cambridge professor, Dr. Eldred Corner concurs. "I still think that, to the unprejudiced, the fossil record of plants is in favor of special creation ... Can you imagine how an orchid, a duckweed and a palm tree have come from the same ancestry, and have we any evidence for this assumption? ... The evolutionist must be prepared with an answer, but I think that most would break down before an inquisition."[5]

Darwin hoped that one day the fossil record would reveal the numerous intermediate forms predicted by his theory. However, with scientific research making such terrific gains over the last century, Darwin's initial claims appear to be more uncertain than ever due to the failed effort of paleontology to fill the wide gaps between species in the past—the same wide gaps that we observe in the present. The truth is crystal clear; the fossil record reveals only distinct kinds of plants and animals with no evidence of transitional forms leading to them; the detail of their anatomy is virtually indistinguishable from modern life-forms of the same species observed in the present world. Paleontologist and evolutionary biologist Dr. Steven Stanley explains: "Species that were once thought to have turned into others have been found to overlap in time with these alleged descendants ... In fact, the fossil record does not convincingly document a single transition from one species to another."[6] Evolutionary biologist, Dr. Douglas Futuyma of the University of Michigan verifies the accuracy of this claim: "The majority of major groups appear suddenly in the rocks, with virtually no evidence of transition from their ancestors."[7] Gradual change moving from simple to complex organisms, with abundant links connecting species at every stage of evolutionary transition, is wishful thinking on the part of naturalism.

Dr. Colin Patterson, senior paleontologist at the British Museum of Natural History, offered the following response to a direct criticism he received for the noticeable lack of transitional forms illustrated in his book on evolution. He wrote: "I fully agree with your comments on the lack of direct illustration of evolutionary transitions in my book ... If I knew of any, fossil or living, I would certainly have included them. You suggest that an artist should be used to visualize such transformations, but where would he get the information from? I could not, honestly, provide it ... and if I were to leave it to artistic license, would that not mislead the reader? Yet Gould and the American Museum people are hard to contradict when they say there are no transitional fossils. As a paleontologist myself, I am much occupied with the philosophical problems of identifying ancestral forms in the fossil record. You say that 'I should at least show a photo of the fossil from which each type of organism was derived.' I will lay it on the line; there is not one such fossil for which one could make a watertight argument ... The reason is that statements about ancestry and descent are not applicable to the fossil record."[8] What an astonishing admission, considering it came from a world-class evolutionary scientist of the twentieth century! Even if transitional links had been discovered in the fossil record, how could we possibly believe that an intermediate form between two different species would be more fit to survive than the parent organism it evolved from? If a good leg was to slowly evolve into a wing, it would be a bad leg long before it became a good wing. Such an awkward transitional feature would make the poor critter less fit to survive and become an easy target for predation. Any such creature would be quickly wiped out by 'survival of the fittest' adding nothing to the advancement of evolution.

Imagine a reptile that is evolving into a bird. Picture elongated scales on the forelimb of a lizard destined to become transitional feathers? Would this not impose a significant disadvantage for an animal that cannot yet take flight, neither can he run well having now been burdened with an awkward transitional appendage. This imaginary trans-mutating intermediate form would have significantly impaired mobility, and would be greatly disadvantaged in competing for necessary resources and eluding savage predators. As such easy prey, his kind would be routinely wiped out long before any subsequent transitional mutation could be randomly achieved that would make him adequately fit to compete and survive in a hostile, unforgiving world.

Evolution theory is grossly deficient! More than 150 years of rigorous, painstaking research, discovery, and observation has brought the world no closer to validating Darwin's claim.

And instead, a vast accumulation of documented evidence has strongly discredited his strange notion. Late German zoologist, Dr. Albert Fleischmann, explains the problem evolution theory is still facing today: "The theory suffers from grave defects, which are becoming more and more apparent as time advances. It can no longer square with practical scientific knowledge, nor, does it suffice for our theoretical grasp of the facts; no one can demonstrate that the limits of species have ever been passed. These are the Rubicon's, which evolutionists cannot cross ... Darwin ransacked other spheres of practical research work for ideas, but his whole resulting scheme remains, to this day, foreign to scientifically established zoology, since actual changes of species by such means are still unknown." [9] Imagine a century and a half of evidence, which has failed to produce a single undisputed transitional form linking one living kind to another. This fact alone is a devastating indictment against a strange notion that has, unfortunately, received worldwide acclaim. America's leading twentieth-century proponent of evolution, the late Dr. Stephen J. Gould, confessed that: "The absence of fossil evidence for intermediary stages between major transitions and organic design, indeed, our inability, even in our imagination, to construct functional intermediates, in many cases, has been a persistent and nagging problem for gradualistic accounts of evolution."[10] Since gradual change is required of evolution, and can only be found in the imagination of devoted loyalists, proponents will have to look outside the fossil record to find the convincing scientific validation they so desperately seek.

The Descent of Man

It has been over a century and a half since Darwin's classic work, *On the Origin of Species*, was first published. Over that span of time, a well-documented fossil record overwhelmingly denies classical Darwinism; yet this bizarre notion has been taught as factual to most of us Americans through formal public education and mass media bias. The historical footprint of modern man is one area where evolutionists have painstakingly labored to fill in the blanks. I suppose that if they were able to accurately connect the dots, and to paint a clear picture of the emergence of modern man from his 'evolutionary' past, all their other claims about the evolutionary history of planet earth would seem that much more plausible. However, nowhere have we seen greater abuse of power, and more blatant deception, than in the extreme calculated effort of so-called evolutionary science to establish a credible link between modern man and his supposed ape-like ancestors. Based on embarrassingly few fossil fragments discovered by various research paleontology teams at locations around the globe, evolutionists have proclaimed to the world, through a biased scientific community, the discovery of supposed 'intermediate transitions' between man and his so-called primitive primate relatives. Using a few jawbones, teeth, and other fossil remains, evolution-minded researchers have directed artists to construct various models of life-sized primate figures, including some complete human-like specimens fashioned based on only a single bone fragment. They have photographed, published, and have proudly displayed these imaginary models claiming support of an unproven theory. The fraudulent outcome of misinterpreted fossil remains has been used to indoctrinate people into believing what essentially amounts to biased speculation. As you will soon discover, the resulting fallout, from this overzealous reporting, has been extremely deceiving to the unsuspecting public.

A great deal of evolutionary guesswork goes into the construction of primate models. In the following material I expose the work of evolution-minded researchers to interpret the occasional discovery of organic remains to fit their preconceived ideology about the origin of man. If a transitional intermediate between man and apes really did exist, this would be compelling evidence favoring the evolution model. However, revealed facts suggest the dogmatic faith evolutionists have placed in their cherished theory makes it necessary to believe that a single jawbone, tooth, or skullcap had belonged to some brutish transitional creature—a creature that allegedly roamed the earth millions of years ago. When fossil remains have been thoroughly investigated by unbiased researchers utilizing the finest scientific methods available, modern science has discovered that many of these ancient bone fragments have told a very deceptive tale. Such fragments have typically belonged to the skeletal remains of some mammal or man, but never to some bizarre creature described as a transmutation between primate and human. When a determined individual wants to believe something is true, professional, or otherwise, it requires very little evidence to confirm their belief. This human flaw has caused some people to uncover, announce, and widely publish findings that would appear to support a Darwinian worldview, even though such reputed evidence has been grossly misinterpreted.

In 1892, the initial published paper proclaiming the supposed transitional humanoid "Java man" (*Pitheconthropus*), came about when Dutch geologist, Dr. Eugene Dubois discovered a human femur in the same general area as the skullcap of an ape. While failing to report that he also found human skulls in the same area, Dubois proceeded to convince the scientific world that he had discovered a genuine missing link between ape and man. This so-called intermediate life-form was hailed for years as a definitive connection between mankind and his supposed primate ancestors. However, before his death, Dr. Dubois confessed that his Java man was nothing more than a large gibbon (ape).[11] In their eagerness to produce the missing link, some biased researchers resort to dishonesty and deception in making erroneous claims. Unfortunately, the retraction of hasty and overzealous reports never quite receive sufficient publicity to erase the initial impression that fraudulent claims have left on the public mind-set.

Another bizarre case involved bone specimens discovered in England, which resulted in the initial designation of "Piltdown man." In 1912, amateur British archaeologist, Charles Dawson came upon a fossil jawbone and skull at a gravel pit near Piltdown, England. Dawson, and the scientists with whom he consulted, were convinced that these fossil specimens represented a legitimate link to man's evolutionary past. Authorities from around the world became fascinated with Piltdown man, believing it to be a genuine intermediate transition in the evolutionary ascent to modern man. Piltdown man received the official name *Eoanthropus dawsoni* and was called a subhuman ancestor of Homo sapiens. However, after many years of public exposure from the scientific and academic community, suddenly in 1953 Piltdown man was buried once again. In that year an independent investigation had proven this supposed intermediate to be a complete fraud! When the jawbone was tested to determine a relative age, it was discovered to be no fossil at all; researchers reported the sample to be only slightly older than the day it was found. Unbiased investigators determined that someone had purposefully manipulated the jawbone of an ape and the skull of a modern human to create a deliberate deception. Further investigation revealed these specimens had been chemically treated to make them appear old, and scratch marks on the teeth disclosed that they had actually been filed down. Apparently, this was done to make the teeth appear to be more humanlike. After planting these bones in a gravel pit, to enact a charade, some trickster (perhaps Charles Dawson) succeeded in creating a colossal deception that left a lasting impression upon the general public for two generations. This finding turned out to be a complete hoax, but somehow succeeded in deceiving many of the world's so-called 'experts', who apparently were eagerly waiting and ready to embrace and report any possible link to man's prehistoric past! Unfortunately, it was over forty years before this fraud was exposed. In the meantime, the widespread reporting of 'Piltdown man' was to appear in public-school textbooks for some fifty years. In addition, over five hundred books and periodicals were published about this discovery. Why did it take so long to uncover the truth and detect this premeditated forgery? And why did no one notice file marks on the teeth when these specimens were initially examined? Why were investigators unable to discover early on that the dark brown stain that coated the bones was only a thin outer layer? Why were they able to detect what was believed to be some human characteristics in an ape's jawbone and also apelike qualities in a human skull? The obvious answer is because they saw in these bone specimens what they were hoping for, and failed to see the actual truth that was unsupportive of their beloved theory. The late biochemist, Dr. Duane T. Gish, has summed up this disgraceful fraud by stating, "The success of this monumental hoax serves to demonstrate that scientists, just like everyone else, are prone to find what they're looking for, whether it is there or not."[12]

In 1921, Dr. Davidson Black discovered two tooth molars near the village of Choukoutien, located close to what is now modern-day Beijing, China. The original report establishing "Peking man" (*Sinanthropus pekinensis*) as a possible link between humans and apes was, strangely enough, based only on these tooth fragments. The Peking story also made a big public impact; and it too appears to be based on careless and over anxious reporting with little regard to genuine scientific protocol. In 1922, another tooth specimen was discovered, this time in Western Nebraska by a man named Harold Cook. The subsequent publication of this find resulted in the designation of an intermediate known as "Nebraska man" (*Hesperopithecus haroldcookil*). This fraudulent report turned out to be very nearly as embarrassing as Piltdown man. Dr. Henry Fairfield Osborn, a science professor at Columbia University, one of America's foremost fossil experts at the time, was among the first to examine this single tooth discovery. Amazingly enough, Dr. Osborn and his colleagues in the field of fossil interpretation, all agreed that this tooth revealed characteristics of an intermediate link between ape and modern man, even though they knew nothing about the primary characteristics of any such intermediate. Once fully presented to the scientific world as real evidence for the evolution of modern humans, Nebraska man was to take his place in a line of descending links believed to be connecting man to primate ancestors. In that same year, the *Illustrated London Newspaper* published the picture of a life-sized hybrid model constructed from this single tooth fragment. Imagine a full-size humanoid intermediate created from the discovery of a single tooth! That is unbelievable! A few years after the find, additional bones were discovered at the original site, and the identity of the mysterious creature was finally revealed. The infamous tooth of the world-famous Nebraska Man had belonged to an extinct pig. What does this tell us about the blatantly overzealous and far-reaching effort of the world's so-called scientific 'experts' of the twentieth century—experts who have painstakingly, although fraudulently, laid additional groundwork for a false and deceptive worldview adopted by millions today?

Around 1930, jawbone fragments and a few teeth were discovered in India. Again, the experts declared the discovery of a transforming intermediate that was well on the way to becoming human. *Ramapithecus*, considered by experts to be an ancient link connecting man to his ape-like relatives, was thought to walk upright like humans. This discovery, like so many others, was received with great enthusiasm in the scientific community, and with time it would become a household word among evolution proponents and the general public. However, after more thorough, unbiased investigation was completed, *Ramapithecus* turned out to be apelike for sure, but was in no way related to humans. The experts were finally forced to admit that this supposed evolutionary link was completely unrelated to man, having subsequently been identified as the primate mammal we know as an "orangutan." Again, desperation ruled the day and once more false and misleading information was presented to the unwary public, adding more fuel to the mounting deception caused by other fraudulent reports. By jumping so quickly to erroneous conclusions these investigators, were guilty of deception by prematurely publishing ambiguous information. They failed to conduct a thorough and adequate investigation well before enthusiastically announcing their research findings to the public at large.

You may remember hearing of an intermediate hominid specimen known as "Cro-Magnon man" (Homo sapiens). The Frenchman, Louis Laret, discovered this skull specimen in 1868. This brings an image to mind of some brutish-looking forerunner to modern man. However, experts now agree that this proposed link to man's evolutionary past was actually fully human, and for all practical purposes identical to today's modern European.

Around 1982, a skullcap was discovered in southern Spain, and declared by Spanish experts to be the oldest fossil link, connected to modern man, ever discovered on the European continent. Initially referred to as "Orce man," experts were again determined to nominate another ancient specimen as a possible link in a long chain of ape-to-man evolutionary transition. However, this Orce man did not amount to much. French experts, upon thoroughly examining the specimen in question, discovered that this skullcap had belonged to a four-month-old donkey! Thoroughly embarrassed Spanish investigators quickly sent out five hundred letters canceling invitations to a major international symposium, scheduled for May of 1984, that would have announced the discovery, to the scientific world, of yet another intermediate link to man's evolutionary past. This was just another bizarre example of how anxious and overly enthusiastic investigators have sought to uncover what they are so desperately hoping for and wanting to find.

In 1974, Dr. Donald Johanson, an American anthropologist, discovered the remains of several ape-like specimens in the Afar Triangle region of Hadar, Ethiopia. Johnson believed this find to be the unveiling of a new hominid species to which he proceeded to assign the official name *Australopithecus afarensis*. What was thought to be the best-preserved fossil skeleton in the group was determined to be a small female, whom he affectionately named "Lucy". Johanson reported a distinctively apelike creature who was thought to walk upright, and was well on the way to becoming fully human. Lucy was subsequently dated at 3.2 million years old, and eager anthropologists the world over hailed this discovery as a definitive intermediate link connecting man to his primitive, prehistoric past. However, this affectionate identification story does not end there. Research on *Australopithecus* remains was conducted by Dr. Solly Zuckerman, a British zoologist and anatomist, who graduated from Yale University and taught extensively at Oxford before becoming a professor of anatomy at Birmingham University in the UK. He, along with a select team of well-qualified scientists, carefully studied the fossil remains of *Australopithecus* for some fifteen years, using the most sophisticated methods of anatomical analysis available at the time. After many years of painstaking research, Lord Zuckerman's team of experts finally concluded that the subjects of their long-term investigation were definitely not transitional between apes and human.[13] Although these 'Australopithecus' remains were found to be somewhat different from modern apes, in the final analysis, unbiased research conducted by a well-qualified team of scientists, determined these creatures were nothing more than another primitive ape species with no confirmed human characteristics. Lord Zuckerman's final report is summarized and documented in a book he subsequently published, entitled *Beyond the Ivory Tower.*[14] According to Dr. Zuckerman, who does not profess to be a creationist per se, there is simply no reliable evidence revealed by the fossil record that adequately supports the strange notion that man evolved from apelike ancestors. African born, British paleoanthropologist, Dr. Richard Leakey, the director of the National Museums of Kenya, offered his professional opinion regarding the 'Lucy' controversy. He wrote, "Echoing the criticism made of his father's *habilis* skulls, he added that Lucy's skull was so incomplete that most of it was 'imagination made of plaster of Paris,' thus making it impossible to draw any firm conclusions about what species she belonged to."[15] Dr. Charles Oxnard, a student of Lord Zuckerman, became a professor of anatomy at the University of Chicago and USC. He studied in detail the post-cranial skeletal remains of Australopithecus. Dr. Oxnard, who employed the most sophisticated techniques in completing his research, determined that Lucy was neither human-like nor ape-like, but more closely resembled an orangutan in terms of physical characteristics. Even though he was not known to be a creationist, his thorough investigation confirmed that these primate creatures did not walk upright like humans, and were not intermediate between apes and man.[16]

"Neanderthal man" (*Homo neanderthalensis*) was first

discovered in Germany in the 1800s and is a familiar name often associated man's evolutionary past. However, this specimen, like so many others, is also a profound misconception. German pathologist and anthropologist, Dr. Rudolph Virchow, who carefully examined Neanderthal remains over a century ago reported that the somewhat primitive-looking features he examined were not the result of Neanderthal's subhuman status, but were, in fact, due to diseased pathology, namely arthritis and rickets. For many years, determined evolutionists stubbornly refused to accept Dr. Virchow's expert finding until other Neanderthal skeletons were examined. This research finally confirmed Dr. Virchow's authoritative conclusion. Today, there is no longer any doubt among unbiased professional researchers that the world-famous 'Neanderthal man' is, and always has been, fully human.

American biochemist, Dr. Duane T. Gish, formerly of the Institute for Creation Research, has summarized the research findings of Neanderthal man. Dr. Gish, an expert on the fossil record, offered this, on Neanderthal status: "It is now known that Neanderthal man was fully erect and in most details was indistinguishable from modern man, his cranial capacity even exceeding that of modern man."[17] It is now believed that the stooped posture observed in this famous case was, as Dr. Virchow had suggested years ago, the result of some crippling disease, probably arthritis, rickets, or both. When you look closely at the fossil record to examine the history of *Homo sapiens*, the evidence of unbiased research findings has revealed the following facts: man has always been human, and apes have always been primate! They are clearly unrelated!

Dr. Timothy White, paleoanthropologist and professor at UC Berkley, has reported another incident involving wild speculation and overzealous reporting that has led evolution to further embarrassment in the scientific community. "A 5-million-year-old piece of bone that was thought to be a collar bone of a human-like creature was actually part of a dolphin rib, ... He (Dr. T. White) puts the incident on a par with two other embarrassing faux pas by fossil hunters: *Hesperopithecus*, the fossil pig's tooth that was cited as evidence of very early man in North America, and *Eoanthropus* or 'Piltdown Man', the jaw of an orangutan, and the skull of a modern human claimed to represent the 'earliest Englishman.' The problem with a lot of anthropologists is that they want so much to find a hominid that any scrap of bone becomes a hominid bone."[18] Listen to the comments of Dr. Greg Kirby, a senior lecturer and teaching professor at Flinders University in Australia. He is suggesting that some field scientists are very tempted to interpret various bone fragments in a unique way that favors what they are hoping to find. "Not being a paleontologist, I don't want to pour too much scorn on paleontologists, but if you were to spend your life picking up bones and finding little fragments of head and little fragments of jaw, there's a very strong desire to exaggerate the importance of these fragments."[19]

The fossil record has revealed distinctive gaps with no conclusive evidence of intermediate or transitional links leading to modern man from apelike ancestors. Many well-publicized, but unconfirmed reports, believed to be connecting modern man with animal primates, have been thoroughly discredited by unbiased scientific investigation. It now appears virtually certain that the hypothetical common ancestor between ape and man is only a figment of the imagination. Again, we hear from Harvard's Dr. Stephen J. Gould: “Most hominid fossils, even though they serve as a basis for endless speculation and elaborate storytelling, are fragments of jaws and scraps of skulls.”[20] He is apparently not impressed by the misleading speculation of those who impose their biased views with endless storytelling, based on misunderstood bone fragments. The missing links in man's descent are imagined creatures conceived in the minds of those who are determined to find any bit of evidence that may help to validate their reliance on evolution as the key to understanding the emergence of life on earth. Dr. Henry M. Morris, who served as the acting director of the Institute for Creation Research, has accurately summed up the current dismal state of evolution theory: “No one has ever seen anything evolve, no one knows how evolution works, the fossil record shows no evolutionary transition taking place, and the basic laws of science show it to be impossible; yet evolutionists insist that this is 'science' and should be taught as proven fact to school children.”[21]

Recapitulation Theory

For many years, science textbooks have taught that the early stages of human embryological development support the theory of evolution. They point out that the human embryo has gill slits like a fish, thus proving common ancestry with aquatic life-forms from hundreds of millions of years ago. Innocent school children have been taught that the growing human embryo goes through stages reflecting evolutionary transition from fish to reptile to ape and finally to man. This supposed evidence for evolution was popular during Darwin's time, and broadcast the world over as indisputable proof for an evolutionary pathway to human life. This false and misleading information has been taught in the schools of developing nations for well over one hundred years and still appears in many textbooks yet today. About the time Darwin's theory was taking hold in England, a German scientist named Ernst Haeckel was working on, and would eventually publish, some biology drawings that he had designed. These controversial drawings supposedly illustrated the various stages through which a human embryo passes before becoming distinctly human. Haeckel's drawings received worldwide attention, were published extensively, and had introduced a concept referred to as the "biogenetic law." Many believed that Haeckel had provided the proof for evolution that the scientific world of his day was eagerly anticipating, and ready to embrace. He used the slogan "ontogeny recapitulates phylogeny," which simply means a developing human embryo passes through stages reflecting major steps in evolutionary transition, moving from aquatic toward human life, suggesting common ancestry with lower life-forms. As it turns out, the human embryo does not have these imagined gill slits. This was a clever misrepresentation based on a serious deficit of knowledge at that time! Remembering back to my embryology class in college, I seem to recall a series of bars and grooves running along the neck region of the developing human embryo, which early evolutionists conveniently mistook for gill slits. These simple structures are called "pharyngeal pouches." During gestation, these formations will eventually develop into the lower jaw, the middle ear, and certain glands over time. In 1875, it was reported that Ernst Haeckel had been convicted of fraud by a group of his peers! He intentionally manipulated his biological drawings to make them appear to resemble the evolutionary stages of human development he so strongly believed in. Even though a nineteenth-century German biologist was rightfully discredited for his deceitful drawings, this publicly circulated fraud left a powerful impression that became deeply ingrained in the mind-set of the developing Western nations. The lasting, but terribly deceptive impression that was left by Ernst Haeckel's controversial charts and drawings has, only in recent years, been exposed publicly for what it really is: a fraudulent misrepresentation of the truth. Today, modern embryologists no longer believe in a recapitulation theory. In fact, some years ago, anthropologist, Dr. Ashley Montague, who was considered one of the world's foremost evolutionists, having taught and lectured often at Harvard, Princeton, and Rutgers—publicly admitted that Haeckel's theory regarding the early development of the human embryo was thoroughly discredited, and should certainly no longer be cited to promote the evolution model. Unfortunately, these misleading and fraudulent illustrations remained in many of our public-school textbooks long after Haeckel's deception was exposed within the scientific community. Sadly, these drawings were used extensively in the Western world during the most influential period of worldwide evolutionary propaganda—throughout the twentieth century.

What Are Vestigial Organs?

Regarding human anatomy and physiology, there is another topic I would like to call to your attention where evolutionists are guilty of reporting false and misleading information based on presumption and a lack of knowledge. For many years, evolutionists claimed that the presence of so-called 'vestigial organs' in the human body proved modern man evolved from lower life-forms. They define "vestigial organs" as the useless vestiges of various organs that were necessary at some previous stage in man's evolutionary history. Evolutionists claim that certain structures and appendages are unnecessary in the human body, or may have a somewhat diminished purpose today when compared to their vital function at some previous stage of evolution. These structures were believed to be remnants that may have become totally obsolete to modern man, but were essential at some time during his evolutionary past.

In 1893, German anatomist, Robert Wiedersheim, had published a list of 86 vestigial structures in man.[22] Through an expanded interpretation, the original list was eventually enlarged to 180. Wiedersheim had picked up on the concept of "rudimentary" organs mentioned by Darwin in his second book, *The Descent of Man*. The zoology expert, Horatio Newman, formally testified, in a written statement that was read into evidence in the famous 1925 Scopes trial, explaining that "there are, according to Wiedersheim, no less than 180 vestigial structures in the human body, sufficient to make of a man a veritable museum of antiquities."[23] Thanks to modern anatomical study and research into human physiology, that list has now been reduced to virtually zero. Many of the structures on the list are now known to be essential to human health, while some others support and enhance the body's function in various ways. If you were to come across some object that was mysteriously unknown to you and, therefore, decided to pronounce it of little or no value, this may simply mean that you are ignorant of its beneficial utility.

It should be obvious to us that "the list" of 1893 represented the physiological deficit-of-knowledge and ignorance of the day. To validate evolution, scientists of the nineteenth century were simply guessing at what they did not understand and, in the process, laid a public foundation for the monstrous deception of evolution—a deception that today has thoroughly permeated Western culture and greatly influences the worldview decision-making of a misinformed public.

It seems hard to imagine, but among the essential organs and structures that were once claimed as vestigial, by overzealous evolutionists of the past, are the pituitary gland, thymus gland, pineal gland, tonsils, appendix, and the coccyx (tailbone). Vestigial organs are one more example of how over one hundred years of misinformation, lack of knowledge, and at times fraudulent reporting have left a dark cloud of deception over Western culture. The bogus claim of vestigial organs in human anatomy have been used to indoctrinate millions of people in favor of an evolutionary view for the origin of human life. This damaging deception, however, has now been discarded on the trash heap of history by the advancing knowledge of modern science. The study of human anatomy has clearly identified the useful purpose of many human organ systems and structures that were once terribly misunderstood.

Closing Thoughts

We live in an age when empirical science and theoretical science are discussed in great detail. However, to be accurate, scientific theories born out of observation and experimentation should always carry more weight and authority than hypotheses that are not. There has always been the danger that biased presupposition may tempt man to endow weak hypotheses with the same authority as empirically proven scientific discovery. This has certainly been the case with evolution theory–a speculative collection of hypothetical ideas that, over time, have now deeply permeated every facet of Western culture. Many scientists bring with them preconceived notions that influence their decision-making in every situation they encounter. Despite a growing number of experts that deny naturalism, the impression left by evolution theory on Western societies has been overwhelming. Even some of the more conservative, creation-minded scientists have felt inclined to bow down to a relentless pressure to compromise their views in order to accommodate the accepted scientific paradigm. Some people have even attempted to introduce "naturalism" into the biblical narrative of creation. This denial of scriptural integrity has led to much confusion, doubt, and even dissent in the Christian community—and to a false perception of God's magnificent power displayed in his creative works. Brilliant scientist, Dr. Henry M. Morris, an advocate of "special creation," speaks about accommodation of evolution by some church organizations: "The evolutionary scenario with its billion-year spectacle of random variation, evolutionary meandering, struggle for existence, suffering and death and extinction, utterly contradicts not only the plain teaching of Scripture, but also the very nature of God. Evolution is the most cruel and inefficient process that could be devised for populating a world and creating man, and God should not be held responsible for any such system!" [24]

Every finite effect requires a sufficient cause, and no effect can be greater than its cause. A house requires a builder, a watch requires a watchmaker, and a painting requires an artist. It is illogical to assume that a remarkable universe and the beautiful planet earth, teeming with abundant life and complex design, could somehow be explained by an infinite number of randomly occurring accidents. Evolution theory makes no sense from a logical standpoint as spoken by Dr. Wolfgang Smith: "And the salient fact is this: if by evolution, we mean macro-evolution (as we henceforth shall). then it can be said with the utmost rigor that the doctrine is totally bereft of scientific sanction... Now, to be sure, given the multitude of extravagant claims about evolution promulgated by evolutionists within an air of scientific infallibility, this may indeed sound strange ... And yet the fact remains that there exists to this day not a shred of bona fide scientific evidence in support of the thesis that macro-evolutionary transformations have ever occurred."[25] This alone should cause us to abandon the evolution model and forsake the thought of "materialistic naturalism" in favor of a more reasonable belief system—a historical system that relies on an initial 'intelligent cause' of a remarkable universe. In his written work, *The Battle for Truth*, the philosopher, religious leader, and writer David A. Nobel states what should be obvious to all who are determined to follow the evidence wherever it leads. He once wrote: "Reason requires the biologist to abandon evolution and to embrace the more rational explanation (for life): Creation! Of course, creationism is untenable for all atheists; therefore, even if the atheist recognizes the irrationality of traditional evolutionary theory, he must postulate an equally indefensible theory to circumvent the notion of God."[26]

We will pause here to consider a personal confession offered by atheist-turned-Christian, legal expert, writer, and speaker Mr. Lee Strobel. Lee Strobel is the author of several popular books including *The Case for Christ*, which became a full-length feature film released in public theaters in the spring of 2017. As an atheist, Strobel was very skeptical of the strange idea that genuine faith in an unseen God was a position that could, in fact, provide him with needed intellectual satisfaction and fulfillment. However, after thoroughly examining the evidence as a trained investigative journalist, he discovered that he could, in no way, refute the claim for the historical life of Jesus Christ, the fact of his brutal death by crucifixion at the hand of the Romans, and the monumental truth of his confirmed resurrection from the grave several days after burial. In the following candid remarks, Strobel speaks openly about a common philosophical worldview that is upheld by so many today who continue to believe in an unproven theory, and are convinced that it offers them the freedom to live a self-directed life without moral restraint and personal accountability. "I was more than happy to latch onto Darwinism as an excuse to jettison the idea of God, so I could unabashedly pursue my own agenda in life without moral constraints ... And try as I might, I couldn't turn my back on nagging inconsistencies that were undermining the foundation of Darwin's theory."[27]

CHAPTER FOUR

* * *

NOT BY CHANCE

PROBABILITY ANALYSIS

A recent frontier in science and mathematics is a discipline known as "probability analysis." This field allows today's scientists to analyze the mathematical probability that certain events could occur solely because of blind, random chance alone, without any aid or guidance. When we consider the evidence of probability analysis, the unlikelihood of 'random chance' as a model to explain the appearance of biological systems becomes very clear to us. When evolutionists speak in terms of billions or millions of years, the average, person will accept the word of the so-called experts. After all, who would dare to question the scientific opinion of the university PhD, wearing a white coat? The average person will simply conclude that the experts must know what they are talking about—after all, haven't they studied this topic all their lives? However, we need to know how accurate the dogmatic opinions of these self-proclaimed experts really are. Could it be that an erroneous worldview has led many experts to the wrong conclusions? Is it then possible to be a misinformed and misguided expert because of preconceived bias? We owe it to ourselves and to our impressionable children and grandchildren to seriously consider whether evolution is really a logical, evidence-based concept supported by science and history. Is this popular theory worthy of our trust, confidence, and allegiance? Should a purely 'chance' explanation of life be the guiding principle that determines our own personal worldview?

Complex biological systems are composed of billions of intricate proteins, which are specialized molecules interacting with each other to direct organic function within living cells. Proteins are long chains of amino acids in very specific arrangements directed by DNA genetic code. In each case, the sequential order of amino acids must be exact to produce functioning metabolism. Each one must network with other proteins to perform a vast array of cellular functions. These amino acids are the building blocks of life. If evolution were true, the theory would predict that millions of complex protein molecules have been generated purely by random chance. A major problem for evolution is that many thousands of complex proteins had to appear all at the same time just to assemble the molecular machinery required to maintain a single living cell. Of the many amino acids found in nature, only twenty of them are used as the essential components of organic proteins. These twenty essential amino acid molecules must line up in a precise linear sequence in order to generate a viable protein called a "polypeptide." This requirement makes the prospect of self-generated polypeptides completely unsupported by probability science. We must also keep in mind that in nature, amino acids occur in mirror-image pairs called "isomers" (in chemistry we call it a "racemic mixture"). For the sake of this study, however, we will think of amino acids as having right-handed or left-handed configuration. The production of proteins from amino acids creates another big problem for evolution as we discover that only the left-handed isomers are used to make them. This 'handedness' is called chirality. For evolution to work, the random forces of nature would have to select the correct twenty amino acids, place them in the exact sequential order, and differentiate between all the right and left-handed ones at the same time. For the sake of argument, however, let us assume that unguided, random events of nature were accidentally able to select only the organically approved, left-hand configured amino acids (a big stretch indeed) to spontaneously generate a biological protein. Let us also assume that nature is building a very small protein structure of only hundred amino acids (the average protein is about three to four hundred amino acids in length). The odds that nature could assemble, by blind chance, this small structure based on the mathematical law of probability, has been established at one chance in 10^{190}. Now, to begin with, the remote chance of getting all one hundred amino acids to appear in the L-form is 1 chance in 10^{30}.

When a biologically active protein is produced, its folded shape appears in a precise three-dimensional geometric configuration upon which specific biochemical activity depends. This folding allows the host cell to recognize the function for each protein, and exactly where it fits into the grand scheme of cellular physiology (somewhat like a lock and a key). The precise sequence of amino acids allowing the correct three-dimensional shape must be linked together in "peptide bonds." However, the purely random, undirected binding of amino acids will produce the required peptide bond at only a 50 percent rate. Because of this, there is only a 1 in 2 chance probability of forming a peptide bond between every two chemically linked amino acids. Again, this leaves us with a probability for peptide bonds in all one hundred amino acids at 1 chance in 10^{30}. Thus, in order to randomly achieve a series of one hundred levoform amino acids that are all strung together with peptide bonds, we have an initial probability of only 1 chance in 10^{60}.[1] This remote scenario alone, by chance, is considered to be an impossible occurrence based on probability science.

In the organic world, the generation of functional protein polypeptides proceeds with uncanny accuracy because the mandatory chirality and peptide bonding of the amino acids is controlled and maintained by cellular DNA and its genetically coded guidance. This difficulty for random chance, however, is minor compared to the mandatory sequence by which these one hundred amino acids must be linked. The accurate order in which every amino acid component must appear is because protein molecules exhibit a very high level of molecular specificity. Even a single misplaced amino acid would likely result in catastrophic failure. Randomly selecting one hundred acid base molecules in the correct order from this pool of twenty amino acids would generate a chance probability of 1 in 20^{100}, which equates to 1 in 10^{130}.[2] If we then combine these odds with our initial chirality and peptide bonding of 1 in 10^{30}, there is no question that it would be scientifically inconceivable to believe that mother nature could form a one hundred amino acid polypeptide by random chance processes alone. Scientists have concluded that the random chance occurrence of an event with a probability exceeding 1 in 10^{50} is scientifically impossible. Regarding our simple protein and the precise sequential order of amino acids alone, consider the following scenario. If there were only twenty letters in the alphabet and they were all mixed together and placed in a hat, what would happen if the letters were selected randomly, one at a time, by a blindfolded man until a total of one hundred selections have been made? By placing the letters in the exact order selected, what is the chance of correctly spelling a predetermined sentence of one hundred letters in length? Could we expect to accidentally spell, with exact precision, the following short sentence by random chance alone: "We hold these truths to be self-evident, that all men are created equal, that they are endowed by their Creator with rights!" All one hundred characters in this simple statement must be arranged in a precise sequential order; otherwise, the resulting message would be left partially unintelligible. The same is true of biological systems when required proteins are generated in the cell. The sequence of amino acids must be precise for the cell to recognize the intended purpose of a given polypeptide polymer. Even if our simple experiment of randomly generating a short written message was to be repeated twenty billion times, do you really think this precise sentence could ever be spelled out correctly by chance alone? If you believe this to be impossible, as I do, let us consider why.

DNA is the language of life. Without exception, any written message or coded communication that is intelligible will always require a mind to originate. You can be certain that each word in this book was selected by a mindful process—not the result of an accidental explosion in the printing shop! For someone to combine the letters of a formal alphabet in a manner that would produce a legible message is inconceivable apart from intelligent agency.

Dr. Gerald Schroeder is a Jewish physicist who graduated from MIT, receiving his PhD in nuclear physics and earth and planetary science in 1965. He completed significant work in probability analysis, in which he has refuted the idea that life arose by chance in a godless universe. He reports on an experiment using a group of monkeys banging away on computer keyboards, which demonstrates the futility of attempting to explain the emergence of life as a colossal super accident. In a bold experiment conducted by the British National Council of Arts, computers were placed inside a cage with a group of monkeys to demonstrate the possibility of creating meaningful words by random chance. Dr. Schroeder reported that after one month of randomly hammering away on the keyboards (along with using them for a urinal) six monkeys had produced fifty typed pages, but not a single word—not even a simple one-letter word such as "a" or "I" was found as a result. According to Dr. Schroeder, the likelihood of producing, by accident, a single one-letter word using this primate investigational tool would be one chance out of twenty-seven-thousand. Using the science of probability analysis, Dr. Schroeder has also calculated the extremely remote chance that a monkey could accidentally type a complete Shakespearean sonnet (14 lines totaling 488 letters) at only one chance in 10^{690}. Schroeder compared these incomprehensible odds with the total number of atoms in the known universe, which has been estimated to be 10^{80}. Gerald Schroeder's conclusion is that you could never produce a Shakespearean sonnet by chance alone, even if the entire universe was filled with computer chips that are randomly producing letters at the rate of one million per second from the beginning of time until now. It is obvious that nature, by random chance alone, could never fulfill the requirements to produce life. It would clearly require the direction of intelligence!

Naturally occurring random chance events, acting totally independent of any significant guiding influence or intelligence, would never provide an adequate explanation for the basic production of complex cellular design. Microbiologist, Michael Behe, compares the accidental probability of linking one hundred amino acids in the correct sequential order to a blindfolded man locating a single marked grain of sand somewhere in the vastness of the Sahara Desert, and doing it, not just once, but three times.[3] In his book *Darwin Was Wrong*, I. L. Cohen confirms the limit of probability: "Mathematicians agree that any requisite number beyond 10^{50} has, statistically, a zero probability of occurring."[4] In other words, it is mathematically impossible to form even a simple protein molecule through chance processes alone, even if all correct amino acids are available, and each one appears in the mandatory left-handed configuration. This is why many researchers are beginning to abandon any chance-based approach to explaining life's origin.

Keep in mind that the unique chemistry of a single cell is now known to be vastly more complex than this simple polypeptide example might suggest. And in fact, generating all the necessary chemistry for a complete living cell would require the spontaneous generation of thousands of specialized proteins, each averaging perhaps three to four hundred amino acids, and all coming together randomly at the same time by pure chance. These complex organic molecules would have to be accompanied by the accidental appearance of DNA and RNA (the most complex chemical polymers found in organic systems), followed very closely by the energy required to produce biologically active physiology.

The famous British astrophysicist Sir Frederick Hoyle made a colorful comparison by equating the odds of the random chance generation of a simple living cell with this interesting remark: "The chance that higher life-forms might have emerged in this way is comparable to the chance that 'a tornado sweeping through a junkyard might assemble a Boeing 747 from the materials therein." [5] Hoyle goes on to explain his personal point of view based on a lifetime of science. "A common-sense interpretation of the facts suggests that a 'super intellect' has monkeyed with physics, as well as with chemistry and biology ... the likelihood of the formation of life from inanimate matter is one to a number of 10, with 40,000 naughts (zeros) after it. It is enough to bury Darwin and the whole theory of Evolution ... There was no primeval soup, neither on this planet, nor on any other ... and if the beginnings of life were not random, they must therefore have been the product of a purposeful intelligence."[6] Nobel Prizewinning cellular biochemist, Christian de Duve, also described the enormous improbability of the chance generation of a living cell with the following concise remarks: "If you equate the probability of the birth of a bacterial cell to chance assembly of its atoms, eternity will not suffice to produce one ... faced with the enormous sum of lucky draws behind the success of the 'evolutionary game,' one may legitimately wonder to what extent this success is actually written the fabric of the universe."[7] As a lifelong educator and conference speaker, Kenneth Ham, the founding director of Answers in Genesis, has also commented about the vast complexity of living cells. He wrote, "Life is built upon information ... In fact, in just one of the trillions of cells that make up the human body, the amount of information in its genes has been estimated to fill at least 1,000 books of 500 pages of typewritten information."[8] People today are forced to admit that the spontaneous generation of basic proteins for life is now known to be mathematically and scientifically incomprehensible.

This undeniable scientific conclusion that unbiased investigators have accurately confirmed after many years of exhaustive cellular research, also became a personal reality to a retired Apollo 14 astronaut by the name of Edgar Mitchell. Commander Mitchell commented openly about a vivid revelation and personal experience that had impressed him so greatly on his heroic flight into outer space: "When I went to the moon, I was a pragmatic test pilot. But when I saw the planet Earth floating in the vastness of space, the presence of divinity became almost palpable, and I knew that life in the universe was not just an accident."[9]

DNA: The Language of Life

Francis Crick, James Watson and Maurice Wilkins all shared the Nobel Prize in 1962 for their discovery of the double helix structure of DNA, which seems to resemble a spiral ladder, and is composed of long continuous chains of sugar-based molecules called "nucleotides." The rungs of the ladder are formed by the base-pair bonding of four specific organic components: adenine, thymine, cytosine, and guanine (ATCG). The precise sequence of rungs on the spiral DNA ladder carries specific, detailed instructions encoded with this four-letter chemical alphabet. A gene consists of a grouping of these bonded chemical letters, specifying the production of certain protein molecules. The "genome" of an organism includes a complete set of genes encoded within the DNA matrix. Within each organic cell is found all the information that is necessary to build and maintain a given organism. The remarkable human genome, which is neatly compressed within the nucleus of every one of our twenty trillion cells, constitutes a remarkable information-bearing text containing over 3.5 billion chemical letters of coded instruction—enough information to fill an entire library. The remarkable human DNA is so long and detailed that it must be tightly coiled within the nucleus of each cell. Stretching it out in a straight linear form would require approximately two meters (about six feet) in length. DNA (deoxyribonucleic acid) is clearly the most complex chemistry known to mankind. The genetic material within each living cell is a blueprint for life that precisely spells out the characteristics of every living organism. DNA is classified as an organic "polymer," which means it is a concise conglomeration of many smaller molecular compounds. Each nucleotide link is distinguished from all others by the base-pair chemical arrangement. The sequencing of the base pairs determines a hereditary blueprint. This repeating sequence of chemical bases comprises a genetic master plan that codes for all organic design and activity. According to Dr. Francis Collins, who served as the acting director of the Human Genome Project, "A live reading of that code at a rate of three letters per second would take 31 years, even if reading continued day and night. Printing these letters out in regular font size on normal bond paper and binding them together would result in a tower the height of the Washington Monument."[10] The entire human genetic data-base is duplicated with every cellular division resulting in a perfect copy of the information-bearing content within this complex DNA polymer.

The DNA instructs organic cells in the complicated art of manufacturing proteins. As previously stated, proteins are composed of amino acids, of which there are twenty specific ones that make up all such molecules within living things. The order of the specific amino acids in each polypeptide molecule determines its role and purpose in overall organic design. Some protein molecules are referred to as enzymes, which describe their function as a catalyst in biochemical reactions. During DNA replication, in order to limit unwanted copying errors, cellular physiology is equipped with a proofreading system that detects and corrects any mistakes in the duplication process. Because of this highly specialized cellular proofreading, the copying-error rate can be as low as one single mistake per every one hundred billion bits of information. This accuracy rate can be compared to only one misplaced letter in fifty million typed pages of transcript.

Multiple millions of genes making up these base pair segments are organized into what we know as "chromosomes." All human cells contain forty-six chromosomes (twenty-three pairs) with offspring having received twenty-three chromosomes from each parent. The different species vary significantly in the number of chromosomes that make up the genome of the particular organism. Chromosomes are simply nature's way of packaging DNA within the cell's nucleus, and the number of chromosomes of a given species is certainly not an indicator of the complexity of the organism. A dog, for example, has thirty-nine pairs of chromosomes, and a carp has a whopping fifty-two pairs. A specific DNA polymer may consist of several hundred thousand to as many as several billion links or subunits connected to one another. Instruction within each subunit encodes for the specific genetic information required for the unique growth, development and functioning of a living organism.

Think of the DNA polymer as a long spiral ladder upon which are written chemical messages recording detailed instructions spelled out by nature's unique four-letter chemical alphabet. The chemical letters in DNA contain coding commands that instruct cells to make specific proteins. With the work of another genetic polymer called RNA (ribonucleic acid) the information of a single gene is transcribed directing the manufacturing process. Many scientists, who are committed exclusively to a materialistic view of life, have insisted that the information-bearing properties of the DNA polymer and the transcribing properties of RNA have emerged purely by chance from raw, chemical elements by some mindless, unguided process. With a complete genetic blueprint residing within the nucleus of every living cell, from bacteria to humans, the vast amount of information found in DNA and transcribed by RNA is the same apparatus of genetically mediated guidance that is now known to be active in all living organisms. Complexity reigns in all cellular chemistry from an amoeba to a man. Perhaps God designed it this way so that our elementary understanding of His grand design would inevitably point us back to Him.

According to evolutionary biologist Richard Dawkins, "What has happened is that genetics has become a branch of information technology ... the genetic code is truly digital in exactly the same sense as computer codes. This is not some vague analogy; it is the literal truth."[11] Do you suppose it takes intelligence to write computer code? Of course! This is surely something we all agree on. The chemical, computer-like code found in DNA spells out biological function, structural design, and molecular activity controlling every facet of organic chemistry within the host organism. As aptly stated by educator Nancy Pearcey, a senior fellow at the Discovery Institute, "We can now explain why all experiments to create life in a test tube have failed, because we have tried to build life from the bottom up, by assembling the right materials to form a DNA molecule. But life is not about matter, it's about information."[12] Her same basic thought continues with this obvious fact: "Information does not arise from natural forces within matter, but it has to be imposed upon matter from outside by an intelligent agent."[13]

When we personally observe an example of a codified communication, or written message, it would be irrational to think of its origin in terms of a random chance process. Like computer code, chemical letters spell out the details of a digital message, making biological function and structural differentiation possible. This remarkable genetic chemistry within living cellular design is an extremely complex mechanism that has caused many great men of science to conclude that the evolution model is useless when it comes to explaining how it came about. According to well-known geneticist, Dr. Francis Collins, DNA transmits instructions in live cells similar to the way one might program a cellphone to beep for specific reasons. He explains this further: "DNA is a three-billion-lettered program telling the cell to act in a certain way. It is a full instruction manual."[14] Referring to the progress made in DNA research, Dr. Collins has also commented: "We have caught the first glimpses of our instruction book, previously known only to God."[15] How can we account for a unique language within living cells, if not by intelligent cause? Australian microbiologist, Dr. Michael Denton, informs us that "the capacity of DNA to store information vastly exceeds that of any other known system; it is so efficient that all the information needed to specify an organism as complex as a man weighs less than a few thousand millionths of a gram. The information necessary to specify the design of all the species of organisms which have ever existed on the planet ... could be held in a teaspoon ... and there would still be room left over for all the information in every book ever written ..."[16]

In his book entitled *In the Beginning Was Information*, Dr. Werner Gitt, who formerly served as director at the German Institute of Physics and Technology, made a claim that it is scientifically impossible for intelligible information to arise from disorder by chance. His explains that "a code system is always the result of a mental process (it requires an intelligent origin or inventor) ... It should be emphasized that matter, as such, is unable to generate any code ... All experiences indicate that a thinking being voluntarily exercising his free will, cognition, and creativity is required."[17] He continues: "There is no known natural law through which matter can give rise to information, neither is there any physical process or material phenomenon known to do this."[18] Whenever information is traced to a cause, it leads to an intelligent source. Information is the central characteristic of all living things. The information they contain, which regulates all of life's processes, can ultimately be traced back to an intelligent source, a creative mind.

According to Microsoft founder Mr. Bill Gates, "DNA is like a computer program, but far, far more advanced than any software we've ever created."[19] The obvious question so many researchers ask after carefully examining the evidence is, how in the world did the DNA genetic coding system and its translation apparatus originate? The following facts we know for certain:

1. All living things require a complex DNA database incorporating a chemically mediated language system, which constitutes a complete blueprint for life.

2. The only logical source of language-based complexity requires the input of intelligence.

3. Probability science demonstrates that random, unguided events are entirely incapable of generating such complexity.

If generating DNA without intelligent guidance sounds impossible, consider this: the chemical commands required to assemble complex proteins comes from the DNA. The obvious question then is, where did DNA come from? DNA/RNA chemistry presents a massive problem for evolution theory because these polymers are vastly more complicated than other organic structures. The mathematical odds against assembling the right material for DNA, by random chance, are so astronomical that very few researchers today even think in these terms. We must keep in mind that about twenty different protein enzymes are required in the manufacture and duplication of DNA (replication). DNA controls RNA synthesis by which the information encoded therein is transcribed (transcription). RNA in turn controls all protein synthesis as directed by the genetic information within DNA (translation). This unique process constitutes a perfect example of the 'irreducible complexity' previously referenced, in which the proteins are required to produce DNA—DNA is required to produce RNA—while RNA is required to produce the proteins that are essential to the production DNA. This critical circle-of-life cell chemistry can never be disrupted without certain death as the outcome. Because each of these critical phases of genetic physiology has required the others in a continuous irreducible process since the very beginning of all carbon-based life, the incomprehensible chance that each cellular component of this manufacturing system appeared spontaneously, all at the same time, is unthinkable.

Consider the single DNA marker that codes for blood clotting factor VIII. This scenario may help to illustrate just how complex genetic chemistry really is. This specific gene contains 186,000 nucleotides linked in a complex molecular chain, which tells the human body how to produce an essential blood clotting protein containing about 2400 amino acids. Without this protein human blood clots very poorly resulting in a genetic disease called hemophilia. If for instance, while RNA is engaged in transcribing the chemical sequence that codes for blood clotting and one nucleotide is somehow misplaced during reproductive cell duplication, the offspring will inherit the defective gene and we have a child incapable of producing factor eight protein. This child would be facing the reality of a disadvantaged life as a hemophiliac. The rare mutation would greatly affect that person's quality of life. Almost without exception, mutations produce lethal, harmful, or useless outcomes. The Achilles heel of evolution theory is its absolute dependence on the existence of imaginary beneficial mutations which have become impossible to identify. The DNA polymer, providing life's genetic instruction, is critical in directing all of life's physical processes, yet the evolution model is completely powerless to explain how it came to be. Some scientists suggest that DNA is the physical world's most obvious indicator of a 'designer' and among the most powerful arguments ever presented to countermand evolutionary theory.

American engineer and young-earth creationist, Dr. Walter T. Brown, offers us a fascinating indication of why this reasoning may, in fact, be true. He reminds us that: "DNA can only be produced with the help of at least twenty different proteins, but these proteins can only be produced at the direction of DNA. Since each requires the other, a satisfactory explanation of the origin of one must also explain the origin of the other .... Apparently, this entire manufacturing system came into existence simultaneously. This implies creation!"[20]

DNA chemistry also presents yet another devastating problem for evolution: chirality!"[21] Chirality refers to a molecular structure that has, or which produces, a non-superimposable mirror image of itself. In nature, chiral compounds typically appear equally in both right and left-handed configurations called "optical isomers." The simplest example of chirality is found in the human hands. The right hand cannot be superimposed on top of the left hand. However, when the palms of the right and left hand are facing each other, they appear to be identical; they are perfect mirror images. In organic systems, chirality implies a biological requirement regarding DNA and protein synthesis. DNA molecules are formed by very long sequences of nucleotides (sugars) that must be of a certain molecular orientation to produce useful function. As we have previously discussed, in nature, all amino acids occur randomly in a racemic mixture (equal numbers of right- and left-handed forms). Remember, however, that only the left-handed (levoform) amino acids are useful in the construction of cellular proteins in organic systems; not one wrong component is allowed or the entire manufacturing process fails. Again, the chiral requirement for protein synthesis must be met in every case. Just as amino acids must be left-handed to produce functioning proteins, the nucleotides, which are required to produce DNA (also occurring as a racemic mixture) can only be used in the right-handed configuration (dextroform). Can you begin to imagine the obstacle this presents for the random process of evolution? Because DNA is made of only right-handed nucleotides, and organic proteins are produced with only left-handed amino acids, a major problem for evolutionary thought is that chemical reactions producing either nucleotides or amino acids will always result in a racemic mixture of both right-handed and left-handed molecules. For example, a small DNA segment of one hundred thousand nucleotides means that all the sugars would have to appear with the right-handed configuration to produce useful code. Our finite minds cannot begin to grasp the monumental odds against the spontaneous generation of a complex DNA polymer. However, just to give you a little perspective of the chiral requirement alone, consider the following: to get all the required sugars to show up in the right-handed configuration by random chance would be the same odds as flipping a coin one hundred thousand times, while correctly guessing heads every time. This fact alone presents a colossal problem for evolutionary theory.

Abiogenesis (the spontaneous generation of life from nonliving material) means that the first living cell required the perfect combination of all necessary right-handed nucleotides and left-handed amino acids, in the perfect order to produce many thousands of proteins along with a complete complement of chromosomes, all at the same time. The infinitely impossible odds against accidentally producing these outrageously complex cellular structures belongs light-years beyond the comprehension of mortal man, no matter how many billions of years might be spent waiting for this miracle to happen by chance. Regarding the chance formation of complex DNA from isolated chemical elements in a primeval world, Dr. Francis Collins of the Human Genome Research Project had this to say: "How could a self-replicating information-carrying molecule assemble spontaneously from these compounds? DNA with its phosphate-sugar backbone, and intricately arranged organic bases, stacked neatly on top of one another, and paired together at each rung of the twisted double helix, seems an utterly improbable molecule to have 'just happened.'"[22] With this in mind, the only reasonable conclusion we can draw from today's research into molecular biochemistry is that the magnificent biological machinery found within the living cell is the unique and marvelous design of an ingenious, purpose-driven Intelligence with a superb imagination and an unbelievably complex mind.

There's No Room for Error

When American astronauts first walked upon the moon, nobody was claiming that this outstanding and dramatic feat had resulted from a long series of remarkable accidents. Thousands of scientists, along with gifted mathematicians and engineers worked diligently for many years to plan and prepare for this world-changing event. Complicated inventions of mechanical genius were conceived, planned, designed, constructed, and tested, resulting in millions of components integrated together with extreme precision to produce a working plan that would accomplish the goal of landing a man on the moon. Why were we successful in our effort to place man on the moon? There can be only one reason: because absolutely nothing was left to chance! This, the greatest engineering achievement of the twentieth century was attributed to the utilization of meticulous design and planning. And yet, by comparison, the simplest living cell, thanks to modern bio-molecular research, is understood to be vastly more complicated than any spacecraft or machinery ever designed and built by human engineering. Still many evolutionists refuse to entertain the possibility of supernatural intelligence, and yet, they do not hesitate to explain the alarming complexity of living things in the simplest terms of blind chance. You and I are purely accidental by their way of thinking, leaving them to conclude that our lives have no more value or purpose than a snail or a slug. One thing we can be sure of is that wonderful things are not the result of chance, but originate by planning and intelligence. Diehard Darwinian intellectuals have essentially backed themselves into a corner by refusing to believe that a master design engineer is required to explain a vast, virtually incomprehensible universe. Over the last couple hundred years, worldwide scientific research and investigation has uncovered immeasurably more data regarding the physical universe than could ever have been imagined by our ancestors. The Hubble telescope, for example, has revealed that there are at least one hundred billion or more galaxies with, perhaps, one hundred billion or more stars per galaxy. It seems an act of pride and human arrogance to explain away a need for a transcendent creator God in light of such a vast, and virtually unknowable universe.

Research Says "Abandon Ship"

Can we really trust in 'spontaneous generation' as the correct model to explain the origin of biological systems? As we consider where the evidence is taking us in our search for the truth about the origin of life, let us examine an authoritative statement made by molecular biologist Dr. Michael Denton, who has authored the book *Evolution: A Theory in Crisis*. Speaking from years of professional research in cellular physiology, Dr. Denton provides his assessment in light of the monumental complexity of the living cell: "Although the tiniest bacterial cells are incredibly small, weighing less than 10 (-12) grams [10 to the minus12], each is in fact, a veritable micro-miniaturized factory containing thousands of exquisitely designed pieces of intricate molecular machinery, made up altogether of 100,000 million atoms, far more complicated than any machinery ever built by man, and absolutely without parallel in the nonliving world."[23] The study of bacterial cells reveals a complexity of design unparalleled in the nonorganic world, and yet these tiny, although extremely multifaceted organisms, are so minute that you could fit a trillion of them in a single teaspoon.

Proponents of evolution still insist on random chance to explain the unique complexities of life. However, today's most sophisticated scientists, conducting up-to-date origin-of-life research, are, by virtue of the latest scientific findings, rapidly abandoning and discarding all chance-based theories. Imagine every living cell in the human body containing much more information than the world's most advanced computer hard drive. Then try to imagine all this information-bearing physiology coming together by random chance, all at the same time. As I have already emphasized, DNA alone demonstrates the monumental futility of scientists who insist they can explain the origin of life in the terms of simple, natural processes. Many others believe it is now time to look seriously at a supernatural explanation for the origin of life! Eminent scientist Dr. Walter L. Bradley explains it this way: "What I've found is absolutely overwhelming evidence that points toward an intelligent designer."[24] Because of more fair and honest research, the evidence is indeed compelling great men of science to look beyond nature for the answers they seek. Even Oxford evolutionary biologist Richard Dawkins was compelled to concede on this point. "The more statistically improbable a thing is, the less can we believe that it just happened by blind chance. Superficially, the obvious alternative to chance is an Intelligent Designer."[25]

In the early 1900s, evolutionists were forced to admit that Darwin's theory was beginning to appear rather inadequate considering advancing scientific achievement, which was identifying insurmountable problems his outdated theory could never overcome. This problem prompted a transformation in evolutionary thought over the next forty years culminating in what is now commonly referred to as the "Neo-Darwinian theory." This update has combined Darwin's theory of biological evolution by natural selection with current ideas from the field of population genetics initially championed by Gregor Mendel. This remodel placed a greater emphasis on random mutations as the sole source of evolutionary variation, suggesting that the accumulation of small genetic changes over time has eventually led to so-called 'species variation'. Although the newest concepts in evolution theory further explained adaptive change and microevolution (variation within a kind), the updated theory fails to address the two biggest complaints about evolutionary thought that were presented with powerful arguments by Darwin's critics: 'spontaneous generation' and 'macroevolution'.

No credible evidence has ever been presented to confirm that life arose spontaneously from nonliving matter, nor has any data ever explained how new information can be added to the genetic composition of a living organism. The revisions in Darwin's theory were a desperate attempt to answer the many difficulties that new discoveries have presented. Not long after modified evolutionary ideas were adopted, the advance of scientific knowledge began to accelerate at light speed, and has continued to move forward at a blistering pace for the last seventy years. The desperate attempt to satisfy the early critics of Darwin has now fallen on hard times as well, and can no longer square itself with current scientific data. The biggest problem facing evolution today, even if the infinitely impossible assumption of spontaneous generation was a given, is that multiple, mutation-driven changes would have to occur at the same time and become complementary for major improvements in a species to occur. For example, suppose an animal experienced a rare mutation resulting in longer lower leg bones. This single mutation would clearly result in diminished fitness, and greater predation, unless this rare anomaly was accompanied by a series of other beneficial and complementary mutations. This creature would also require longer upper leg bones (for balance and equal distribution of weight); larger joints; longer muscles, tendons, and ligaments, along with more skin. These vital changes all working together might allow for more explosive land speed in pursuing prey and evading predators. Yet, without multiple coincidental, complementary mutations, no physically adaptive advantage would be achieved and the creature would be wiped by predation.

Remember that the vast majority of mutations are injurious, if not lethal. The evolutionary hypothesis relies on the hope of multiple "beneficial mutations" occurring one after another, resulting in a cumulative advantage. Dr. Stephen J Gould of Harvard understood this problem and attempted to get around it by suggesting a new concept called "punctuated equilibrium." This pipe dream suggested that isolated episodes of rapid speciation may have occurred, followed by extremely long periods of little or no change. Accelerated speciation would have required multiple beneficial mutations in short order causing sudden species transformation. However, it is obvious that mutated creatures are greatly disadvantaged and would be quickly wiped out from survival of the fittest. The impossible odds of a rapid succession of favorable mutations occurring within the germ cells of the same organism at the same time creates an infinitesimally remote probability that is far too outrageous to be given serious consideration. These so-called beneficial mutations would each have to contribute positive information to the genetic blueprint. The extreme degree of desperation that is found in this kind of wishful thinking has never been adequately confirmed by any documented observations within natural science.

Darwinism, new or old, has never revealed an adequate, practical mechanism that could explain how the genome of an organism could have accumulated vast amounts of additional information in its genetic programming, and the extremely remote and exceptionally rare occurrence of new copying errors in germ cells, during DNA replication, is still a major obstacle for neo-Darwinian theory. Dr. Stephen C. Meyer, who holds a PhD in history and the philosophy of science from Cambridge University, said: "While many outside origin-of-life biology may still invoke chance as a causal explanation for the origin of biological information, few serious researchers still do."[26] Today's serious investigators in scientific disciplines such as biology, biochemistry, and genetics are raising serious objections to the evolutionary hypothesis, claiming that its implications are based on incomplete and flawed data. On close examination, many so-called factual elements of the theory are unraveling based on current research findings. Scientists are beginning to concede that an 'intelligent designer' must have been at work in the initial appearance of the earth's complex living creatures. Dr. Michael Behe of the Lehigh University department of biochemistry wrote, "The conclusion of intelligent design flows naturally from the data itself--not from sacred books or sectarian beliefs ... The reluctance of science to embrace the conclusion of intelligent design ... has no justifiable foundation .... Many people, including many important and well-respected scientists, just don't want there to be anything beyond nature."[27] Microbiologist, Dr. Michael Denton, adds this comment: "[Evolutionary theory] is still, as it was in Darwin's time, a highly speculative hypothesis entirely without direct factual support, and very far from the self-evident axiom some of its more aggressive advocates would have us believe."[28]

The technological world of scientific research has shown a bright and revealing light upon the evolution model, exposing numerous problems that, many are convinced, cannot be overcome. Consequently, the most cherished hypotheses of 'Neo-Darwinism' are beginning to unravel in light of continuing scientific investigation. Dr. Denton offered a very serious accusation against evolutionary thought, by stating: "Nowhere was Darwin able to point to one bona fide case of natural selection having actually generated evolutionary change in nature ... ultimately, the Darwinian theory of evolution is no more nor less than the great cosmogenic myth of the 20th century."[29] Because of abundant and glaring flaws, when compared with observable facts, many of today's renowned scientists and researchers are beginning to abandon ship, and who can blame them? When you examine the worldviews of today's top scientists, you will discover very few atheists among them. This is because they have experienced firsthand the futility of attempting to explain natural phenomena, and especially organic life in terms of 'blind chance'. The many challenges to Darwinian theory that cannot be overcome have now led more than 1000 PhD scientists, from major universities around the world, to sign and endorse a document entitled: "A Scientific Dissent from Darwinism". The document reads in part: "We [the signatories listed here] are skeptical of claims for the ability of random mutation and natural selection to account for the complexity of Life......" Darwinism is looking more and more like a blind leap that well-educated men and women of science and philosophy are unwilling to take. Coming to the conclusion of a 'Designer' is not akin to committing intellectual suicide. It is now looking like the most logical and rational decision one can make based on the evidence. And today, the world's most devoted atheists, who desperately cling to the rotting corpse of evolution, are most likely to be found among educators in the darkened hallways of liberal, ivory-tower universities where they instruct their students in the social sciences and humanities.

Life in the Primordial Soup

There are two immutable truths in the field of biology that have never changed. No evidence discovered through empirical science has even come close to contradicting these fundamental laws of nature. I would challenge anyone to show sufficient evidence to the contrary.

1. Only the living can give rise to new life! Dead things cannot produce life. Raw chemicals and the basic elements found in the crust of the earth cannot produce life in and of themselves.

Life comes forth only through the reproductive mechanism that was designed for this purpose within the world's vast array of complex living organisms.

2. Plants and animals can only reproduce after their own kind! This factual truth has never been violated. However, a wide variation of traits has been written into the genetic code of every living kind; thus, the ability to express a variety of complementary features within each basic plant and animal group in order to successfully adapt to the environment. This adaptation allows a survival of the species. Even with the many variations we observe and enjoy within the basic kinds, we can be certain that dogs will always reproduce dogs, and corn will always reproduce corn.

Dr. Michael Denton says the incomprehensibly drastic jump from a nonliving material world to the living, breathing world of organic design "represents the most dramatic and fundamental of all the discontinuities of nature. Between a living cell and the most highly ordered non-biological systems, such as a natural crystal or a snowflake, there is a chasm as vast and absolute as it is possible to conceive."[30] Denton has further explained that "molecular biology has also shown us that the basic design of the cell system is essentially the same in all living systems on the Earth, from bacteria to mammals. In all organisms, the roles of DNA, RNA, and proteins are identical. The meaning of the genetic code is also virtually identical in all cells ... In terms of their basic biochemical design, therefore, no living system can be thought of as being primitive or ancestral with respect to any other system, nor is there the slightest empirical hint of an evolutionary sequence among all the incredibly diverse cells on Earth."[31] The 1965 Nobel Prize-winning biochemist Dr. Jacques Monod concurs: "We have no idea what the structure of a primitive cell might have been. The simplest living system known to us, the bacterial cell... in its overall chemical plan is the same as that of all other living beings. It employs the same genetic code and the same mechanism of translation as do, for example, human cells. Thus, the simplest cells available to us for study have nothing 'primitive' about them ... No vestiges of truly primitive structures are discernible."[32] Therefore, we must never underestimate the vast complexity of a living cell, no matter what organism it comes from.

The intricate interlocking physiology within all integrated cells are abounding with irreducibly complex molecular machinery. They operate with sophisticated, precision-coordinated moving parts capable of manufacturing protein polypeptides with uncanny accuracy and stunning efficiency. According to biochemist Michael Behe, flagellum (the complex propeller-like structure that enables a bacterium to swim) are so miniscule that if you laid thirty-five thousand of them end to end their total length would occupy approximately 1 millimeter of space (0.04 inch).[33] Dr. Behe argues that the existence of irreducibly complex, micro-molecular machinery within the living cell can only point toward the intelligent design of these complex organic features. He stated that: "To a person who does not feel obliged to restrict his search to unintelligent causes, the straightforward conclusion is that many biochemical systems were designed. They were designed, not by the laws of nature, nor by chance and necessity; rather, they were planned ... The designer knew what the systems would look like when they were completed, then took steps to bring the systems about. Life on earth at its most fundamental level, in its most critical components, is the product of intelligent activity."[34]

It must be emphasized that no team of scientists anywhere in the world has been able of create something remotely close to organic life from nonliving chemicals. However, many evolutionists remain convinced that what cannot be accomplished by specially trained scientists, with the most up-to-date equipment, in the artificially controlled environment of a modern laboratory, somehow occurred randomly, by blind chance several billion years ago. Men and women who tenaciously cling to this monumental fantasy have invested great faith in a protracted series of mathematically impossible biological accidents, which contemporary mankind, with his most sophisticated and technologically advanced surroundings, has not come remotely close to reproducing today. In 1864, Dr. Louis Pasteur addressed a major university in Paris where he predicted that the strange notion of the "spontaneous generation" of a living cell from dead, lifeless chemicals in some ancient primordial state would never sufficiently recover from the lethal blow that was inflicted upon it by his various painstaking experiments and observations. Those profound words have indeed stood the test of time as science continues to struggle in providing a reliable answer to explain the original emergence of carbon-based life. Consider the commentary of famous British astrophysicist Sir Fredrick Hoyle, who stated: "If there were a principle of matter, which somehow drove organic systems toward life ... its existence should easily be demonstrable in the laboratory ... One could, for instance, take a swimming bath to represent the primordial soup. Fill it with any chemicals of a non-biological nature you please. Pump any gases over it, or through it, you please, and shine any kind of radiation on it that takes your fancy. Let the experiment proceed for a year and see how many of the 2,000 enzymes (proteins produced by living cells] have appeared in the bath ... I will give you the answer, and so save the time and trouble and expense of actually doing the experiment. You would find nothing at all, except possibly for a tarry sludge, composed of amino acids, and other simple organic chemicals. How can I be so confident of this statement? Well, if it were otherwise, the experiment would long since have been done and would be well known and famous throughout the world. The cost of it would be trivial compared to the cost of landing a man on the moon; in short, there is not one shred of objective evidence to support the hypothesis that life began in an organic soup here on the Earth."[35]

Imagine belief in a materialistic myth to explain the origin of organic life against insurmountable odds. Though evolution has been taught as factual throughout America and elsewhere, the evidence does not support this hypothetical conclusion. Dr. Wolfgang Smith is a well-respected mathematician, physicist, and philosopher of science, who taught at MIT, UCLA, and Oregon State University. Dr. Smith has offered this professional assessment from the current trends in science: "Today, 128 years after it was first promulgated, the Darwinian theory of evolution stands under attack as never before .. the fact is that in recent times there has been increasing dissent on the issue within academic and professional ranks, and that a growing number of respectable scientists are defecting from the evolutionist camp. It is interesting, moreover, that for the most part these 'experts' have abandoned Darwinism, not on the basis of religious faith or biblical persuasions, but on strictly scientific grounds, and in some instances, regretfully, as one could say ... We are told dogmatically that evolution is an established fact; but we are never told who has established it, and by what means. We are told, often enough, that the 'doctrine is founded upon evidence', and that indeed this evidence is henceforward, above all verification, as well as being immune from any subsequent contradiction by experience; but we are left entirely in the dark on the crucial question wherein, precisely, this evidence consists."[36]

Charles Darwin's own personal letters reveal a troubled soul whose confidence became shaken by the direction life had taken him. "Often a cold shudder has run through me, and I have asked myself whether I may have not devoted myself to a fantasy."[37] Darwin was forced to recognize that evidence from the fossil record did not and perhaps never would support his theory, and for this, and other reasons, he remained skeptical of his own conclusions. If he was having such disturbing second thoughts during his lifetime, imagine what he would be thinking today after 150 years of investigations, revealing countless scientific breakthroughs, that have noticeably shaken his worn-out theory. The proposal that a primitive hypothesis modeled in the early 1800s could possibly account for the origin of every organic structure in the irreducibly complex world of molecular microbiology is laughable. An obsolete theory has become a religious dogma so deeply ingrained in its adherents that they have become totally oblivious to its many absurdities. Recent decades of cellular research have dramatically widened the broad gulf that separates the realities of empirical science from the myth of Neo-Darwinism, a theory that is hopelessly riddled with unfounded assumptions. If Charles Darwin was alive today, he would likely join thousands of other scientists and philosophers who have abandoned an obsolete theory.

Dr. Ilya Prigogine was a Belgian chemist who was awarded a Nobel Prize in 1977. When he was questioned about the probability of the chance generation of carbon-based life by an unguided series of millions of chemical accidents. He said this: "The statistical probability that organic structures, and the most precisely harmonized reactions that typify living organisms, would be generated by accident is zero!"[38]

Beyond Reason—Strange Logic

Logic alone teaches us that every design requires a designer. Everything created required intelligent planning, designed purpose, and deliberate intent. Consider the fashioning of a magnificent arrowhead as compared to a crude piece of obsidian, or perhaps Mount Rushmore appearing next to a rugged granite mountaintop; do these vivid images speak to us of design? How about a sandcastle along the seashore, a beautiful painting, a massive skyscraper, or even a simple paper airplane? Intelligent design is always employed when basic raw materials are assembled into something that is useful, intelligible, or in some way pleasurable to the senses. We may observe the majestic grandeur of a snowcapped peak, or walk along the seashore as wave's pound forcefully against the sand; and perhaps we have observed a myriad of stars on a moonless night. To say that these remarkable sights are the product of blind chance is a grave departure from sound reason. Every effect demands an adequate cause. Consider the words of contemporary physicist Paul Davies. "Every advance in fundamental physics seems to uncover yet another facet of order."[39] He continues with this statement: "The laws [of physics] ... seem to be the product of exceedingly ingenious design. The universe must have a purpose."[40] These authoritative statements, coming from a world-class, non-Christian scientist, are a tremendous blow to the materialistic view of blind chance.

It is entirely unreasonable to believe that the unique complexity and order we have observed in the universe is merely the accidental product of freakish, random events. Creation, on the other hand, refers to the purposeful acts by which God, by his unlimited power, has called forth time, space, matter, energy, and life all interacting in a perfect world designed specifically around his plan and purpose for mankind. Everything we have learned from origin science tells us, without question, that the universe had a beginning. This suggests that someone is responsible as the initial cause; to believe otherwise is to infer that time, space, and matter have always existed, or have created themselves, which is irrational. When you break it down to the basic hypothesis, the argument for evolution amounts to this bizarre and unreasonable claim: no one (no God) plus nothing (no matter) times blind chance (freak accidents) equals everything (the entire universe). Do you have enough faith to believe such a strange and fantastic notion as this? I certainly do not!

As a student of both science and history for over thirty years, and one who has examined the question of origins during that time, I have personally come to the following conclusion—the thoroughly and aggressively promoted concept of Darwinian evolution is complete and utter nonsense. Do not be fooled by it! It is designed to exalt man and to diminish the need for biblical faith in a world that seems eager to deny the Creator. This vast, remarkable universe is, at best, barely comprehended by the combined historical achievements of a fragile human race. So why would people cling to a philosophical dogma that cannot stand up against the scrutiny imposed on it by today's technological world of scientific research? The answer can only be found in man's free volition—the right to exercise one's God-given will in attempting to control the circumstances related to personal choice and lifestyle decision-making. Once people make a firm commitment to a philosophical point of view, very little evidence is required to confirm what they already have decided to believe. It is simply a matter of choice! Consider the following statement from evolutionist Richard Dawkins. "Even if there were no actual evidence in favor of the Darwinian theory ... we should still be justified in preferring it to all rival theories."[41]

As a well-known personality within the scientific circles how can Dawkins justify making such a bold statement? It is because he has long since ruled out any option that does not conform to naturalism; and naturalism, to the unaided human mind and self-centered human heart sits easier on the post-modern intellectual mind than a firm commitment of believing faith in the supernatural. There are those who believe that religion, and especially the Christian worldview, has significantly impeded the progress of science. Is it true? Have doctrines of faith hindered the acquisition of scientific knowledge as many naturalist proponents are claiming? A close look at the historical footprint of Western civilization would seem to indicate otherwise. I would invite them to examine the historical roots of the scientific revolution—for it was Christianity and nothing less that gave rise to the advent of modern science as nearly all the early scientific pioneers held a Christian worldview. Perhaps it is atheism, and not religious faith, that is hindering the progress of scientific achievement.

How can we explain the reluctance of secular science to accept the idea of a creator when evidence for a designer is simply overwhelming? Four hundred years ago a courageous innovator named Galileo sought knowledge of the physical world based solely on evidence derived by observation. Galileo's inquiry met with objection as many believed scientific facts were best understood through traditions embraced by Aristotle. Despite great opposition and even imprisonment by the mainstream establishment of his day, the perseverance of Galileo and his contemporaries successfully forged a new scientific paradigm, and had ushered in the enlightenment of a new scientific revolution. Men became scientists because they believed in the laws of nature, and they believed in the laws of nature because they believed in a lawgiver. Their fact-finding pursuit of empirical scientific knowledge was based on the firm belief in a Creator God, and the goal of such investigative inquiry was to discover a small measure of His supernatural plan in designing a physical world that generated so much irresistible curiosity.

Because the advent of evolution theory has fanned the flames of "materialism" in modern cultures the world over, many within scientific circles seek to redefine science in a way that conforms with naturalism and excludes any mention of supernatural causation for a remarkable universe. With this short-sided point of view becoming common-place, we can understand why many would view religion as an obstacle to the advancement of scientific achievement. If, however, evolution theory has led to a false view of the world around us, as reliable evidence overwhelmingly suggests, then it is this theory, pointing toward a strict materialistic worldview, that has become the greatest hindrance to a deeper understanding of the universe we observe.

We hear secular scientists referring to the "god of the gaps." This is the kind of god atheists say religious people believe in. To those who deny any mention of deity in the grand scheme of things, this is a type of god created out of necessity and ignorance. They will suggest that weak people need a crutch to lean on. If we cannot understand it, then god did it. For example, early humans did not understand about thunder and lightning. In Greek culture, this phenomenon was attributed to the deity Zeus (the god of the sky). However, a single class in atmospheric science today and the god Zeus quickly disappears. The gods of Greek mythology were imagined by those who could not understand certain characteristics of the natural world. Once science can explain a given phenomenon in the public discourse, the god attributed to it disappears. If this were true, then of course you would have to choose between god and science. The so-called 'god of the gaps' is not, however, the God of the Bible. The God of the Bible is the author of the whole show, regardless of whether we can understand a given physical phenomenon from a scientific point of view. Science exists because there is a sovereign creator God, and we seek to know his grand design by using the intelligence and curiosity he has given us.

Many humanist philosophers completely disregard the factual evidence that points to a creator. A Kansas State University professor wrote this comment in the journal *Nature.* "Even if all the data point to an intelligent designer, such an hypothesis is excluded from science, because it is not naturalistic."[42] Richard Dawkins said, "Biology is the study of complicated things that give the appearance of having been designed for a purpose."[43] Many scientists are just afraid of the implications of design. The determination of naturalists, who hold tight to a prior commitment with materialism, was expressed by another Nobel Prize winner, Dr. George Wald. "There are only two possibilities as to how life arose. One is spontaneous generation arising to evolution; the other is a supernatural creative act of God ... There is no third possibility. Spontaneous generation, that life arose from nonliving matter, was scientifically disproved 120 years ago by Louis Pasteur, and others ... That leaves us with the only possible conclusion--that life arose as a supernatural creative act of God. I will not accept that philosophically, because I do not want to believe in God. Therefore, I choose to believe in that which I know is scientifically impossible; spontaneous generation arising to evolution."[44] Amazing! Because of their devoted commitment to a worldview that excludes God, the proponents of evolution theory are forced to interpret the evidence in a way that conforms with their emotionally preferred philosophical point of view, even though known scientific facts clearly dictate otherwise. Geneticist and evolutionary biologist Theodosius Dobyhansky has offered a strange remark that seems to explain for us why so many scientists have willingly ignored valid empirical evidence and continue to cling ever so tightly to their preferential worldview. "No evidence is powerful enough to force acceptance of the conclusion that is emotionally distasteful."[45]

Writer and educator Nancy Pearcey sums up these irrational conclusions with the following: "Even if there is no evidence in favor of Darwinism, and all the evidence favors intelligent design, still we are not allowed to consider that in science. Clearly, the issue is not fundamentally a matter of evidence at all, but of a prior philosophical commitment."[46] Many other people have simply avoided the creation-evolution controversy altogether, having fallen victim to a modern-day secular culture that is moving further and further away from belief in a personal, loving God. Perhaps they are no longer interested in where the evidence is leading science, but are simply content with the ideology that an increasingly secularized culture is demanding. The message is this—believe what makes you comfortable whether the evidence supports it or not.

Many of these individuals will continue to maintain the following unfortunate disclaimer: "My mind is made up do not confuse me with the facts!"

CHAPTER FIVE

* * *

BEYOND THE IVORY TOWER

OLD MOTHER EARTH?

How old is the earth? Do we have any idea? Is there any information that can help us determine our planet's relative age? Scientists love to invoke the magic of colossal age. The mystique generated by outrageous dating makes it sound as though anything is possible, even if it's not. Scientists insist that our planet is between 4.6 and 5.3 billion years old. This age is based on radiometric dating—an undependable dating system that is riddled with assumptions. If we take their word for it, we are merely buying into a statistic that is required for the false assumption of evolution to appear tenable. We are told that life appeared spontaneously on this planet in a primordial soup of chemicals about 3 billion years ago, and that modern man began to emerge from his long evolutionary past around 1 million years ago. Since literally billions of years are required to accommodate the evolution model (if it were even possible), the relative age of planet earth has become an important factor that deserves our consideration. The real question we need to address is whether the earth is relatively young or extremely old as measured by thousands, millions, or billions of years. Because standard evidence for the evolution model is so weak, the theory must rely on an extremely ancient earth to make it seem at all plausible.

Again, we discover that the evolutionist is relying heavily on speculation and guesswork rather than factual data that can be observed and verified. The truth is that we cannot determine an accurate age for planet earth; however, what we can do is point to verifiable physical evidence that may indicate to us whether our planet is relatively young or exceedingly old. Since there are so many technical problems associated with radiometric dating, for better understanding see the reference material noted here.[1] As previously mentioned, scientists rely heavily on the presence of fossils to date rock formations. This unscientific method uses "circular reasoning." (the dating of rocks by the fossils they contain, and the dating of the fossils by the rocks they are found in) Consider this frank admission from British geologist Robert H. Rastall of Cambridge University: "It cannot be denied that from a strictly philosophical standpoint, geologists are here arguing in a circle....... The succession of organisms has been determined by the study of their remains embedded in the rocks, and the relative ages of the rocks are determined by the remains of the organisms they contain."[2] Many have considered such dates to be both reliable and scientific, even though their dating method is totally dependent on a theory that is bankrupt of evidential support. As a professor of paleobiology at Kansas State University, Dr. Ronald R. West has commented on the unscientific practice of circular reasoning. "Contrary to what most scientists write, the fossil record does not support the Darwinian theory of evolution because it is this theory which we use to interpret the fossil record. By doing so, we are guilty of circular reasoning if we then say the fossil record supports this theory."[3] Paleo-evolutionist and American Museum curator Dr. Niles Eldredge confirms a terrible weakness in evolutionary thinking. "Paleontologists cannot operate this way. There is simply no way to look at a fossil and say how old it is unless you know the age of the rocks it comes from. And this poses something of a problem: if we date the rocks by their fossils, how can we then turn around and talk about the patterns of evolutionary time in the fossil record?"[4] It is obvious that there is no justification to validate a vast geological timescale as proposed by the evolution model when using such blatantly biased and unscientific methods. Consider this professional opinion of world-renowned British organic chemist, Dr. A. E. Wilder-Smith: "So firmly does the modern geologist believe in evolution, up from simple organisms to complex ones over huge time spans, that he is perfectly willing to use the theory of evolution to prove the theory of evolution ... applying the theory of evolution to prove the correctness of evolution. For we are assuming that the oldest formations contain only the most primitive and least complex organisms, which is the base assumption of Darwinism ... If we now assume that only simple organisms will occur in old formations, we are assuming the basic premise of Darwinism to be correct. To use, therefore, for dating purposes, the assumption that only simple organisms will be present in old formations is to thoroughly beg the whole question. It is arguing in a circle."[5]

I am always amazed at the poor dating methods that are used to support the evolution model. The only evidence favoring evolution is the assumption it is true, based on the overwhelming degree of indoctrination modern cultures have expounded on its behalf. When considering the age of the earth, we soon discover that the primary evidence for an extremely ancient earth history turns out to be its absolute necessity to validate evolutionary thought. A consistent drumbeat of "billions of years" is simply necessary to create the mystique required to attract the unwary public to a theory that cannot possibly rely on known facts of true science. No scientific method could ever confirm the accuracy of a five-billion-year-old earth, a number that is based solely on unscientific assumption. In fact, looking at data that can and has been measured, we see a very different story entirely. For example, according to geology expert Dr. Stuart Nevins, who has carefully studied and documented the rate of sedimentary deposition within the ocean basins, approximately 27.5 billion tons of sediment is currently being washed into the oceans by river systems every year. Based on the present rate of sedimentary erosion over the surface of the planet, he concluded that all the earth's continents would have been completely reduced to sea level in about 14 million years.[6] How could one possibly explain the rate of sedimentary erosion in terms of a 5.3-billion-year-old earth? According to evolutionary theory, comets are believed to be the same age as the solar system, yet each time they orbit near the sun large amounts of debris is lost. This rapid comet disintegration within our solar system is very telling, and according to Dr. Henry M. Morris, the comets are burning up at a rate suggesting that they are only thousands and not billions of years old, as evolutionists have claimed.[7]

Decay-rate data from measuring the earth's magnetic field may also be an indicator of our planet's relative age, at least with regard to its ability to support life. Scientific research has indicated that the earth's magnetism is decaying at a rate of approximately five percent every one hundred years. If we can assume the rate of decay of our planet's magnetism has been constant throughout the historical past, then, by going back around ten thousand years we would soon discover that the earth's magnetic field was 128 times stronger than it is today.[8] The great amount of heat generated by a magnetic field of that intensity would have prevented life's existence on the earth at that time. Dr. Thomas G. Barnes, a former professor of physics at the University of Texas in El Paso, has aptly pointed out that the strength of our planet's magnetism has been studied, and carefully measured with empirical data recorded for over 130 years. He has reported a current rate of decay in the magnetic field, which equates to a half-life of 1,400 years. This means that every 1,400 years the strength of the earth's magnetism has diminished by one-half. At such a rapid rate of decay, according to Dr. Barnes, it is inconceivable that a starting point for the earth's magnetism could have produced a life supporting planet with an age greater than ten to twenty thousand years. This data amounts to powerful evidence in support of an earth that could be much younger than the so-called 'experts' insist upon!

Every day an influx of metallic dust is settling to the planet's surface resulting from 'fine particle cosmic drift,' as debris from outer space is constantly entering the earth's atmosphere. The best measurements of this vast accumulation of particulate matter inbound from outer space were made by German mathematician

Hans Pettersson.[9] He has calculated that cosmic debris is reaching the earth's surface at a consistent rate of over 14 million tons per year. If we multiply this large number by 5 billion years (age of the earth by evolutionary standards), according to Pettersson, we could estimate an enormous stockpile of cosmic dust at a current level of 182 feet in depth covering the entire earth. However, there is no indication of this vast predicted accumulation of cosmic debris covering the planet. Even erosion has failed to account for the noticeable absence of this massive buildup of foreign dust that should be present for us to observe. This truth has been confirmed because the "missing" cosmic dust is known to have a composition that is very distinctive and quite different from the native raw material found in the earth's crust, especially regarding its comparably high content of nickel and iron. The current concentration of these minerals in the earth's oceans does not even come close to identifying the enormous influx of space particles that should be accounted for with a five-billion-year-old earth. Since the presence of nickel is relatively uncommon as a naturally occurring element on the earth, and therefore, this mineral's existence is largely due to the vast accumulation of cosmic dust, it is very interesting to note that researchers are reporting a very small amount of nickel concentration in ocean basins as a result of the river-wash deposition of space debris. According to earth scientist Dr. Henry Morris, when considering today's current level of nickel concentration found in the oceans, this measurement, due to river-fed seawater accumulation, can only account for about nine thousand years of earth history.[10]

What about the current salt concentration in the oceans of our planet? Could this measurement indicate a relative age of the earth? The water cycle is well established with rain falling from the clouds, running off the dry land back to the sea, only to evaporate once again by solar radiation to form new clouds. If today's water cycle has existed on earth for billions of years, the oceans should be much saltier than they are today. The mineral salts deposited in the ocean basins are constantly adding to the saltiness of seawater. A significant portion of the earth's precipitation falling from the sky eventually returns to the sea. Returning river water that reaches the oceans has leached minerals salts from exposed soil and river basins, which are then deposited into seawater. Evaporation from solar radiation removes fresh water from the oceans, leaving the accumulation of mineral salts behind. Because of this, the oceans are gaining salt content every day. Wikipedia reports that seawater is estimated to have a current salt concentration of about 3.5 percent. By using predictable rates of mineral deposition in seawater, it has been calculated that the oceans could have become noticeably salty within only about five thousand years.

What about population statistics? If modern man began to descend from primate ancestors approximately a million years ago, doesn't it seem strange that overpopulation has only been discussed in recent years? It is estimated that the world's population is growing at a rate of two percent per year. If we can assume that human descendants have been populating the earth for nearly a million years, we can easily calculate the reproduction rate that would be necessary to produce today's population of around seven billion people. We would soon discover this growth rate to be only 0.002 percent.[11] This means that the world's current growth rate is one thousand times greater than would be necessary to validate evolutionary teaching about the emergence and occupation of human life on this planet. On the other hand, if the population growth began with eight people, some five thousand years ago as the Bible teaches, we find ourselves well within range of the world's current population. If, for example, we are to consider a very modest family size of say 2.5 children (a very conservative estimate on a worldwide scale), this would produce a population of about 3.5 billion people in only four thousand years. If we apply this growth rate to the typical evolution model using one million years as the beginning of mankind's reproductive occupation, it makes no sense that twenty-five thousand generations of human history would only produce a current population size of around seven billion? The obvious reality is that 2.5 children per family over twenty-five thousand generations of human reproductive life would produce a worldwide population so enormous that it would be far beyond the realm of reason. In fact, this utterly impossible number would result in a world that could not possibly sustain its own human population.[12] If we could somehow imagine the number of people that would have lived and died throughout twenty-five thousand generations of human history—another question we would have to ask ourselves is--where are all the bones? The whole earth could not contain the enormous pile of bones that you and I would be living on top of. Many years ago, a Canadian professor, Dr. Harold Armstrong, conducted a very interesting study on human population in which he calculated that the average ethnic group would double in number every 150 to 175 years. With this statistic in mind, Dr. Armstrong began his study with the current worldwide population. By reversing the data projection, he was able to estimate the starting point for the earth's initial population growth at around 4,800 years ago, which coincides closely with a worldwide flood account at the time of Noah as recorded in the book of Genesis.

Dr. Russell Humphreys PhD is a brilliant American physicist who has studied earth/planetary science and cosmology throughout a long and productive career. In regard to the earth dating methods that have been examined in recent years, he reports that there are 101 such methods that support an earth age of thousands of years—not millions and not billions as conventional scientific sources have claimed for decades. In fact, of the many dating methods currently given serious consideration, the vast majority favor a young earth compared to the billions of years that are required to make evolution theory seem feasible. These billions of years are constantly disappearing today based on an ever-increasing catalog of research data to the contrary.

Is It Science or Religion?

Even though he remained a committed evolutionist, paleontologist Colin Patterson has offered this caution: "Karl Popper warns of a danger: 'a theory', even a scientific theory, may become an intellectual fashion, a substitute for religion, an entrenched dogma. This has certainly been true of evolutionary theory."[13] The teaching of evolution theory has become deeply ingrained in the education process among many teachers and professors who have never gained any first-hand knowledge through research experience, but have adopted a philosophy of life consistent with intellectual demands that have increasingly excluded God from the mainstream of public life. Many have earnestly embraced this godless worldview because it fits neatly into their personal disdain for the idea of biblical moral absolutes. They have convinced themselves that evolution is scientific and explains the emergence of life on the earth without the need for supernatural agency. Listen to the words of British zoologist L.H. Matthews of Cambridge: "The fact of evolution is the backbone of biology, and biology is thus in the peculiar position of being a science founded on an unproven theory. Is it then a science or a faith? Belief in the theory of evolution is thus exactly parallel to belief in 'special creation'. Both are concepts, which the believers know to be true, but neither, up to the present, has been capable of proof."[14]

People generally believe evolution to be factual by virtue of its remarkable prevalence in our culture. This unproven and ungodly philosophy of life has been offered as truth to young people through every subject in the learning process, and especially in the many disciplines of science. No other explanation for the existence of carbon-based life is considered or even tolerated in modern academia's relentless effort to shape the emerging worldview of today's eager young minds. Nevertheless, world-renowned French zoologist Pierre-Paul Grasse openly ridiculed Neo-Darwinism as a religious philosophy, and nothing more, when he wrote, "Directed by all-powerful (natural) selection, 'chance' becomes a sort of providence, which, under the cover of atheism, is not named but which is secretly worshiped."[15] Cal. State University history professor and author Dr. Theodore Roszak became best known for his 1969 publication, *The Making of a Counter Culture*, in which he thoroughly chronicled and publicized a drastic shift in Western culture during the 1960s. He has offered a disturbing insight regarding the profound effect of evolution theory on the American mind-set: "The irony is devastating. The main purpose of Darwinism was to drive every last trace of an incredible God from biology ... but the 'theory' replaces God with an even more incredible deity: omnipotent chance."[16] The extreme biased indoctrination within academia is not what we find when we get 'beyond the ivory tower.'

Serious and credible research conducted with the most sophisticated technology has all but cast evolution theory onto the scrap heap of history. Many impartial and open-minded scientists have concluded, not only that evolutionary theory is completely barren of factually based support, but given what we now understand, the basic tenets of the entire theory would be mathematically impossible. For this and other compelling reasons, unbiased researchers, armed with today's first-hand scientific knowledge, are rapidly discarding chance-based theories, and are beginning to look well beyond nature for the answers they seek. Dr. Dean Kenyon is a professor emeritus of biology at San Francisco State University. After weighing all the facts from many years of study in his chosen field, he came to the following conclusion: "When all the relevant lines of evidence are taken into account and all the problems squarely faced, I think we must conclude that life owes its inception to a source outside of nature."[17] Scientists can speculate about the past and guess about the future, but they can only observe the present.

Both creation and evolution are philosophical positions requiring faith. They each examine the same evidence but from opposing worldviews. When it comes to matters of faith, we need faith that is consistent with reason not in conflict with it. We need faith that is grounded in reality and supported by physical evidence, not contrary to it. The impossibility of spontaneous generation and macro-evolution can only be embraced as factual by those who have sufficient faith to steadfastly deny the possibility of a supernatural power at work behind the forces of nature. Science has never observed evolution at work, and there is no indication of a remarkable series of improvement-bearing mutations leading to an abundance of transitional forms in the fossil record. Since no one has ever observed this bizarre process at work in nature, we are forced to conclude by the obvious lack of evidence that the strange notion of evolution does not qualify as a valid scientific theory. Listen once again to the expert judgment of brilliant Austrian scientist, Wolfgang Smith. "I am convinced, moreover, that Darwinism, in whatever form, is not in fact a scientific theory, but a pseudo-metaphysical hypothesis, decked out in scientific garb. In reality, the theory derives its support, not from empirical data or logical deductions of a scientific kind, but from the circumstance that it happens to be the only doctrine of biological origins that can be conceived with the constricted worldview to which a majority of scientists, no doubt subscribe."[18]

Any proposed theory that is thought to be scientific must first be tested against the known physical laws of nature. When the creation model is held up against the natural laws of science and physics, we discover that it stands up very well to the scrutiny, and bears out the truth of what is recorded about the physical world in the Word of God. Evolution theory, on the other hand, requires its devoted adherents to imagine things that are against the laws of science, against logical reasoning, and even against vast historical knowledge. Evolution, therefore, repeatedly fails the rigorous testing that it has been subjected to. Many people are concluding, and rightfully so, that the creation model is actually more scientific than evolution theory. Evolution certainly is not empirical science, but perhaps it would qualify as science fiction! Speaking of fiction, consider this definition of the term, which may help us understand what we are up against with evolution theory. The word fiction constitutes 'a deliberate or unintentional fantasy without any basis in reality' ... 'an imagined creation or pretense, without a solid factual basis' ... 'something (an idea or concept) accepted as factual, but without adequate justification, and believed merely for the sake of preference'. Webster's New World Dictionary defines fiction as 'a counterfeit ... something made up or imagined ... something that is accepted as fact for the sake of convenience, although not necessarily true'. Evolution theory is certainly a convenient ideology among those who choose to reject the concept of a supernatural creator. Since the loyal adherents of naturalism cannot rely on documented physical evidence to support their case, they can only hold fast to this ideology as their preferential worldview in order to satisfy intellectual demands for a humanly gratifying "natural" explanation of life.

The basic position of the evolutionist begins and ends with the premise: there is no God! This narrow-minded view leaves only the possibility that random, blind chance events are all that is necessary to explain the complexity and diversity of life, as we know it. Consequently, this has left evolution theory in a very precarious position. Why? Because its adherents are unable to provide relevant evidence that would establish a reasonable degree of credibility to their core belief (that there is no God, and chance is all that is needed to explain life's existence). This glaring lack of real evidence makes evolution extremely religious and not very scientific at all, because it is not based on verifiable facts, but unproven assumptions demanding great faith for one to maintain. Consider the words of prominent preservationist and author, Louis T. Moore: "The more one studies paleontology, the more certain one becomes that evolution is based on faith alone; exactly the same sort of faith which is necessary to have when one encounters the great mysteries of religion."[19]

This reveals that evolution is nothing more than a philosophical dogma. Just like any other creed, evolution requires serious and exceptional faith. As a religious ideology, it must be accepted by personal choice apart from any direct empirical scientific support. Attempting to prove the nonexistence of God is certainly an impossible thing for mankind, but it remains the centerpiece of the Darwinian worldview as we know it today. Adherence to evolution is the founding principal of a dangerous and widely emerging religious philosophy called "secular humanism." Hitler, Stalin, Pol Pot, and Mao Tse-tung were the ruthless dictators of the most ungodly and violent political regimes in human history. Each one of these twentieth century political figures possessed an evolutionary driven humanistic view of life that led to the sadistic extermination of multiple millions of innocent lives. Listen to a description of humanism by the well-known atheist, Julian Huxley, who was himself convinced that mankind controls his own destiny and answers to no authority greater than himself: "I used the word 'humanist' to mean someone who believes that man is just as much a natural phenomenon as an animal or plant; that his body, mind, and soul were not supernaturally created, but are products of evolution, and that he is not under the control or guidance of any supernatural being or beings, but has to rely on himself and his own powers."[20]

Our federal government is supposed to be restricted by the First Amendment with regard to actively promoting any religious philosophy over another. Consider a Supreme Court ruling by Justice Abe Fortas: "Government in our democracy, state and federal, must be neutral in matters of religious theory. It may not foster or promote one religious theory as against another."[21] This is clearly not the case in America today, where a purely religious ideology of 'evolutionary humanism' is so often publicly endorsed, by government authority, at taxpayers' expense. Evolution is a strange worldview that people have adopted out of choice, not because the evidence favors this ideology, but because it has become a popular, media and education-driven philosophy with favorable public and didactic support. The theory is easy to accept by those who are not interested enough to examine the evidence for themselves. Belief in evolution is widely accepted among the populace who are opposed to the idea of a just and moral God interfering with their personal lifestyle choices and decision making. The 'secular' approach to explaining life's origin on earth will always demand a natural explanation for every observed phenomenon, and it strongly objects to any suggestion of divine origination, authority, and judgment. The intelligent design model is openly ridiculed and rejected because it infers the existence of divine supernatural agency that the mind of man cannot begin to comprehend.

A mistake that many people make in their search for understanding is attempting to find the truth about life through human intellect, and refusing to consider what their hearts are telling them, based on the evidence. Many have rejected the belief that faith and science can and do coexist. Western culture is saturated with the ridiculous notion that science alone provides the answers we need to explain the many mysteries of life. However, it is becoming more apparent than ever that the most incredible hoax of the twentieth century has been the common misconception that biological evolution is a reliable scientific fact. Obviously, neither philosophical view on the origin of life can be tested, demonstrated, or repeated before the watchful eyes of qualified witnesses, and therefore, remains within the belief system of the individual. What we can and must do, however, is to examine all the best available evidence we can gather to determine which philosophical worldview is most strongly upheld by the known, verifiably facts.

According to American philosopher David A. Noebel, "The belief that God created all things, including man in his own image, requires faith. But evolution requires more faith, because it runs very much contrary to reason, science, and history. Still, many people hold on desperately to this so-called theory, simply because it is the only explanation of origins that excludes God."[22] After very much soul-searching about the true nature of the evolutionary hypothesis, science philosopher, Michael Ruse, of Florida State University openly admits that "evolution came into being as a kind of secular ideology, an explicit substitute for Christianity; even today, it is promulgated as an ideology, a secular religion, a full-fledged alternative to Christianity with meaning and morality."[23] Still an evolutionist who maintains an unwavering commitment to naturalism: Ruse was forced to yield the following point in response to the argument of well-known creationist Dr. Duane T. Gish: "I must admit that in this one complaint ... the (biblical) literalists are absolutely right. 'Evolution is a religion'. It was true of evolution in the beginning, and it is true of evolution still today." [24] This powerful statement is a serious admission coming from an evolutionist, and should be cause for great concern by today's parents, whose children are being systematically indoctrinated into a religious worldview at government expense.

Consider the words of Mr. R. Kirk published in *National Review* in 1983: "Darwinism is a creed not only with scientists committed to document the all-purpose role of 'natural selection'. It is a creed with masses of people who have, at best, a vague notion of the mechanism of evolution theory as proposed by Darwin, let alone, as further complicated by his successors ... Clearly, the appeal cannot be that of a scientific truth, but of a philosophical belief, which is not difficult to identify. Darwinism is the belief in the meaninglessness of existence."[25] The evolutionary creed is a religious dogma enabling people to justify a lifestyle in conflict with traditional moral values. After discarding God's established moral laws, man can decide for himself what is right and wrong. With no divine authority to govern his decision making, morality vacillates with the norms of society, as determined by public opinion, and swayed by intellectual bullying. Evolutionary ideology is a nonscientific religious dogma, and nothing more. Promoting this ideology as factual amounts to proselytizing the religion of 'secular humanism', which crowns man supreme, and stubbornly denies that he was created in the image of a personal, loving God.

To offer the reader evidence of just how far evolutionary thinking has driven modern man away from his traditional moral framework, keep in mind that Adolf Hitler's hateful extermination of over six million Jews was justified in his evolutionary thinking. His bizarre hatred for the Jewish people was conceived out of some twisted, self-righteous delusion leading him to believe that his actions were somehow speeding up the evolutionary process—a process that strongly favored his Aryan race. The widespread dissemination of evolutionary thought throughout the world has led to disastrous and deadly consequences like that of Joseph Stalin in communist Russia. The Soviets adamantly believed they could reject God and retain value and dignity for human life, but they are now admitting they were terribly wrong and this mistake cost the lives of, some estimate, upward of fifty-million human souls.

Consider the wide impact of American eugenicist Margaret Sanger, the founder of Planned Parenthood. (The leading provider of abortions in America today) Sanger was profoundly influenced by Darwinism, which led her to conclude that by controlling the human population through eugenics, achieving a superior race of humanity was possible. Eugenics to Margaret Sanger was the idea of advancing human evolution by discouraging reproduction among the "less fit" or "inferior races." This concept led to profound racism, as she believed that inferior races included Negroes, Jews, and Italians, among others. She believed the brain capacity of the Australian Aborigine was only slightly more developed than a chimpanzee. Sanger was the leading advocate for birth control distribution and strongly encouraged its use in populations she considered to be inferior. However, in her sick desire to control population and reproduction, she also became a strong advocate of euthanasia, abortion, and human sterilization for those who should not be allowed to procreate. In addition to this, her books encouraged sex outside of marriage and were instrumental in the sexual revolution that has radically changed our culture over recent generations leading to rampant sexual immorality, sexually transmitted disease and a rate of unwed pregnancy that remains out of control. The influence of evolutionary thinking has led mankind down a pathway that is littered with disastrous consequences in our world.

The Great Debate Rages On

The proponents of evolution theory have been hiding behind the cloak of science for over 150 years, but now that false perception has been exposed. Consider this direct statement from Dr. Marjorie Grene on the philosophy of science: "Today the tables have turned. The modified, but still characteristically Darwinian theory, has itself become an orthodoxy, preached by its adherents with religious fervor, and doubted, they feel, only by a few muddlers imperfect in scientific faith."[26] Anatomical structures examined in public school biology labs are typically interpreted based on the misleading claims of evolution, as thousands of unwary students are being indoctrinated by polished professors who hold tightly to a preconceived bias, in spite of where all the evidence is leading. NASA senior systems analyst and computer specialist, David Coppedge' has elaborated about this unfortunate reality. "Darwin is liked by evolutionists, because he liberated science from the straitjacket of observation and opened the door to storytellers. This gave professional evolutionists job security so that they can wander through biology labs as if they belong there."[27] As an interesting side note, David Coppedge, who had a successful career with the Jet Propulsion Laboratory, became the victim of severe persecution. He was fired from NASA for his personal views on intelligent design. His case, and that of many others who have been 'blackballed' in their careers, demonstrates, once again, that freedom of speech and thought in the public domain is mostly reserved for those who advocate for godless materialism. Paleontologist, Colin Patterson, reflected on the numerous storytellers associated with evolution theory when he said: "It is easy enough to make up stories of how one form gave rise to another, and to find reasons why the stages should be favored by natural selection. But such stories are not part of science, for there is no way of putting them to the test."[28]

Although arduous disagreement still exists among evolutionists today, those in the mainstream agree that life on planet Earth came about through blind, random chance events with no divine architect to guide it. On the other hand, Christians generally agree that the known universe and everything in it is the marvelous creative handiwork of an unseen, personal God. At the heart of this debate over origins is the question of whether an "intelligent agent" exists. Was a living God at work in preparing the design and complexity of the vast and profoundly diverse world that we observe around us? Unfortunately, many Western societies have been saturated with a bogus theory, which lies in stark contrast with many revealed facts of science– facts that have been observed in the present and painstakingly uncovered from the past. Why then is this unproven 'philosophical dogma' so prevalent in American society and elsewhere?

Evolution theory is the central issue in a 'cultural war' between two opposing worldviews that can best be described as biblical Christianity and secular humanism. Evolution is a religious philosophy cleverly disguised as science, and taught as factual to naive school children with government sanction. It is the core belief system establishing a secular, state-funded, and media-driven religion that is taught, preached, and promoted from public-school classrooms, instead of a pulpit. Humanism routinely seeks to indoctrinate people (especially young people) with the false claim that man has descended from simple life-forms by way of apelike ancestors millions of years ago. However, it is certain that evolution theory is not just about fish becoming frogs and monkeys becoming human. This popular ideology is a profoundly religious philosophy that simply tries to explain everything in the cosmos without God. Although many dedicated evolutionists make the grave mistake of believing that all creationists are ignorant, gullible dupes, who believe in myths, legends, and fanciful stories, these evolutionists maintain that scriptures penned several millennia ago should have little or no relevance to the scientifically enlightened men and women of the twenty-first century. By continually mocking the Word of God, they have failed to understand that creationists do not live in blind faith, as they do. The hard, factual evidence repeatedly validates the traditional Judeo-Christian worldview.

Dr. Richard Bliss, professor of biological science at Christian Heritage College, stated: "The miracles required to make evolution feasible are far greater in number and far harder to believe in than the miracle of creation."[29] If it requires miracles either way, it is more logical to believe in a miracle-worker with intelligence and power to create with purpose, than placing one's faith in blind chance events with no guiding hand to direct them. Esteemed biochemist, Dr. Klaus Dose, provides additional insight into the many difficulties and immense challenges evolutionists are up against. He commented that: "more than thirty years of experimentation on the origin of life in the fields of chemical and molecular evolution have led to a better perception of the immensity of the problem of the origin of life on earth rather than its solution ... At present, all discussions on principle theories and experiments in the field either end in stalemate, or in a confession of ignorance."[30] Evolution elevates humanity as supreme, seeking to eliminate any accountability to divine authority. The cultural battle for hearts and minds, in the war over the souls of men, basically comes down to two all-important questions. Is there a supreme creator God behind the scenes who is responsible for the appearance of life and the existence of a material world that we are privileged to observe and experience? And, if there is a supreme being working out his purpose, how should we respond to him as the subjects of his creative power and authority? There is so much at stake here that we cannot afford to underestimate the importance of these fundamental questions. The answers we embrace will determine both how we live our lives today and how we spend eternity tomorrow. If evolution is a false and erroneous claim that's offered by those who detest the idea of a creator, then we have been handed the most colossal deception the world will ever know—a deception that will bear eternal consequences for millions of human souls. Humanists remain defiantly dogmatic as they attempt to explain life without God. While many have preferred to bury their proverbial heads in the sand and ignore a great body of evidence, which strongly opposes their preferential, secular worldview.

Dr. Walter L. Bradley, scientist, author, and professor at Baylor University, said this about the current state of origin-of-life research: "There isn't any doubt that science, for the moment at least, is at a dead end. The optimism of the 1950s is gone. The mood at the 1999 international conference on the origin-of-life was described as grim--full of frustration, pessimism, and desperation. Nobody pretends that any alternative provides a reasonable path of how life went unguided from simple chemicals to proteins to basic life-forms."[31] Despite such reporting, the Darwinists have offered the world a religious alternative to biblical Christianity, so they and others may continue to pursue lifestyle choices thought to be free of limiting moral constraints, personal accountability, and, ultimately, judgment. Many years ago, when asked in an interview why the scientific community had so readily embraced Darwin's ideas, evolutionist Julian Huxley offered this response: "I suppose the reason why we leapt at the Origin of Species was that the idea of God interfered with our sexual mores."[32] At least he offered an honest and I believe accurate response; the truth be known, God is oftentimes viewed as an unwelcome intruder regarding the moral choices and lifestyle decision making of carnal man.

In Defining Science, the Stakes Are High

Considering what is at stake in the great debate after a century of indoctrination in Darwinism, most people still reject this deceiving ideology. In American today, polls reveal that those who maintain a creation worldview outnumber those who believe in evolution by a wide margin of nearly two to one. However, humanist believers have attempted to change the rules to generate greater compliance with their anti-God sentiment. Nancy Pearcey reminds us that: "historically speaking, it was Darwinism, more than anything else that barred the door on any consideration of Christianity as objective truth."[33] Even so, Mr. Darwin's ideas hold little credibility considering today's technical, research-minded, scientific revolution. The progress of empirical science has found his crude ideas to be completely impractical. By attempting to redefine true science in the broad terms of naturalism, atheist-minded evolutionists have excluded any view that does not comply with their materialistic framework. The scientific reluctance toward religious faith is quite evident in the following statements by Harvard evolutionary biologist, Dr. Richard Lewontin. He writes: "Our willingness to accept scientific claims that are against common sense is the key to an understanding of the real struggle between science and the supernatural. We take the side of science in spite of the patent absurdity of some of its constructs ... in spite of its failure to fulfill many of its extravagant promises of health and life ... in spite of the tolerance of the scientific community for unsubstantiated ... 'just so stories', because we have a prior commitment, a commitment to materialism... It is not that the methods and institutions of science somehow compel us to accept a material explanation for the phenomenal world, but, on the contrary, that we are forced by our priori adherence to material causes to create an apparatus of investigation, and a set of concepts that produces material explanations, no matter how counterintuitive, no matter how mystifying to the uninitiated. Moreover, that materialism is absolute, for we cannot allow a Divine foot in the door."[34] Dr. Richard Lewontin admits that a steadfast commitment to materialism and a firm denial of the supernatural has forced evolutionists to believe in absurd hypotheses that are scientifically flimsy at best and completely nonsensical at worst. A major concern for diehard evolutionists is that a widespread invasion of supernaturalism into the world of modern science, as they have conveniently defined it, would provide a firm foundation for the entire orthodoxy within the Christian worldview, complete with its vast theology and biblical moral absolutes. However, in the final analysis, people should not allow themselves to become indoctrinated by elite scholars, even if their absurd dogma is convincingly supported by mass media and academic bias. If we fail to consider the impartial evidence available today, we are essentially permitting the so-called 'enlightened elite' to shape our personal worldview for us, and doing so without valid evidence to verify and authenticate their materialistic claims.

Whatever happened to open-minded scientists who depended on objective evidence to point them down the correct path? With heads firmly buried in the sand, many have attempted to redefine science so it fits neatly into their preferred philosophical point of view. Julian Huxley made this very clear when he said: "Modern science must rule out special creation or divine guidance."[35] One thing, however, is certain, as pointed out by creation scientist and former atheist, Dr. Gary Parker: "If evolutionists had to prove their case in court, evolution would be thrown out for lack of evidence."[36] This stands as the professional judgment of two well-known attorneys, Norman Macbeth[37] (author of Darwin Retired) and Philip Johnson[38] (author of Darwin on Trial). Neither of them argue in favor of the Bible, but are simply writing legal opinions from years of experience in the rules of evidence and logic. Macbeth concluded that: "This does not mean that the profession is about to abandon Darwin forever or endorse my views publicly. The situation remains much as it was: the inner circles are full of doubt, but the public utterances are confident. The doubts may be greater now and the confidence less serene, but it will be a long time before the public is given the full dark picture ... There is still need for a dissenting voice, a devil's advocate, a skeptical whistle-blower."[39] Yet, their determined attempts to manipulate the evidence and to shroud the history of science in a cloak of theoretical jargon are certain to fail.

As we consider an historical appraisal of man's great search for knowledge and understanding, it soon becomes clear that modern science, as we know it today, sprang forth from a Christian cradle. The compelling drive of sixteenth century man to understand the design and purpose of the Creator was the initial spark that ignited a major scientific revolution—a revolution that exploded on the scene in Western Europe out of a commonly held biblical worldview. Even those who aggressively oppose Christian faith, like Richard Dawkins, are now admitting this is true. You might even say modern science was a gift to the world courtesy of a Judeo-Christian heritage. Early Christian pioneers who engaged in the scientific endeavor were compelled by the desire to understand why their Creator had designed things as they are. Up to the present day, modern science has done nothing at all to diminish the impression that the universe is not self-existing but 'caused'. The scientific revolution sprang forth out of Western culture because those early investigators held in common a unifying concept of a Creator, and they set out to discover what the Creator had done in establishing the governing laws of nature. As spoken emphatically by professor John Lennox of Oxford, 'men loved science because they believed in the laws of nature, and they believed in the physical laws of nature because they believed in a "lawgiver"'. Remarkable historical figures such as Galileo, Kepler, Boyle, Faraday, and Newton, along with their contemporaries, believed in an eternal Creator. The great mathematician and astronomer, Johannes Kepler, declared that, "the chief aim of all investigations of the external world should be to discover the rational order and harmony, which has been imposed on it by God, and which he revealed to us in the language of mathematics."[40] Unlike Lewontin's firm claim that 'godless materialism' was needed to develop an apparatus of investigation in order to disallow 'a divine foot in the door', these early pioneers understood that they were investigating the master plan of a divine architect, and they never questioned his existence and authority as the Creator. Galileo had a similar thought to Kepler's when he wrote, "The laws of nature are written by the hand of God in the language of mathematics."[41] Sixteenth-century scientist Sir Francis Bacon, founder of the scientific method, spoke of two books from which he received inspiration. He spoke of 'the book of nature' and 'the book of God'. Sir Isaac Newton, a major contributor to the scientific revolution, expressed openly that the more he discovered about God's grand design, the greater became his commitment and belief in the 'Grand Designer'.

The pioneer spirit modeled by early experimental inquiry, is still often found in some of today's most brilliant minds—men who have essentially allowed the evidence to speak on behalf of the Creator. Keith Ward is a philosopher, theologian, and scholar, who speaks of the mathematical connection between the physical universe and the mind of God. "The continuing conformity of physical particles to precise mathematical relationships is something that is much more likely to exist if there is an ordering cosmic mathematician who sets up the correlation in the requisite way. The existence of laws of physics strongly implies that there is a God who formulates such laws and ensures that the physical realm conforms to them."[42] Dr. Henry F. Schaefer, professor, and director of the Center of Computational Quantum Chemistry at the University of Georgia commented that: "The significance and joy of my science comes in those occasional moments of discovering something new and saying to myself, so that's how God did it ... My goal is to understand a little corner of God's plan."[43] Dr. Kurt Godel, an Austrian mathematician, logician, and science philosopher has impacted the revived 'theistic revolution' that began in the latter part of the twentieth century. His professional career won him the Albert Einstein Award in 1951. After examining the evidence using his unique expertise in logic and mathematics, Dr. Godel concluded that, "the order of the world reflects the order of the supreme mind governing it."[44]

The absurd idea that religion is at war with science, and has been for hundreds of years, is nothing but a psychological myth! We should never have to make a choice between God and science; faith and knowledge are not mutually exclusive concepts by any means. Science has had remarkable success at probing the nature of the material world and discovering many of the mechanisms by which the physical laws operate, but true science has done a very poor job of explaining why the laws of nature exist. We must always distinguish between mechanism and agency. When Sir Isaac Newton discovered the universal law of gravitation, allowing man to account for planetary motion in the solar system, he understood this law existed because of a 'lawgiver'. Through investigation of how this law operates, Newton became overwhelmed with admiration for the magnificent, intelligent designer who created it that way. In an outstanding publication entitled *God's Undertaker ... Has Science Buried God*? British mathematician John Lennox of Oxford thoroughly discusses the effect that science has had on the concept of faith in God in a world governed by the fundamental laws of physics. Lennox remarked that, "far from science abolishing God, it would seem that there is a substantial case for asserting that it is the existence of a Creator that gives to science its fundamental intellectual justification."[45] Lennox continues with this thought about the rigid view of materialism: "The world of strict naturalism in which clever mathematical laws all by themselves bring the universe and life into existence, is pure fiction .... Theories and laws simply do not bring anything into existence."[46] I like the argument he will often use when he addressing the common complaint proposed by atheists regarding a perceived war on science. Professor Lennox will offer the following illustration to solidify his rebuttal: "To say that God is competing with science as an explanation for life is like saying Mr. Henry Ford was competing with the laws of the internal combustion and mechanical engineering as a sound explanation for the extraordinary innovation of the Ford Motor car."[47] Henry Ford was the designer of the motor vehicle developed for everyday use by the common man. He was the innovative mind behind this historical achievement. However, in order to produce truly innovative transportation, it was necessary that Ford understood the laws of internal combustion, and was able to utilize the basic principles of mechanical engineering. For mankind to arrive at such a remarkable end point, we needed both the creative mind and the physical mechanism. To deify science, claiming it to be the 'innovative force' that introduced carbon-based life on earth is, in my view, terribly misleading and intellectually dishonest. British astrophysicist, Stephen Hawking, who held the academic chair at Cambridge once belonging to Sir Isaac Newton, but did not share Newton's Christian faith, admitted in a television interview: "It is difficult to discuss the beginning of the universe without mentioning the concept of God. My work on the origin of the universe is on the borderline between science and religion, but I try to stay on the scientific side of the border. It is quite possible that God acts in ways that cannot be described by scientific laws."[48]

Long before the classical Greek philosophers, ancient Hebrews recorded that time was linear, and expressed a belief that the physical elements had their beginning at the time of God's initial creative works. This profound worldview held a powerful allegiance, dominating the civilized world's intellectual landscape for centuries. The conviction that the universe had a beginning is once again the predominant view in the contemporary world, where we are seeing a major consensus favoring this position. Many, however, still object to the idea of a beginning because it implies a first cause, which points, philosophically, toward a Creator. The concept of a creator who designed the universe with purpose in mind, is gaining stronger credibility as the most rational explanation for why we exist. According to the Nobel Prize winner, Dr. Arno Penzias, "Some people are uncomfortable with the purposefully created world. To come up with things that contradict purpose, they tend to speculate about things they have not seen."[49] Another Nobel Prize winner, physicist and educator Dr. Charles Townes said: "In my view, the question of origins seems to be left unanswered if we explore it from a scientific point of view. Thus, I believe there is a need for some religious or metaphysical explanation ...... I believe in the concept of God and in his existence."[50] No matter the evidence against the claims of evolution, a firm commitment by many in scientific circles has already been made. According to them, only theories that imply naturalism, to the exclusion of God, will be considered worthy of discussion. Sadly though, the contemporary view of science has gone mostly unchallenged throughout the American public school system, becoming impressed upon the minds, and thoroughly engrained in the lives of young students who lack the knowledge, maturity, and boldness to challenge it.

Teaching Children: "Age of Enlightenment"

The goal of humanism is to capture the minds of young people at an early age; this is why humanists have pressed so hard to take over public education with their narrow-minded, anti-God philosophy. Let us consider another aggressive statement by the late, atheist/evolutionist Julian Huxley, who wrote this:

"It is essential for evolution to become the central core of any education system, because it is evolution, in the broad sense, that links inorganic nature with life, and the stars with Earth, and matter with mind, and animals with man. Human history is a continuation of biological evolution in a different form."[51] Subversive ideas from Huxley, and other atheists like John Dewey, have aggressively infiltrated the American public education system for decades with a blatant anti-God philosophy. Sadly, humanists have had their way with the education methods that are used today to the peril of millions of unsuspecting and naive young people, who, without knowing it, have become thoroughly indoctrinated in a religious worldview that denies God. I did not know how invasive evolutionary teaching had become in our public schools until I came across a junior-high-school science textbook first published in 1968.[52] In this and other publications, disseminated to impressionable young minds while a tidal wave of evolutionary propaganda swept our nation in the 21st century, two entire chapters are devoted to expounding on the evolutionary history of planet Earth. The text offers no evidence to verify claims that are made, but simply proceeds with a narrative suggesting to the reader that the principles of evolutionary thought are undisputed scientific facts.

Consider a sample quotation from a typical high-school textbook: "Many people believe that a supernatural force or deity created life. That explanation is not within the scope of science."[53] How about this statement from another textbook: "By attributing the diversity of life to natural causes, rather than to supernatural creation, Darwin gave biology a sound scientific basis."[54] This narrow-minded view comes from a college textbook: "Biological phenomena, including those seemingly designed, can be explained by ... 'purely material causes', rather than by divine creation."[55] And this aggressive statement comes from a university textbook on evolution, by Monroe Strickberger (Museum of Vertebrate Zoology, Berkeley, California): "The fear that Darwinism was an attempt to displace God in the sphere of creation was therefore justified. To the question: is there a divine purpose for the creation of humans? Evolution answers No! ... According to evolution, the adaptations of species, and the adaptations of humans come from natural selection and not from design."[56] In yet another university textbook entitled, *Evolutionary Biology* by Dr. Douglas Futuyma, we read: "By coupling undirected purposeless variation to the blind uncaring process of natural selection, Darwin made theological or spiritual explanations of the life processes superfluous."[57] Harvard evolutionary biologist, Richard Lewontin, adds this bold commentary about evolutionary thought: "It is time ... to state clearly that evolution is fact, not theory ... birds arose from nonbirds and humans from nonhumans. No person who pretends to any understanding of the natural world can deny these facts any more than he or she can deny that the Earth is round, rotates on its axis, and revolves around the sun."[58]

This indoctrination is being directed at our children. Imagine the impression these bold claims have left upon their eager young minds. And worst of all, no valid evidence is ever presented to back them up. The profound influence of men like Richard Lewontin have led to a commonly held impression among university professors, students, and the general public that evolution theory has buried God under an avalanche of widespread scientific and didactic revelation on the topic of origins, making Him appear obsolete and irrelevant. Many people have become convinced that a designer is unnecessary when mindless, unguided processes are able to explain the emergence of life in the physical dimension. It is no stretch of the imagination to believe that the theory of evolution has had the devastating impact of a tsunami upon the human quest for meaning and significance in a vast, virtually unknowable universe.

However, if we choose to consider the evidence, a different story quickly emerges. We find lurking behind the cloak of science a sinister dogma whose implications have led to a drastic cultural shift in American life. We might even say that humanist indoctrination has led to a state-sponsored religion that is systematically forced on children and young adults in public schools, colleges, and universities across the nation. In public life we have all but forsaken the God of our fathers, and have replaced Him with a brand of pseudoscience birthed in the darkened halls of academia throughout the land. Education in America used to enrich children with the facts concerning the world they were growing up in; but it has become increasingly more about the spreading of a philosophical worldview without a factual basis. Children are being routinely deprived of newly discovered scientific data and verifiable historical details for the purpose of propping up a theory that cannot stand on its own merit. Schools are teaching religion today more than ever, but the religion taught amounts to blind faith in random chance processes.

Humanist leaders are surprisingly open about using the public schools as the principal means of proselytizing their godless religion. I was initially shocked at a statement that first appeared in a copy of *The Humanist* magazine in 1983. This frightening scenario is exactly what the twentieth-century humanist movement has secretly planned and thoroughly implemented throughout America for decades with minimal resistance. This ambitious effort to reshape America into a secularized society, ripe for socialism, has been underway for over seventy years. Imagine the implications of this horrific quotation. If only every conservative parent of school-age children was aware of it. "I am convinced that the battle for humankind's future must be waged and won in the public-school classroom by teachers who correctly perceive their role as the proselytizers of a new faith: a religion of humanity that recognizes and respects the spark of what theologians call divinity in every human being. These teachers must embody the same selfless dedication as the most rabid fundamentalist preachers, for they will be ministers of another sort, utilizing a classroom instead of a pulpit to convey humanist values in whatever subject they teach, regardless of the educational level-preschool, day care, or large state university. The classroom must and will become an arena of conflict between the old and the new -the rotting corpse of Christianity, together with all its adjacent evils and misery, and the ... new faith of humanism ... it will undoubtedly be a long, arduous, painful struggle replete with much sorrow and many tears, but humanism will emerge triumphant. It must, if the family of humankind is to survive."[59] This deliberate anti-God religious campaign, which has been deceptively masquerading for years as fundamental science while proselytizing a blatantly materialistic view of life and the cosmos, is designed to divest students of a conceptual knowledge of a Creator. Humanists have declared war on Christian values, principles, and theology.

Evolution is a man-made religion, disguised as science, to convince the unwary public that the deeply philosophical tendencies of man require no special guidance from any unseen God!

CHAPTER SIX

* * *

THE BATTLE FOR TRUTH

EXPELLED—NO INTELLIGENCE ALLOWED

The religion of humanism, firmly rooted in evolution theory, is carefully guarded in the public domain. Teachers and professors who dare to question or challenge this sacred cow of intellectual ideology are swiftly reprimanded and severely reproved with extremely harsh measures. In a highly successful documentary film entitled, *Expelled–No Intelligence Allowed*, hosted by Ben Stein, the tragic loss of academic freedom in America is thoroughly exposed.[1] In an open country that has always prided itself on academic freedom, sadly enough, American instructors in higher education are being denied the opportunity to teach what we know to be true, if it conflicts with the established and accepted views. We see this unfortunate reality appear over and over, especially in the disciplines of science where college and university students are often deprived of factual data and reliable research findings only because such knowledge fails to support a materialistic worldview. With this in mind, a professional critique of American biology textbooks revealed the following conspicuous details about the origin-of-life documentation contained therein: "It should be apparent that the errors, overstatements, and omissions that we have noted in these biological texts ... all tend to enhance the plausibility of hypotheses that are presented. More importantly, the inclusion of outdated material and erroneous discussions is not trivial. The items noted mislead students and impede their acquisition of critical thinking skills. If we fail to teach students to examine data critically, looking for points both favoring an opposing hypotheses, we are selling our youth short and mortgaging the future of scientific inquiry itself."[2]

Reliable scientific findings are often censored and distorted for the purpose of promoting an obsolete theory that is unsupported by known facts. Evidence for the radical claim that all species developed from a single life-form on a primeval earth is completely lacking, and unbiased scientists, doing hands-on origin-of-life research, have made this abundantly clear. Educators in the compromised world of academia (the ivory tower) are forcing their worldview on unwary students at government expense. Consider this statement from writer Timothy Lahaye: "Long before the 55 million children in our American school system are old enough to understand or examine humanism for themselves, already they are convinced it is scientific ... Until educators in the public school system awaken to the fraud they are perpetrating, humanists will continue to dominate them and force them to disseminate unsupported and unscientific dogma"[3] Science education in America has all but destroyed the freedom of thought and inquiry of today's eager young minds as it pertains to the origin of life and the material world. At the same time, we see that brave, honest educators in public institutions, who dare to suggest the obvious conclusion of design and purpose in nature, are browbeaten and stifled at every turn. On most public college and university campuses, any professors who admit to having serious doubts about the factual nature of evolution theory are typically laughed off campus and out of a job. Those who will not be silenced by unrelenting humanist demands risk losing tenure, or being fired and blackballed from the education system.

Consider this blatant example of how university administrators will often deal with a teacher or professor whose own personal philosophical view of life is perceived as a threat to the closely guarded academic position on evolutionary thought. This case involved Mr. Mark Armitage, a research scientist from California State University Northridge. In May of 2012, Armitage was doing research at the Hell Creek formation in the state of Montana. While unearthing fossil remains at this location, his team came upon one of the largest triceratops horn fossils that has been uncovered to date. The Hell Creek location has become legendary for fossil-bearing rock strata, which have produced a variety of well-preserved dinosaur remains. With this fascinating discovery in hand, Mr. Armitage returned to his CSUN laboratory in Los Angeles to complete a thorough examination of the specimen. He used a high-powered microscope to examine the fossil remains in minute detail, and carefully recorded his observations. To his surprise, tissue inside the horn revealed shocking evidence that threatens to alter the standard academic view on paleontology. Armitage discovered preserved layers of soft tissue that were pliable to the touch inside this fossil specimen. Based upon previous laboratory experiments, conventional wisdom suggests that soft tissue and possibly DNA could remain intact for possibly ten-twenty thousand years at most, but certainly not for millions of years.

Members of the scientific community were stunned by this finding, which again suggests that dinosaurs are most likely only thousands, and not millions, of years old. You and I were taught that dinosaurs roamed the earth between sixty-five and two hundred million years ago before becoming extinct. Taken at face value, aside from evolutionary bias, it would seem this Hell Creek discovery casts serious doubt upon the vast ages that are constantly spoken of in the dating of fossils. In this particularly interesting case, biased educators were not dealing with a novice or some fly-by-night rookie who set out to make a name for himself. Mr. Mark Armitage is a well-educated specialist in his field (a doctoral candidate of Liberty University). His work includes an appearance on the cover of scientific journals eleven different times, and is a well-published researcher whose work spans over thirty years. As he did on many other occasions, Armitage published a scientific review revealing his professional observations of the Hell Creek discovery. His research appeared in a peer-reviewed journal called *Acta Histochemica* that was published in February 2013.[4] This scientific journal typically reports on cell and tissue research. His article was peer-reviewed by Mary Schweitzer, a paleontologist at NC State (the same woman who discovered soft tissue in the leg bone of a Montana

T-rex) and was found to be professionally, very well done, without making any religious claims related to the discovery. Armitage simply stated the factual details he observed through a careful scientific analysis.

Two weeks after the article was made public, CSUN officials abruptly pulled back the funding for his laboratory research, silenced the evidence, and immediately terminated Mr. Armitage from his employment with the university, where he successfully managed the biology department's electron and confocal microscopy suite for a number of years. Later, an angry university official made it clear why they were letting him go when he said, "We are not going to tolerate your religion in this department!" An ensuing legal battle began on behalf of Mr. Armitage, who had previously been credited with outstanding job performance. A personal attorney representing him soon issued a statement in which he explained that the act of terminating an employee because of his or her religious views is most definitely illegal. He further explained that the termination of an employee to silence or censor scientific evidence to preserve the university's position on the evolutionary dating of fossil remains, was even more alarming. The attorney suggested that this incident should be a wake-up call to the entire academic world as it was a blatant example of how evolutionary thought is closely guarded and protected by academic institutions even to the point of violating the law to do so.

It should be stated here that the soft tissue deposits revealed by this triceratops horn are not unique. We have already considered the evidence of soft tissue documented from the leg bone of a Montana T-rex. Dr. Kevin Anderson, a molecular biologist and director of the *Van Andel Creation Research Center* in Arizona reported that soft tissue has been discovered in numerous dinosaur fossil remains, including a rib bone and a theropod claw that sat on a museum shelf for one hundred years before being examined and revealing the presence of soft tissue.

The wrongful termination of Mr. Armitage left embarrassed CSUN officials, who did not want to be associated with a well-published creationist, in a difficult position. Not wanting to bring further embarrassment and expense on the university, they chose to settle out of court for a large sum of money. There have been numerous cases of honest researchers and professors being blackballed from the education system because they dared to report on findings that do not support long-standing academic traditions on evolutionary thought. This was the compelling point made by the *Expelled* documentary– that scientific inquiry and curiosity at the university level is welcome and even encouraged as long as results from such inquiry do not conflict with established academic views. In this incident, students of Mr. Armitage were deprived of valid empirically derived scientific data so university officials could continue to dominate their fragile understanding with a traditional evolutionary view of life. Never the less, more courageous researchers are publicly admitting what they have known privately for many years–that the concept of evolution disseminated in America's public schools is based mostly on speculation, and human imagination, and believing in such requires a measure of profound faith.

Very misguided education authorities have blatantly authorized the deceiving misinformation commonly found in public education textbooks. They are convinced, against sound reason, that Darwinian ideology adequately explains how all living systems came to be. Evolutionary thought has erroneously concluded that noble humankind is nothing but an advanced form of animal life. Why should you and I willingly submit to evolutionary indoctrination without putting forth any resistance, whatsoever? Even our nation's highest court has recognized this as a significant abuse of intellectual power. The US Supreme Court ruled that "To promote one faith in the public school system at public expense, while banning the other, is an example of viewpoint discrimination, which the Supreme Court has declared unconstitutional in its variety of cases."[5]

The Media Monster Reigns

During these days of high-tech, digital communication, public media has a profound effect on our lives. Television, radio, magazines, books, and the internet are presenting new ideas and life-styles we were rarely exposed to only a short time ago. Media outlets have succeeded in establishing the concept of evolution as a widely accepted scientific fact in the mindset of the general public. Nothing, however, could be further from the truth! Mass-media bias continually offers evolution as the only possible explanation for life as we know it. Have you noticed that nearly every nature documentary recorded in our generation offers this ideology as factual by direct statements, or through subtle inference. We are constantly hammered with the phrase "millions of years" that is used to re-enforce, in our minds, the false doctrine of evolution. As defined for us in Webster's

Dictionary, a fact is something presented as having objective reality. A fact can be historically or experimentally verified, or perhaps corroborated by eye witnesses. Webster's New World Dictionary describes a fact as 'the state of things as they really are as distinct from fantasy'. Since evolution has never been proven by objective observation, and evidence to support it is still missing, we can quickly eliminate it from the category of scientific fact. People who control the media are, by and large, evolutionist in their thinking, or perhaps fearful of the scorn they would otherwise receive. We know this because the public rarely sees any material supporting an alternate point of view, and strict censorship is constantly employed to filter out any adversarial ideas. Most observers in Western culture will accept unverified statements and misinformation disseminated through the media without question or challenge. Since media outlets have such a powerful influence on all our lives, we owe it to ourselves, and to our children, to be careful and vigilant before accepting suspicious-sounding reports that are delivered to us as factual. Someone once said that if you tell a lie publicly often enough and long enough, the people would begin to believe it. I think this is true. It is called brainwashing, and was a technique that was used and perfected by none other than Adolf Hitler.

There are many unproven ideas that have been presented to the public through media outlets and public education. Unfortunately, these ideas have received wide public acceptance, even though their validity has seldom been seriously challenged. For example, geologists tell us that the earth is 4.6-5.3 billion years old. (It is always changing) We are expected to believe that the beginning of life occurred when mindless elements in some ancient primordial soup randomly assembled themselves by some unguided process to form a complex array of chemical bonds. This imagined colossal accident resulted in the essential proteins required to produce and sustain a living organism. Anthropologists report that dinosaurs became extinct after roaming the earth 65 to 200 million years ago. Evolution-minded scientists state that humans began to evolve from primate ancestors around one million years ago. They also claim to have reliable scientific methods of determining the age of fossils, and even that of geological formations. These so-called 'experts' have also led us to believe that our vast oil and coal deposits resulted from the decay of prehistoric life-forms over multiple millions of years of earth's history, long before humankind appeared. We are also informed that the discovery of organic fossilization within the earth's crust has given us a clear historical record of early life-forms, which, we are told, have largely substantiated the claims of the evolutionary model. However, all these claims are currently being refuted by a vast accumulation of past and current scientific knowledge. The evolutionist is the first to deny the idea of a worldwide flood as recorded in the biblical Genesis account. By the age-long process of mutation and natural selection, we are even expected to believe that somehow one species successfully evolved into another in an endless progression of ever-advancing life-forms from very simple to very complex. Finally, the strange notion of evolution is presented to us as a scientific fact, while the Holy Bible is consistently ridiculed as merely mythological. These bizarre evolutionary principles are continually embellished before us within the public domain as undisputed scientific reality in magazines, newspapers, television broadcasts, textbooks, and even over the internet.

Many people today accept these naturalistic claims with little hesitation because of a public domain that constantly censors any claims to the contrary. In 1982, the Institute for Creation Research requested that *Reader's Digest* (a popular magazine that has been known to give an occasional "nod to God"), publish at least one short research article from their organization defending creationism. This simple request seemed only reasonable considering the undisputed scientific work done by this remarkable organization. Listen to the response from the publishers of *Reader's Digest*: "Because predominant scientific thinking considers evolution as a solid basis for further research, Digest articles focusing on man's origin will continue to refer to evolution as a virtual certainty."[6] It is a sad world that censors scientifically valid adversarial views while also barring free speech and freedom of the press. If we continually allow ourselves to become indoctrinated by popular opinions without challenging the controversial ideas they present, then perhaps we do not deserve freedom of intellectual thought so cherished as an American ideal. Evolutionary theory is all about placing man's opinion above God's Word. The theory of evolution is clearly religious in nature, somewhat like believing any other creed. Why? Because it takes enormous faith to believe that matter, energy, and living organisms came about by blind luck, out of chaos, without God. This cleverly disguised religious philosophy dominates public opinion by the sheer magnitude of the media support it has received.

Now that we better understand the true nature of evolutionary thought, we must decide which origin-of-life model makes the most sense, objectively. Which is reasonable faith supported by scientific and historical evidence, and which is blind faith, designed to exalt man and to negate moral absolutes often viewed as too restrictive with regard to personal lifestyle choices? Many modern-day evolutionists vehemently attacking their creationist counterparts as unenlightened religious fanatics, are themselves promoting a religious dogma that relies on faith alone.

War of the Worldviews

The long-standing dispute between creation and evolution is a fierce battle between differing philosophies of life. Biblical Christianity and secular humanism are the incompatible worldviews that have framed this debate for the last century. These contrary views, at war in our world, form the basis for an ongoing cultural divide that is deeply rooted in the question of origins, and holds the key to determining which basic philosophy of life a person will choose. The public perception that evolution is true science has given naturalists an undeserved advantage when it comes to explaining the origin of life. People tend to think that anyone with a PhD after his or her name must have the right answers. However, when their shortsighted ideology begins with the firm conviction that there is no God, the PhD must become adept at formulating rescue mechanisms in response new scientific data that denies evolutionary claims. A remarkable redirection taking place among many objective-minded scientists is a sure sign that creationism is on the rise while evolutionary dogma is fading. Many scientists are becoming more open to the idea of allowing recently revealed evidence to guide them in a very new direction. Creationist and aerospace engineer, Dr. Luther D. Sunderland has commented on the numerous viewpoint transformations that have taken place on university campuses over recent years. "Hundreds of scientists who once taught their university students that the bottom line on origins had been figured out and settled are today confessing that they were completely wrong. They have discovered that their previous conclusions, once held so fervently, were based on very fragile evidences and suppositions which have since been refuted by new discoveries ...This has necessitated a change in their basic philosophical position on origins. Others are admitting great weaknesses in evolution theory."[7]

The greatest objection to creationism by proponents of evolution involves supernatural events, which cannot be accounted for based of the laws of nature. These physical laws are immutable based on all known human observations. However, the Creator God, who dwells outside of time and space, can circumvent these laws at will whenever it pleases him. As a transcendent being who ordained the laws of physics, God is not subject to them. The miracles described in the Biblical text are extra-ordinary events in which natural laws were suspended for a time, resulting in a unique purpose at God's discretion. A genuine miracle is a supernatural event by the hand of God that cannot be explained by physical laws. For the committed materialist, who subscribes to naturalism in explaining life's origin, no allowance can be made for the possibility of miracles. However, if the Creator is truly all-powerful, as he has proclaimed himself to be, then circumventing the natural laws that he himself ordained would be a small thing indeed. As Creator, the miracle-working God who dwells outside of time and space is never limited by laws He set into motion to govern the physical world. According to Dr. Henry M. Morris, "If God is the creator and sustainer of all natural processes, then he can surely change those processes when and as he wills. Thus, to say that miracles are impossible is simply to deny God."[8]

When it comes to the matter of miracles or supernatural events, the question is not "Are they possible?" but "Have they occurred?" If the weight of historical evidence suggests that many supernatural events have occurred in the past, then we would be mistaken to reject the idea of supernatural intelligence only because we do not understand how or why the miraculous occurs. Years ago, I attended the Yuma Air Show in my home state of Arizona. At this show, I became especially impressed by the performance of one particular airplane, which was entertaining the audience by casting a stream of smoke in a giant circular pattern above our heads. As I gazed at this huge circle in the brilliant afternoon sky, a peculiar thought occurred to me. 'If the vast space inside this circle represented everything there is to know about the physical universe, the combined total of mankind's accumulated knowledge including every generation until now, could be represented inside that circle as a tiny dot the size of a pin prick'. Could an omniscient God be at work within the immense area of a giant circle that is mysteriously unknown to man? It seems profoundly arrogant of men and women who act as though they are far too intelligent, and thoroughly enlightened, to need the help and guiding wisdom of an omniscient Creator.

Years ago, Dr. Albert Einstein, who won the coveted Nobel Prize for physics in 1921, offered a similar thought with this statement: "I'm not an atheist, but I don't think I can call myself a pantheist. We are in a position of a little child entering a huge library filled with books in many languages. The child knows that someone must have written those books. It does not know how ... It does not understand the languages in which they are written. The child dimly suspects a mysterious order in the books, but does not know what it is ..... That, it seems to me, is the attitude of even the most intelligent human being toward God."[9] Dr. Antony Flew was an award-winning British philosopher and well-known atheist whose final major publication was a book with the title: *There Is a God: How the World's Most Notorious Atheist Changed His Mind.* Long before Richard Dawkins of Oxford became popular as a contemporary poster-child for neo-Darwinian evolution, world-class philosopher Antony Flew was widely considered the most celebrated atheist in the academic world. Prior to his recent death, however, Flew, renounced atheism, and turned his heart and mind toward a profound belief in a creator. He explained to his critics how modern science had recently uncovered a level of complexity in design that Darwin's theory utterly failed to account for. As a lifelong observer of scientific research, Dr. Flew offered this insight from his own personal experience. "It now seems to me that the findings of more than fifty years of DNA research have provided materials for a new and enormously powerful argument for design."[10] German-born rocket scientist, Dr. Werner von Braun, who became a NASA aerospace engineer, offered this commentary: "I find it as difficult to understand a scientist who does not acknowledge the presence of a superior rationality behind the existence of the universe, as it is to comprehend a theologian who would deny the advances of science."[11] "Nature does not know extinction; all it knows is transformation ... Everything science has taught me and continues to teach me strengthens my belief in the continuity of our spiritual existence after death. Nothing disappears without a trace!"[12]

Living in a vast, virtually unknowable universe, for anyone to exercise faith in the God of the Bible, they must believe in miracles. Unexplained supernatural acts that defy natural laws and human reasoning are characteristic of the God whom the Bible declares to be the Creator of all things. The Bible teaches: "for the invisible things of him from the creation of the world are clearly seen, being understood by the things that are made, even his eternal power and Godhead; so that they are without excuse." (Romans 1:20 KJV) In other words, the Bible says that if we deny God as the creator of the physical world, we will have no excuse when we stand before him in judgment. This is because he has revealed to us everything we need to know in order to exercise faith in Him. It is not that we cannot believe. Believing in a creator God is clearly the most reasonable option. The truth is that many people simply refuse to believe, despite overwhelming evidence that strongly favors a "theistic" universe. As it turns out, however, it is foolish human pride and moral depravity that causes many to deny Him.

Evolutionary theory is designed to give self-centered man the hope of never having to stand before a Holy God in personal judgment. You do not have to see the Creator, or personally observe his powerful acts, to recognize the evidence of design all around you. What need is there for anyone to prove that an artist lives or has lived? His fine works of art will surely validate his existence. When we take an honest look at the factual evidence available today, what we observe clearly and consistently lends commanding support to the traditional, time-tested, creation worldview. Creationist and Christian leader Kenneth Ham, the founding director of Answers in Genesis and a powerful voice in the creation movement, describes it this way: "Only the Bible explains why there is beauty and ugliness; why there is life and death; why there is health and disease; why there is love and hate. Only the Bible gives the true and reliable account of the origin of all basic entities of life and the entire universe."[13]

If we could somehow assembled all the proteins and other structural features to form a perfect, complete organism, you would still not have life because a "life force" must be imposed on the physical elements by a source outside of the material world. The Lord God created Adam from the dust of the earth, and fashioned his body with every chemical structure and physical feature necessary for life. Perhaps, you will recall from a childhood Sunday school class years ago that the first human on the earth did not become a living, breathing creation until, "God breathed into his nostrils the breath of life." (Genesis 2:7 KJV) Then, and only then, did man become a living being. It is very important for us to understand that perfect chemical structure and complete physical form, in and of itself, could never come alive on its own. Life can only come from existing life! The life we see on planet earth could only have come from a live Supreme Being, who alone possessed the power to impose "the breath of life" on the physical elements, which he himself has created. This explains why attempts to create life in the laboratory are certain to fail. Life is top down from a creator, not bottom up by an imagined mechanism dependent on pure chance. We are told about the futility of relying on evolution with the words of Dr. Albert Fleischmann. "The Darwinian theory of descent has not a single fact to confirm it in the realm of nature. It is not the result of scientific research, but purely the product of imagination."[14]

If your belief system is based on what seems intellectually sound, according to the word of so-called 'experts' and the idea of 'naturalism' appeals to your mind, then the concept of evolution is likely to capture your allegiance. If, however, you are more interested in where the physical evidence is leading, then you should conclude that life on earth cannot be rooted in materialism. You and I will be forced, if we are intellectually honest, to consider a worldview that cannot be arrived at by drawing conclusions based on assumption. Can we afford to trust our eternal souls to the wisdom and will of man, whose ever-changing opinions are an insult to the Creator? Or, will we trust our lives and our future to the wisdom of the revealed Word of God?

The truth is that the sovereign God of the Bible has created everything in the physical and spiritual realm. He has purposefully designed a magnificent universe, and he has lovingly fashioned our beautiful home (earth), as He carefully planned the perfect dwelling place for mankind. No single flaw or imperfection existed in the perfect world the Most-High God had created. But man chose to disobey his Maker and fell into sin. Because of this willful rebellion, and disobedience drastic change came over the perfect world God had made. Man's sinful, selfish act invited the curse of death, sorrow, disease, pain, and suffering with no one to blame except himself. "Therefore, just as through one man sin entered the world, and death through sin, and thus death spread to all men, because all have sinned" (Romans 5:12). You and I must live out our days with sin, corruption, greed, lust, and hate all around us. This was certainly not God's original plan! But as the Creator, God retains absolute authority over his creative works! We, as human beings, who were created in the image of Almighty God, are accountable to the sovereign Lord of the universe. As Eternal God with infinite power and authority it was up to Him to establish moral standards that would reflect His image in the earth. Therefore, it is in our best interest to reverently acknowledge and humbly fear the divine authority God has lovingly placed over our lives as the centerpiece of His master plan.

Secular humanists strongly oppose such a worldview, so they vehemently deny the possibility of "God" any credibility in public discourse. They have deceived the public into thinking that by eliminating Christian views from public schools, they have simply left our children in a neutral position with regard to religious faith. This is not at all true! All they have succeeded in doing, which should be of utmost concern for every responsible parent in America, is eliminating Christianity from the public schools in favor of the anti-God religion of 'secular humanism'.

Make no mistake about it—the views of children are radically shaped through public education. Humanists refuse to accept Christianity because they deplore the idea of a creator God with the authority to judge their lives. They claim that evolution theory is scientific fact, but they cannot offer valid evidence to support this claim, and this is, of course, because there is none! Henry Morris PhD, explains their troubling dilemma in these remarks: "No one in all human history has ever observed one species evolve into a more complex and better adapted species by natural selection or any other mechanism ... No one has seen evidence of any mechanism that would make evolution work. In the fossil record of the past, with billions of fossils preserved in the Earth's sedimentary crust, no one has ever found any fossils showing incipient or transitional structures, leading to the evolution of more complex species."[15] He continues: "All the hard data in these life sciences show that evolution is not occurring today, all the real data in the earth sciences show that it did not occur in the past and all the genuine data in the physical sciences show that it is not possible at all. Nevertheless, evolution is almost universally accepted as a fact in all the natural sciences ... all of the real data of these sciences can be understood much better in the context of creationism ... It is the evolutionary framework in which they are taught and the evolutionary premises upon which they are built that ought to be rejected."[16]

The foundational basis for the far-reaching claims of evolution decrees that God does not exist. Therefore, at least to some degree, our influential public institutions have become indoctrination centers for the promotion of atheism. If we examine the physical evidence at face value, we must consider the implications of where this crucial data is directing us. The former atheist and MIT graduate, Dr. Frank Tipler, is a mathematical physicist and cosmologist who converted to Christianity after many years of researching the deep secrets of the universe. Frank Tipler is currently serving a joint professorship in the Department of

Mathematics and Physics at Tulane University. Tipler offers what he believes as the most rational approach to facing the problem of origins when examining physical evidence based on scientific observation alone. "My approach to reality is different. I believe that we have to accept the implications of physical law whatever those implications are. If they imply the existence of God, well, then God exists."[17] I admire his honesty, which is so rare today in the liberal, ivory-tower world of modern academia.

To anyone who is willing to investigate, the evidence is overwhelming that the evolution theory is dead wrong! This grossly inadequate hypothesis amounts to nothing less than the most monumental deception the world has ever known. Despite a devastating lack of evidence, loyal proponents of "the theory" are simply not ready to throw in the towel. These are people who dreadfully fear any connection between God and science! Unfortunately, the most dogmatic defenders of naturalism are often seated in places of authority, having been elevated to lofty positions as professors, scientists, and journalistic editors. In addition, they have received the unequivocal backing of the most influential media distribution network in the world. In spite of this, however, the truth about the physical world is being disseminated more widely today than ever before, by every possible means, to anyone willing to investigate with an open mind.

Once the facts are understood, it becomes apparent that the theory of evolution is long overdue for the scrap heap of history. Like so many species of animals in the past, evolution is destined for extinction. A biomolecular scientist recently told me that he believes within two generations neo-Darwinian evolution will be only a passing memory. Many well-respected scientists are now allowing the facts to speak for themselves, and as a result are defecting from the evolutionist camp in record numbers. The main tactic that is left to the most ardent proponents of evolution theory is to ridicule their philosophical opponents as uninformed simpletons who really do not understand how the theory works. We should deplore the way in which liberal educators have conveniently and consistently censored relevant facts by hiding legitimate scientific data for the purpose of propping up a worldview that continually mocks God and denies his eternal Word.

True Confessions

There is a war against God that is being waged throughout every civilized culture in the world. This radical warfare has been orchestrated against the Almighty by many skeptical men and women who falsely believe they are speaking in the name of science. However, just because a scientist speaks does not mean that what he or she says is scientific or that it should not be questioned and put to the test. The German philosopher Friedrich Nietzsche once declared that "God is dead," inferring that science had buried him under an avalanche of modern intellectual enlightenment. It appears Nietzsche's philosophy has fallen on hard times indeed in a world that is no longer satisfied with wild speculation, biased rhetoric, and unfounded assumptions.

In the popular 2014 full-length feature film entitled *God's Not Dead* by Pure Flix Entertainment, we witness the saga of an atheist college professor whose goal was to coerce his students into denying the existence of God. In this he believed he was doing his unsophisticated students a great service by freeing their minds from the shackles of religious superstition. As the storyline begins to unfold, each character becomes seriously challenged by the difficult circumstances affecting human lives in a broken world. Each one will ultimately face the growing need for a "divine presence" to help them make sense out of personal experiences over which they have no control. This compelling film production reveals an understanding that logic, evidence, and human need will ultimately drive every open-minded soul toward a personal God who loves them. We have plenty of humanists who are telling us that modern science has systematically eliminated the need for God. At the same time, we have a growing number of theist-minded intellectuals who testify it was science that ultimately confirmed their faith in a higher power.

After more than one hundred years of evolutionary propaganda via the public airwaves and taught with religious zeal in public institutions, secularists suggest that science and religion can peacefully coexist only if religion remains private and restricted from the public domain. Even so, there is a new phenomenon taking place in the world of science. A philosophical change seems to be taking place in the evolution camp. The revelation of scientific knowledge that has been observed and documented over the past seventy years or so has now cast a dark shadow of uncertainty over evolutionary claims that were previously thought to be indisputable facts. This new evidence is leading many down a different path than most scientists could have ever imagined. Those within scientific ranks are generally responding in one of three ways to the direction scientific research is now leading. Some are digging in their heels having remained determined to find a way of explaining life without God. In wishing to circumvent the notion of supernatural agency, atheists are rigidly maintaining the drumbeat litany that boldly proclaims: "We know evolution is true, even though we do not know how it works and have never seen it happen." The second wave of conviction suggests that Darwin's theory is all but finished in the minds of those who let the evidence lead wherever it wants to take them. Still these stalwart intellectuals, who are nearly ready to admit Darwin's demise, will often persist by simply stating, "We just don't know; so, we will have to keep on looking for a way to explain the physical realm and the remarkable emergence of organic life." And finally, there are some brave souls who are genuinely ready to acknowledge what the actual evidence is teaching us about the origin of life, even if it means to openly recognize that it is impossible to explain life without a reliance on some principle means of intelligent oversight. Nothing is worth believing, without a reasonable amount of evidence supporting it! William Paley, the eighteenth-century philosopher, wrote the original framework for the following compelling statement: "There is a principle, which is a bar against all information, which gives proof against all argument, and which cannot fail to keep any man in everlasting ignorance. The principle is contempt prior to examination!"[18] This very profound thought reminds us that when a person's mind is already firmly made up, all evidence is to him of little significance. When considering the important question of origins, many people remain very guarded about what they believe, and will often see any attempt to persuade them otherwise as a threat to their intelligence, as though someone were attacking their religion. No surprise, because we now understand that this is exactly what belief in evolution is--a religion! When it comes to debating the questions of life with someone who is deeply committed to an evolutionary worldview, do not be surprised if they turn a deaf ear to sound arguments. Many are hopelessly persuaded, because this is what they want to believe, and any logically derived empirical evidence will bear little consequence to them. "A man persuaded against his will is of the same opinion still!" This firmly engrained posture is commonly observed when people are considering creation versus evolution. However, after more than one hundred-fifty years of revealed facts that have flatly denied Darwin's claims, the scientific tables are beginning to turn. We are witnessing a resurgence of the traditional Judeo-Christian worldview. This trend is not from biblical persuasion, but seems based on the laws of physics and the evidence of design.

We need to consider a few of the revised opinions, conclusions, and statements that some of today's evolutionists are now inclined to share publicly because of their admitted frustration over the lack of evidence to support their preferred worldview. Paleo-evolutionist, Dr. Niles Eldredge, is a curator with the American Museum of Natural History. As such, he has evaluated a vast array of fossil evidence over the years, and now, he is publicly revealing the dark secrets of paleontology. He wrote, "We have proffered a collective tacit acceptance of the story of gradual adaptive change, a story that strengthened and became even more entrenched as the synthesis took hold.

We paleontologists have said that the history of life supports that interpretation, all the while really knowing that it does not."[19] Another frank acknowledgement comes from a Scottish anthropologist, Sir Arthur Keith: "Evolution is unproved and unprovable ... We believe it only because the only alternative is 'special creation', which is unthinkable."[20] When the famous molecular biologist Francis Crick, and his colleague James Watson, discovered the remarkable double helix structure of the DNA polymer and both were awarded the Nobel Prize in 1962, their work popularized DNA research as never before. Later, as a research professor at the Salk Institute for Biological Studies in La Jolla, California, Francis Crick remained on the cutting edge of molecular biological research for many years. After a lifetime of devoted investigation at the molecular level, Francis Crick ultimately came to the following conclusion: "An honest man, armed with all the knowledge available to us now, could only state that, in some sense, the origin of life appears, at the moment, to be almost a miracle ... so many are the conditions which would have had to have been satisfied to get it going."[21] Another evolutionist, Roger Haines Jr. has openly confessed: "The arguments for macroevolution fail at every significant level when confronted by the facts."[22] Again and again, we see startling admissions coming from the evolutionist camp. Professor Edmund J. Ambrose of the University of London has offered this comment: "At the present stage of geological research, we have to admit that there is nothing in the geological records that runs contrary to the view of conservative creationists, that God created every species separately, presumably from the dust of the earth."[23] Now, consider a remarkable confession by evolutionary paleontologist Mark Czarnecki as he highlights a difficult obstacle for naturalism: "A major problem in proving the theory (of evolution) has been the fossil record; the imprints of vanished species preserved in the Earth's geological formations. This record has never revealed traces of Darwin's hypothetical intermediate variants; instead species appear and disappear abruptly, and this anomaly has fueled the creationist argument that each species was created by God."[24] Gifted astronomer, physicist, and highly regarded cosmologist Dr. Robert Jastrow reveals the current dilemma that many scientists are now facing: "Scientists have no proof that life was not the result of an act of creation, but they are driven by the nature of their profession to seek explanations for the origin of life that lie within the boundaries of natural law. They asked themselves, 'how did life arise out of inanimate matter'? And what is the probability of that happening?' ...And to their chagrin, they have no clear-cut answer, because chemists have never succeeded in reproducing nature's experiments on the creation of life out of nonliving matter"[25] Harvard's evolutionary biologist and paleontologist, Dr. Stephen J. Gould, was undoubtedly the leading evolutionary voice in America during the late twentieth century. He laments over his own personal comments on the obvious incompleteness of the fossil record: "The extreme rarity of transitional forms in the fossil record persists as the trade secret of paleontology. The evolutionary trees that adorn our textbooks have data only at the tips and nodes of their branches; ... the rest is inference, however reasonable, not the evidence of fossils ... We fancy ourselves as the only true students of life's history; yet to preserve our favored account of evolution by natural selection, we view our data as so bad that we never see the very process we profess to study."[26] Award-winning astronomer, Dr. Allan Sandage, who is widely regarded as the modern world's greatest living cosmologist, made this candid remark after having explored the many mysteries of astronomy for a lifetime: "I find it quite improbable that such order came out of chaos. There has to be some organizing principle. God to me is a mystery, but is the explanation for the miracle of existence, why there is something instead of nothing."[27] H. J. Lipson, a professor of physics, asks the question on the minds of so many evolutionists today who have repeatedly failed to discovered what they so earnestly expected to find. "In fact [subsequent to the publication of Darwin's book, Origin of Species], evolution became, in a sense, a scientific religion; almost all scientists have accepted it, and many are prepared to bend their observations to fit with it ... To my mind the 'theory' does not stand up at all ... If living matter is not, then, caused by the interplay of atoms, natural forces, and radiation, how has it come into being? ... I think, however, that we must go further than this and admit that the only acceptable explanation is creation ...I know that this is anathema to physicists, as indeed it is to me ... but we must not reject a theory that we do not like if the experimental evidence supports it."[28] Listen to these remarks from lifelong physicist and mathematician Dr. Wolfgang Smith: "A growing number of respectable scientists are defecting from the evolutionist camp ... moreover, for the most part, these 'experts' have abandoned Darwinism, not on the basis of religious faith or biblical persuasions, but on scientific grounds, and in some cases, regretfully."[29] It is now easy to understand why esteemed, world-class figures like Sir Frederick Hoyle, Antony Flew, and many others have decided to abandon their entrenched philosophical worldview in recent years and turn to a more realistic conclusion about the origin of life.

Many open-minded seekers of truth and knowledge have recently come around to agree that the majestic universe, as we know it, could never reasonably be explained as a colossal cosmic accident. Highly regarded science historian Frederick B. Burnham, who appeared briefly on ABC's Nightline with Ted Koppel, publicly confirmed the current openness of modern science to the intelligent design model with this remark: "The scientific community is prepared to consider the idea that 'God created the universe' a more respectable hypothesis today than at any time in the last 100 years."[30] Established science does not have the faintest idea how life began on planet Earth, and this admission should be confessed openly and honestly throughout the scientific community— and to the public at large. Because of this well-established fact, scientists and biased professors must refrain from confusing the minds of students with statements and ideas that are based solely on philosophical beliefs and nothing more. The last seventy years of scientific inquiry has clearly denied the traditional evolutionary model. Men of true science are beginning to consider, and in many cases accept, the only alternative: that a supernatural agent must be acknowledged as the chief architect and design engineer of a remarkable universe.

Final Thoughts

Canadian geologist, Sir William Dawson, helped to bring additional clarity to the argument when he summarized for us the plight of evolution theory. "This evolutionist doctrine is itself one of the strangest phenomena of humanity--a system destitute of any shadow of proof, and supported merely by vague analogies and figures of speech … Now, no one pretends that they rest on facts actually observed, for no one has ever observed the production of even one species ... let the reader take up either of Darwin's great books, or Spencer's Biology, and merely ask himself ... as he reads each paragraph, 'What is assumed here and what is proved?" ... and he will find the whole fabric melt away like a vision .... We thus see that evolution as an hypothesis has no basis in experience or in scientific fact, and that its imagined series of transmutations has breaks which cannot be filled."[31]

After many years of study and research into the question of origins, I have encountered a vast amount of material from both the modern world of science and the past world of natural history. This catalog of physical evidence is known to lend much greater credibility to the creation model of origins than to an evolutionary account. After 500 years of scientific inquiry and investigation, it is safe to say that the vast preponderance of evidence gathered throughout the scientific age most strongly favors 'theism' as opposed to naturalism; so many are the unexplained anomalies that have greatly strained the fragile confidence of objective scientists. Some were even willing to manipulate evidence in order to promote the worldview they have so desperately believed in against insurmountable odds. Evolutionists have a very serious problem on their hands because they have been forced into the uncomfortable position of having to logically account for the remarkably complex designs found in nature without the creative power of intelligent design. They have insisted that our children should be taught such things in school as if they were undeniable facts. There was a time in America when observable facts would identify truth and knowledge, but I suppose that time was long before the current wave of humanistic pretense. Today's cultural climate compels the American public to embrace and believe outlandish claims that are grounded far more in guesswork and wishful thinking than empirical, observable, and testable scientific methodology. Many aspects of the vast physical domain in which we live cannot be understood when God is excluded from the equation. The basic flaw we find in the evolution model has been the obvious inability to explain where the information came from that is essential for life to exist. The critical information contained in genetic DNA was produced by a cognitive mental process. Evolutionary thought is bound by the unthinkable proposition that the vast information for spelling out life required no intelligence, but simply owes its truly miraculous existence to a mindless, unguided process. Dr. Henry M Morris was perhaps one of the greatest scientific philosophers and educators of the twentieth century. This man devoted his entire adult life to the interpretation of observed and applied science in the present world, and the world of the past. He offers the following expert analysis of the major cultural and intellectual battle that is facing America today. "The modern creation/evolution conflict is more than a mere scientific controversy, or even a battle between science and religion, as many evolutionists pretend. It is nothing less than a new and critical phase in the age-long conflict between the only two basic worldviews ... One is centered on the creator of the world, and his redemptive work on behalf of that lost world; the other is centered in the creatures of that world, not only man with his self-oriented goals, but also the devil himself, who is ultimately behind all rebellion against God."[32]

German-born, British biochemist Ernst Chain was among the recipients of the 1945 Nobel Prize. After the remarkable discovery of penicillin by Alexander Fleming, Dr. Chain worked with both Fleming and Howard Florey on penicillin research for which they all shared the Nobel Prize. Chain was successful in identifying the structure of penicillin, and able to isolate the active substance in his laboratory research. For this, Chain is considered a founding pioneer in the medical field of antibiotic therapy. Regarding his views on Darwin's theory of evolution, Dr. Chain found the theory to be "only a very feeble attempt to explain the origin of species based on presumptions so flimsy that it can hardly be called a theory." Chain viewed the evolution model's reliance on chance mutations as a "hypothesis based on no evidence and irreconcilable with the facts." He wrote: "These classic evolutionary theories are a gross oversimplification of an immensely complex and intricate mass of facts, and it amazes me that they were swallowed so uncritically and readily, and for such a long time, by so many scientists without a murmur of protest."[33] Dr. Chain finally concluded that he "would rather believe in fairies than in such wild speculation" as Darwinism. These powerful statements are the personal reflections of a world-class, Nobel-Prize-winning, scientist whose dazzling research successfully pioneered a new scientific field of investigation by altering the future of medicine with groundbreaking discoveries.

Dr. Robert Jastrow, a decorated astronomer, physicist, and cosmologist, became the founding director of NASA's Goddard Institute for Space Studies in 1961. He offers another interesting perspective. Although a self-proclaimed agnostic, Dr. Jastrow has provided a very revealing conclusion after finally reaching the end point of scientific limitation where he, and many others made a very startling discovery. Consider the implications that are suggested by his profound observations: "A sound explanation may exist for the explosive birth of our universe; but if it does, science cannot find out what the explanation is. The scientist's pursuit of the past ends in the moment of creation ... This is an exceedingly strange development, unexpected by all but the theologians. They have always accepted the word of the Bible: In the beginning God created heaven and earth ... At this moment it seems as though science will never be able to raise the curtain on the mystery of creation. For the scientist who has lived by his faith in the power of reason, the story ends like a bad dream. He has scaled the mountains of ignorance; he is about to conquer the highest peak; as he pulls himself over the final rock, he is greeted by a band of theologians who have been sitting there for centuries."[34]

In 1929, astronomer Edwin Hubble discovered that the galaxies are receding from one another as the universe appeared to be continually expanding. This observation eventually led to a firm conclusion by the vast majority of physicists and cosmologists that our universe began at a single moment of time– a profound concept that is now commonly referred to as the "big bang." Physicists are now virtually all in agreement that the present universe began as a dense and unexplainably dimensionless point of pure concentrated energy. The scientific concept of an explosion of energy "in the beginning" is currently being considered, by many, to be perfectly compatible with the Judeo-Christian tradition as stated in the opening words of Genesis, "In the beginning God ...Then God said: Let there be light; and there was light" (Genesis 1:1, 3). Dr. Arno Penzias, who won a Nobel Prize for physics in 1978, came to a similar conclusion based on a lifetime of scientific achievement: "The best data we have [concerning the big bang] are exactly what I would have predicted, had I nothing to go on but the five books of Moses, the Psalms, and the Bible as a whole."[35] Many of today's scientists claim to have discovered answers that would circumvent what theologians have believed for centuries; however, their bold and often dogmatic claims have not always turned out as they expected. The proven scientific facts gathered from many years of disciplined technical research continue to refute these claims. The Bible is not in conflict with science, as evolutionists believe. In fact, it can be stated, categorically, that not even a single known contradiction exists between the biblical record and any proven or otherwise observed fact of science. They insist that God was not the initial cause, so their scientific conclusion is that 'nothing exploded and became everything'.

Swedish embryologist and biochemist, Dr. Soren Lovtrup, took a parting shot at the evolution model that is still reverberating throughout the scientific community. "Micro-mutations do occur, but the theory that these alone can account for evolutionary change is either falsified, or else it is an un-falsified, hence meta-physical theory. I suppose that nobody will deny that it is a great misfortune if an entire branch of science becomes addicted to a false theory ... But this is what has happened in biology ... I believe that one day the Darwinian myth will be ranked the greatest deceit in the history of science ... When this happens, many people will pose the question: How did this ever happen?"[36] The more advanced man's revelation becomes, the more evident it appears that creation is the correct explanation for the origin of life on earth. The amazing degree of complexity and order that man has discovered cannot possibly be the product of blind chance events. Because scientific observation clearly supports the Genesis account, it is most unfortunate that so many intelligent people are still clinging to the unproven and historically repudiated philosophy of evolution. Dr. Louis Bounoure is a former biology professor at the University of Strasbourg and served as the director of the Strasbourg Zoological Museum. He put it rather bluntly when he said, "Evolutionism is a fairy tale for grownups. This theory has helped nothing in the progress of science. It is useless!"[37] Dr. T. N. Tahmisian of the U.S. Atomic Energy Commission elaborates further: "Scientists who go about teaching that evolution is a fact of life are great con-men, and the story they are telling may be the greatest hoax ever ... In explaining evolution, we do not have one iota of fact."[38]

Ultimately, belief in evolution leads mankind to a form of self-worship or humanism. This false belief maintains that there never has been an intelligent power greater than man. The Bible even refers to this type of thinking in Romans chapter 1, beginning with verse 22, where we read this powerful reproof: "Professing themselves to be wise, they became fools, and changed the glory of the un-corruptible God into an image made like corruptible man ...who exchanged the truth of God for the lie [evolutionary humanism] and worshipped and served the creature rather than the Creator ... and even as they did not like to retain God in their knowledge, God gave them over to a reprobate mind." (Romans 1:22, 25, 28) People are eager to latch onto evolution because they believe it somehow justifies a self-directed lifestyle by offering mankind the imagined hope of never having to face personal accountability and judgment. However, the evolution model provides its followers with only a false sense of security— suggesting that at the end of life they will have no one to answer to. This 'false hope' is the ultimate lie of evolution! It is a monumental deception wrought with deadly consequences. The most significant danger is that evolution leads mankind to a rejection of the authority of the Bible, resulting in a failure to appropriate the promises therein, leading to salvation and eternal life. The Creator had a definite purpose in his undeniable creation, and it His purpose was centered primarily in mankind. It is terribly foolish to deny God. To deny his divine power and authority is a dangerous choice. "Who are you to reply against God? Will the thing formed say to him who forms it, 'why have you made me like this'? Does not the potter have power over the clay?" (Romans 9:20, 21).

When it comes to the origin of life and the universe, we can now separate the truth from fantasy. Evolution theory is the consummation of man's stubborn refusal to submit his personal will to a powerful intelligence greater than himself— a submission that, once offered, will prove to be truly life-altering toward the prideful independence and arrogance of the self-centered human heart. How should we answer the ultimate questions suggested by the direction scientific discovery is pointing us in? Which existed first, mind or matter? Was there intelligence before the physical domain came to be? Do we owe our existence to a power greater than random chance? After all these arguments have been thoroughly examined, either we are the accidental outcome of billions of random events in an impersonal universe, or there is a living God in heaven that has planned for us to be here. As Oxford Professor, John Lennox, has aptly stated, "Either human intelligence ultimately owes its origin to mindless matter; or there is a Creator."[39] In spite of over one hundred years of evolutionary propaganda, the historic Judeo-Christian tradition remains remarkably strong as revealed by the many expert testimonies we have encountered here.

1. The physical elements are clearly not eternal but had a beginning.

2. The universe had its beginning through an 'initial cause'.

3. The causal agent is the God of the Bible, Creator of heaven and earth.

4. A transcendent God existed before the universe was and is entirely independent of it.

5. The physical dimension was designed and created with humanity in mind.

6. This eternal Creator God has thoroughly revealed himself to his creation!

Naturalism has been exhaustively weighed in the balance and found wanting! Now that 'the theory' has been thoroughly discredited and exposed for the monstrous deception that it is, how should we respond to the powerful evidence that strongly supports a 'theistic universe' and the creation worldview? The religious ideology of 'evolutionism' is the most colossal fabrication ever imposed upon human lives? Now that we know the truth, we must share it with those who don't.

"And you shall know the truth, and the truth shall make you free." (John 8:32)

"But without faith it is impossible to please Him; for he who comes to God must believe that he is [that he exists], and that he is the rewarder of those who diligently seek Him." (Hebrews 11:6)

"In the beginning, O Lord, you laid the foundations of the earth, and the heavens are the work of your hands." (Hebrews 1:10 NIV)

"I have made the Earth, and created man upon it: I, even my hands, have stretched out the heavens, and all their host have I commanded." (Isaiah 45:12 KJV)

"The god of this age has blinded the minds of unbelievers, so that they cannot see the light of the Gospel of the glory of Christ, who is the image of God." (2 Corinthians 4:4 NIV)

"For thus saith the Lord that created the heavens; God himself that formed the Earth and made it; he hath established it, he created it not in vain, he formed it to be in the inhabited: I am the Lord; and there is none else." (Isaiah 45:18 KJV)

It has been said that around the beginning of the twentieth century an unknown scientist of that time had discovered and subsequently reported what he sincerely believed to be the five basic properties defining the physical realm. These five essential properties of the physical dimension were subsequently identified as time, force, action, space, and matter. I would submit to you that this early-twentieth-century scientific thinker, whoever he was, had simply stumbled upon what was the most basic knowledge that a sovereign creator God had revealed to his creation thousands of years before! "In the beginning [time] God [force] created [action] the heavens [space] and the earth [matter]" (Genesis 1:1 KJV, [with amendments]. The jury is back on the question of scientific skepticism and a verdict has been reached. To the intellectually honest seeker of truth, divine providence cannot be denied.

So let him, the only sovereign God of all creation and the supreme Lord of the universe, who alone is found to be worthy of our worship, praise, and allegiance, be glorified by the thoughts we are thinking, the lives we are living, and the God-ordained posterity we are producing!

CHAPTER SEVEN

* * *

ALMIGHTY OR ALL MYSTERY

THE ALL-IMPORTANT QUESTION

On the one hundredth anniversary of Darwin's revelation to the scientific world, atheist Julian Huxley said this, "In the evolutionary scheme of thought there is no longer either need or room for the supernatural. The Earth was not created, it evolved. So did all the animals and plants that inhabit it, including our human selves, mind, and soul as well as brain and body... "[1] Mr. Huxley was certainly living proof that atheism is the logical outcome and ultimate consequence of evolutionary thinking, even though his dogmatic, and provocative statements are now proven to be totally without merit. I hope you will agree that the material from the previous chapters is more than enough to convince any interested person that the existence of the universe and everything in it had nothing to do with unguided, random chance events.

We have discovered the evolution model as terribly lacking and grossly inadequate to explain the remarkable diversity and complexity of life on planet Earth. Therefore, the most logical conclusion we can make, based on the evidence, is to realize that there must be a living intelligence with supernatural power who is responsible for the incredibly unique beauty and design that surrounds us. In this chapter, we will examine, more closely, the evidence for a purposeful God behind the scenes– the one who planned, designed, created, and sustains the universe in which you and I are privileged to live. The skepticism about God is widespread, so we will examine compelling evidence that will help answer this all-important question. Was there a personal, purposeful creator God at work initiating the appearance of the physical world? By answering this one critical question, we should find that all of the other deeply philosophical mysteries of life will become much easier for us to understand.

If God is real, we must know this for sure, because an accurate worldview and promising future depend on it. How do we know if there is a divine, unseen force governing the universe? Since we cannot detect the unseen God directly through our physical senses, we can only experience him through faith based on what we observe, and what He has chosen to reveal to us. Who is this Almighty God of the spirit world that desires a relationship with mankind? What is his divine claim as the chief architect of the material world? Has he revealed himself through his creation? Has he taken satisfactory measures to make himself personally known to us? In this chapter, I offer the reader sufficient reason to accept the reality of God's existence. I hope you will agree that discovering the truth about Him is not nearly as difficult as many people would like to think.

If there truly is a sovereign God beyond time and space, in a reality scarcely understood by mortal man, then it is entirely up to the Creator to reveal himself to us; and he would have to do this in a way that will satisfy the deep longing within humanity for knowledge and understanding about his Maker. You and I can only view reality from the remote perspective of a mundane three-dimensional world. This leaves us with only two basic ways by which we can discover the truth about God, if we are looking for him with an open heart and mind. The first method is found through "observation": we can infer what is true about God by observing the physical features on display on earth and in the vast universe that surrounds us with all its fascinating complexity and design. The second means of discovering the truth about God is found through "revelation": We can know more specifically the many details about God and his nature only if he is willing to reveal them to us. The truth about God, and his undeniable role in designing the physical reality, is of the utmost importance in our quest to fully understand "The Mystery of Life." How could a powerful, just, and loving God, beyond time and space, reveal himself and his character to finite man in a way that addresses the question about his eternal existence, as well as his great love and undeniable devotion to those he has created in his own image? We shall examine some reliable evidence as we begin a thorough investigation into numerous, compelling reasons for believing in the existence of the Almighty.

Observation: The Anthropic Principle

Through investigative research we find that both secular and Christian scientists have reported detailed data about the sensitive physical characteristics of planet Earth, in what is now often referred to as the "anthropic principle." Scientists report that it appears the earth's most delicate physical parameters are finely tuned and perfectly balanced for the purpose of allowing organic life to exist. This reveals that the features of our planet, with respect to the solar system and the galaxies, are dialed in to exactly where they must be to support life. According to author and conference speaker Nancy R. Pearcey, a senior fellow at the Discovery Institute, when you examine the unique physical characteristics of the universe with so many primary constants set with amazing precision: "What you discover is that each of the thousands of dials just happen to be set to exactly the right value for life to exist. Even the slightest tweak of one of the cosmic knobs would produce a universe where life was impossible."[2] She tells us the slightest variation in one of these critical physical parameters would result in our rapid demise. World-renowned scientist Stephen Hawking, the British astrophysicist sought after the knowledge that has eluded mankind— for when we find it he states: "Then we shall be able to take part in the discussion of the question of why it is that we and the universe exist ... If we find the answer to that, it would be the ultimate triumph of human reason for then we would know the mind of God."[3]

Many believe that the current evidence points to only one possible conclusion. Contemporary astrophysicist, Dr. Paul Davies has explained: "Why is nature so ingeniously, one might even say suspiciously, friendly to life? ... It's almost as if a Grand Designer had it all figured out."[4] "Everywhere we look, says Davies, from the far-flung galaxies to the deepest recesses of the atom, we encounter regularity and intricate organization ...

moreover, the behavior of physical systems is not haphazard, but lawful and systematic."[5] He concludes with this statement after many years of researching the physical laws: "The equations of physics have in them incredible simplicity, elegance, and beauty. That, in itself, is sufficient to prove to me that there must be a God who is responsible for these laws and responsible for the universe."[6] Consider a similar thought from NASA planetary scientist, John O'Keefe, who writes: "We are, by astronomical standards, a pampered, cosseted, cherished group of creatures ... If the universe had not been made with the most exacting precision, we could never have come into existence ... It is my view that these circumstances indicate the universe was created for man to live in."[7]

Cambridge University physicist John Polkinghorne states: "When you realize that the laws of nature must be incredibly finely tuned to produce the universe we see, that conspires to plant the idea that the universe did not just happen, but that there must be a purpose behind it."[8] British astronomer and world-renowned mathematician, Sir Frederick Hoyle describes earth's life supporting parameters with these comments: "Such properties seem to run through the fabric of the natural world like a thread of happy coincidences. But there are so many odd coincidences essential to life that some explanation seems required to account for them."[9]

In identifying a cause of the physical world, we have two choices to pick from, faith in a living supreme creator or faith in the accidental appearance of all living and nonliving matter in an impersonal, godless universe. Can we believe that the physical world was accidental, or should we consider it the ingenious work of a divine architect–one who planned and prepared a perfect world, with mankind as his primary motivation? Theoretical physicist, Dr. Steven Weinberg, the winner of the Nobel Prize for physics in 1979, stated: "How surprising it is that the laws of nature and the initial conditions of the universe should allow for the existence of beings who could observe it. Life, as we know it, would be impossible if any one of several physical quantities had slightly different values."[10] The list below includes a few of the many critical physical constants of planet earth that, if varied only slightly, could impose dire consequences on the existence of life.

1. Our planet is the ideal size for the atmosphere we need to survive. We know of no other planet with an atmosphere containing the perfect mixture of gases uniquely suited to support plant, animal, and human populations.

2. The earth is perfectly positioned within our solar system; it is exactly where it needs to be with regard to its perfect distance from the sun and other planets. If positioned closer to the sun, we could not survive the heat; if farther from the sun, life would succumb to the frigid temperatures. Even the slightest variation in this critical parameter would be catastrophic.

3. The exact speed of the earth's rotation, and its precise orbit around the sun, are perfectly suited to allow for the warming and cooling of the continents, creating an excellent environment for living organisms. A slowdown of our planet's velocity would result in temperature fluctuations between day and night that would be far too extreme to support life.

An increase in rotational speed would produce devastating wind velocities.

4. The moon is the perfect size and distance from the earth, providing a gravitational pull that allows for critical tides and currents. This feature keeps the oceans from becoming stagnant, and maintains a healthy environment for the marine life so essential to our food supply.

5. An abundance of water on earth is necessary for the hydration cycle; precipitation provides the moisture offering life-giving support to all living organisms. Nothing can survive without it.

6. The earth's gravitational constant is also finely tuned to support life.

7. The ratio of electrons to protons cannot vary by more than one part in 10^{37}. Otherwise, electromagnetic interactions would prevent chemical reactions from occurring in nature.

8. The earth's magnetism is finely tuned to support living things.

9. The ozone layer above us is a critical characteristic of our planet that protects us from a dangerous overexposure to ultraviolet light, which would be lethal if not adequately deflected away from us in the earth's upper atmosphere. This is a critical planetary feature.

10. The crust of the earth is exactly the thickness it needs to be to support life.

11. The carbon dioxide-to-oxygen ratio is dialed in exactly where it needs to be.

12. The earth's reflectivity of sunlight is also perfectly fine-tuned for life support.

There are dozens of other key parameters dialed in for life support. Do you think it was a random accident that they are all specifically friendly to life? The slightest variation in these primary constants of physics would cause life, as we know it, to cease. Does this physical evidence point to the existence of God? British-born cosmologist, Edward Harrison, certainly seems to think so. "Here is the cosmological proof of the existence of God--the design argument of William Paley updated and refurbished. The fine-tuning of the universe provides prima face evidence of deistic design."[11] Another American astronomer, George Greenstein, concurs: "As we survey all the evidence, the thought incessantly arises that some supernatural agency, or rather 'Agency, must be involved. Is it possible that suddenly, without intending to, we have stumbled upon scientific proof for the existence of a Supreme Being?"[12]

We find even non-Christian scientists have reported that the earth appears to be specifically designed as a suitable home for mankind, which, by the way, is also how the Word of God has described it. "The heaven, even the heavens, are the Lord's; but the Earth has he given to the children of men" (Psalm 115:16).

Consider this remark from a famous Soviet mathematician, Alexander Polyakov: "We know that nature is described by the best of all possible mathematics, because God created it."[13] That the universe must be finely tuned to make carbon-based life possible is a universal concept in science today. A powerful example of what is meant by this fine-tuning is offered by Oxford mathematician, John Lennox, who stated, "The ratio of the electromagnetic force to gravity in the early universe had to be accurate to about one part in 10^{40} in order to have the chemistry of the universe as we now see it."[14] To give you some perspective of the precision commanded by this incredible fine tuning, famous physicist Paul Davies has compared it to the odds of hitting a one-inch target on the opposite side of the universe, some twenty billion light-years away. In regard to these elaborate design arguments, Oxford mathematician John Lennox has offered this candid remark: "The remarkable picture that is gradually emerging from modern physics and cosmology is one of a universe whose fundamental forces are amazingly, intricately, and delicately balanced or fine-tuned in order for the universe to be able to sustain life ... 'the fundamental constants of nature' ... Change any of them just a little, and the universe would become hostile to life and incapable of supporting it. These constants are precision-tuned, and it is this fine-tuning that many scientists (and others) think demands an explanation ... fine-tuning has established itself as an aspect of the universe that merits very serious consideration."[15] This precision-tuning argument has now led many qualified scientists to conclude that the earth and its unique life-supporting parameters were intended with us in mind—we did not just accidentally show up in a world that has randomly turned out to be perfectly designed and exceptionally well suited for us.

Highly esteemed physics Nobel Prize winner, Dr. Arno Penzias, has described the anthropic principle in this manner. "Astronomy leads us to a unique event, a universe which was created out of nothing, one with the very delicate balance needed to provide exactly the conditions required to permit life ... and one which has an underlying (one might say supernatural) plan."[16] Consider an interesting quotation from Dr. Henry Schaefer, a physical scientist who ultimately became a five-time nominee for the Noble Prize: "It is relatively unusual that a physical scientist is truly an atheist. Why is this true? ... Some point to the anthropic constraints, the remarkable fine-tuning of the universe. For example, Freeman Dyson, a Princeton faculty member, made this observation: 'Nature has been kinder to us than we had any right to expect.'"[17]Impressive anthropic evidence suggests that someone designed, and carefully dialed in planet earth's most delicate physical parameters in order to provide the perfect world where many plant, animal, and human populations could flourish and thrive in keen balance and harmony with nature.

Observation: Law of Cause and Effect

Someone eternal was the cause of everything that exists in the material world. If there is a God, why do so many intelligent people remain unconvinced of this truth? The 'law of causality' does not predict that everything requires a cause, but that every finite thing requires a cause. Every effect in the physical world must have a sufficient cause, but by definition, an infinite being does not require a cause. Since we know the universe is finite, we understand that it has had an initial cause. Therefore, we know that someone or something (a necessary being or entity) had to be uncaused (infinite and eternally existing) and would, therefore, qualify as the initial cause agent of the material reality that you and I are so fortunate to observe and experience.

Everyone agrees with the scientific principle of "cause and effect". However, many people are educated to the point of arrogance, and to the degree that God is no longer necessary in their personal lives. Modern societies have built vast education systems that subscribe to cause and effect, but only if the cause conforms with naturalism, and does not invoke supernatural agency. By beginning with the narrow-minded premise of 'no divine cause allowed', educators have generated a strict worldview that excludes any reference to a divine power and authority greater than man. In doing this, they are educating faith in God out of the hearts and minds of young people. This indoctrination scheme will ultimately point students toward atheism. A person is never born an atheist. Atheism occurs when people learn to deny a need for God, resulting from a variety of secular influences within a modern, contemporary culture. The problem is not that there remains insufficient evidence to convince rational individuals of God's existence, but many intelligent and influential people (especially the elite, powerful, and well educated) hold a natural antipathy toward the idea of God, despite the abundant evidence favoring a theistic universe. The notion of God seems to threaten to their autonomy. The very idea of a divine presence is a threat to man's self-centered nature (that compelling desire within the human heart to live life on one's own terms without moral constraint). This may explain why many people remain totally unconvinced of the reality of God; they have willingly chosen to embrace a secular, unyielding worldview that accepts only a natural explanation for every observed effect. This may explain why evolutionary teaching has had such a powerful influence on the most advanced cultures throughout the civilized world. Evolution theory has given self-centered man what he considers to be sufficient reason to deny God as the designing and sustaining influence over all things.

Some will say that belief in God is old-fashioned, and allowing the concept of God any credibility or consideration in the final equation of life would somehow diminish human enlightenment from science and education. By comparing modern discovery with the biblical record, we find that the Bible is, and always has been, in perfect harmony with all the known facts of science. In fact, the ancient Hebrew text has always been far out ahead of science with accurately recorded details about the physical world– knowledge that has taken man thousands of years to verify. Much of what the Bible records about the physical world was not even discovered or validated by mankind's inquisitive search for understanding until only the last couple of centuries. These documented discoveries of biblical truth amount to powerful evidence favoring the existence of God. Sadly, however, many of today's elite scholars and educators are determined to rid their students of "outdated religious superstition" by persuading them that the evolutionary model is trustworthy and scientific. However, nothing could be further from the truth! Empirical science deals only with observable and repeatable events in the present, not with wild conjecture and speculation about numerous unproven, unverifiable events from the past.

Imagine, if you will, a closed empty room with no doors or windows, finished with a smooth, level concrete floor. Consider this room to be completely isolated from any outside influence. In the center of the room, hanging from the ceiling, is a giant bag of marbles. Now, over time, the constant strain of gravity causes the heavy bag to burst open, spilling thousands of marbles onto the polished concrete floor below, where they scatter about in a totally random pattern to remain indefinitely. Now, even if ten billion years were to pass, the marbles in this completely closed system, isolated from any outside influence, would never become organized into an orderly arrangement. To rearrange the marbles (like setting up dominos) in a very specifically designed pattern would require force, energy, planning, and will. These marbles would remain in a maximum state of randomness forever, unless or until an organizing agent from outside the room was able to impose willful design on them.

Likewise, all the dead, lifeless matter in a closed universe would never become spontaneously arranged without willful intelligent agency at work! A cause-driven phenomena can only occur in an 'open universe' that allows for creative will to be imposed on the random physical chaos. For science to believe that all matter was concentrated in an isolated 'singularity' before it randomly exploded to produce the entire universe, in what has become known as 'the big bang theory', would still not constitute a closed universe. The problem is that someone had to hang the big bag of marbles to begin with! An honest atheist, maintaining a strict view of materialism, must maintain that all matter and energy is self-existing and not created, otherwise he or she ceases to be an atheist. If matter had a beginning, as nearly all scientists believe, yet was not created, this leaves the atheist having to imagine that the material dimension appeared randomly, out of nothing, without intelligent cause. Logic would suggest to us that whoever was responsible for the existence of the marbles (stars, planets, galaxies, etc.), which have become highly ordered with remarkable mathematical precision, would clearly be the *uncaused cause* of an intelligently designed universe. I am suggesting that, in order to think logically, we must understand that every material outcome in the finite physical realm demands a sufficient initial cause. Since all atheists are strictly bound to a prior commitment with naturalism, they are unable to explain life without a firm reliance upon freakish, random events that have magically altered the basic raw-materials, whose existence they cannot account for. By insisting that purposeful, supernatural events have never occurred, humanists are confined to a world without cause. Anyone who is open-minded, and willing to follow the evidence wherever it naturally leads, would surely have to seriously consider the logical possibility that intentional design has been creatively imposed on the vast material domain of a magnificent universe that you and I are privileged to enjoy.

When we consider the workmanship of a magnificent handmade clock, or the finely detailed painting of a skilled artist, how is one to believe that the universe with all its amazing detail and complexity could have come into existence merely by chance. The mathematical order in the movement of the stars, planets, and galaxies is so consistent that we can set our clocks by it! Who set the earth at precisely the right distance from the sun? Who set the moon in orbit around the earth? Who caused the earth to revolve around the sun while spinning on its own axis? Who divided the night and the day? The waters of the oceans evaporate to form great clouds, which are pushed by winds across the dry land where they pour out their life-giving moisture on the ground. The melted snow and rain rushes down the mountains to form great rivers, which make their way back to the sea, only to be caught up and delivered across the thirsty dry land once again, providing life-sustaining strength to all the plants, animals, and people of the earth. We observe a very complex world around us, yet everything is in careful balance and harmony, as though it was all diligently planned. We view the majestic splendor of the mountain peaks covered with snow. We may ponder the grandeur of the world's restless seas as their wave's pound forcefully against the sand. We marvel at the fragile beauty of a spring flower, and the magnificent perfection of a single snowflake. We experience four unique seasons as they come and go consistently without ceasing. Standing beneath a star-filled sky on a moonless night, we wonder in amazement at the awe-inspiring scene before our eyes. With the electron microscope we gaze into a single cell, only to find an amazing world about which we know so very little. We are confronted every day with the miracle of life and the mystery of death, and yet the keen balance of nature continues uninterrupted. We endeavor to explain the many mysteries of life with modern science, but still there remains much we do not even begin to understand about the incredible world that surrounds us.

Every observed effect has had a definite cause, and the cause is always greater than the effect. If atheists are right, "nothing" was the initial cause. For the atheist to maintain that the physical world had a spontaneous, unintelligible cause, they have simply created a religion of the much-maligned 'theory of evolution'. The scientific principle of cause and effect would flatly deny this conclusion. It is more logical that the first cause came from beyond the realm of the natural world. The atheist's logic is impossible to understand, because they have come to believe that "*no one, plus nothing, times blind chance, equals everything*." Since there is no real evidence to validate the far-reaching claims of evolution, people must accept this ideology because they cannot tolerate the alternative– that there was a creator God at work, initiating the appearance of the material world. Something could never come from nothing without sufficient cause. This revelation requires that someone very powerful must be eternal (outside the constraints of time and space). This implies God! The initial cause of the physical dimension could not have been part of the time-space, material world, because this transcendent being 'caused' it to appear. Listen again to American astronomer Alan Sandage: "I find it quite improbable that such order came out of chaos. There has to be some organizing principle ...... God to me is a mystery, but is the explanation for the miracle of existence, why there is something instead of nothing."[18]

The first cause had to be both infinitely powerful and eternally transcendent! The scientific principle of cause and effect, or any other natural law, for that matter, would place no limitation on a God that exists outside the wide constraints of the material, time-space reality. Why is it so hard for people to acknowledge life as the work of a divine architect? People have become so eager to explain away the magnificent complexity and diversity of life with a bogus philosophical dogma. After having much time to fully appreciate the ramifications of his wildly popular theory, Charles Darwin offered these revealing remarks. "Reason tells me of the extreme difficulty or rather impossibility of conceiving this immense and wonderful universe, including man and his capability of looking far backward and far into futurity, as a result of blind chance or necessity ... When thus reflecting, I feel compelled to look to a 'First Cause' having an intelligent mind in some degree analogous to that of man; and I deserve to be called a Theist."[19] His tone here reflects the thoughts of a troubled soul, who was unable to find solace in a strange theory that eventually found itself in direct opposition with a vast array of empirical scientific evidence.

"By the Word of the Lord the heavens were made and all the host of them by the breath of his mouth" (Psalm 33:6). The compelling argument of cause and effect demands a credible initial cause for a remarkable universe. This argument points toward a powerful and intelligent being who has revealed Himself to His creation in a number of exceptional ways.

Observation: Principle of Logic and Design

Can you have remarkable design without a designer? Perhaps you have seen the magnificent sculptural masterpiece on Mount Rushmore. Could any rational person conclude that the four faces of former presidents, carved into a mountainside in South Dakota, are the result of millions of years of wind and water erosion? Of course not! That would be totally ludicrous! In response to this astonishing masterpiece of artistic genius, people marvel at the creativity and perseverance of a great professional artisan. Because of the determined efforts of Danish-American sculptor Gutzon Borglum and his son Lincoln, who began construction on this iconic scene in 1927 and ended it in 1941, the world is able to enjoy one of the most extraordinary man-made landmarks of all time. This famous American attraction serves to remind us of the fact that 'design requires a designer'. William Paley made this compelling argument so many years ago, when he suggested that a watch requires a watchmaker.[20] This is a fact that no one, thinking rationally, could possibly deny. However, even though no reasonable person is likely to argue against the need for a watchmaker, a painter, or a sculptor, when the results of their work are so clearly seen, yet, the most complex design ever known, the human body, has been left relegated to the imagined world of blind chance, with no 'design engineer' given well-deserved credit. Consider this interesting thought from Dr. Isaac Asimov, who was a former professor of biochemistry at the Boston University, School of Medicine. He said: "And in man is a three-pound brain, which, as far as we know, is the most complex and orderly arrangement of matter in the universe."[21] For man to declare the human brain as the end-product of a mindless, unguided process in a godless universe seems quite absurd. This kind of thinking makes no rational sense; it defies reason, denying the need for intelligent planning behind the most complex organization of organic chemistry ever known to the physical realm. The truth is that human physiology is an exquisite design of pure genius, and God alone deserves the glory as the chief architect and master design engineer.

The psalmist King David declared the marvel of the human body: "For you formed my inward parts; you covered me in my mother's womb. I will praise you for I am fearfully and wonderfully made; marvelous are your works, and that my soul knows very well." (Psalm 139:13,14)

You would not say to an artist, "I do not believe you painted this; prove it to me." No, it would be up to you or I to prove someone was a fraud. Many believe the greatest evidence for intelligent design in living organisms is DNA. This highly organized and extremely complex chemical library contains a unique language of code-transmitting data. This molecular cipher conveys intelligent information very much like computer code. The human genome contains around 3.5 billion "chemical letters" in every one of our 15 to 20 trillion cells. Every single cell contains the blueprint for life (a complete instruction manual) spelling out precision living commands coded within a four-letter chemical alphabet. How is it possible for one to believe that the remarkable DNA-laden human genome came about by random chance through the unconscious blind forces of nature? We have never discovered a source of information-bearing code that was not the product of an intelligent mind. DNA is no exception; it required a code-maker in the exact same way that written text requires an author. This amazing information storage system has pointed numerous unbiased researchers directly toward a master design engineer, a supreme intelligence, who demonstrated purposeful intent and ingenious planning when he carefully orchestrated his most remarkable design of all— human physiology.

Naturalism, which relies on random chance processes alone, is powerless to explain how programmed information code could be found in the composition of cellular physiology. Nowhere on this earth are we able to locate a single example of written language or code without an intelligent cause behind it. When someone records a detailed instruction manual, they do so with a very definite purpose in mind. Imagine every cell in your body containing more vital information than the world's largest supercomputer hard drive? Dr. Antony Flew's life was forever altered by the magnitude of complexity he discovered and the rational implications presented through DNA research. He said: "The genetic message in DNA is duplicated in replication, and then copied from DNA to RNA in transcription ... following this, there is translation, whereby the message from RNA is conveyed to the amino acids, and finally the amino acids are assembled into proteins."[22] As described by physicist Paul Davies, a gene is nothing but a set of coded instructions with a precise recipe for manufacturing proteins. How foolish to deny God simply because man's finite mind cannot comprehend his infinite power.

The science of mankind cannot possibly explain supernatural events. For this reason, many have concluded there is no creator, but any open-minded observer should recognize the divine fingerprints that have been revealed throughout the universe. This world operates under the authority of natural laws—why is this? Why is the entire cosmos so orderly and dependable? There is certainly no reason for you and I to expect a smart universe that consistently obeys dependable laws of physics and mathematics. American theoretical physicist Dr. Richard Feynman, who won the coveted Nobel Prize in 1965 for his work in quantum electrodynamics, commented in this way: "Why nature is mathematical is a mystery; the fact that there are rules at all is a kind of miracle."[23]

The Holy Bible makes remarkable statements about scientific principles that are at work in nature. Many of these biblical references have been fully verified and confirmed by empirical science. Humans like to make logical decisions based on reason, and display a vast array of complex emotional responses including romantic love. We desire interpersonal relationships, and create music and beautiful works of art. Humans often seek a life of benevolence that gives purpose and meaning to their days. The ideology of naturalism leaves people unfulfilled by not offering them the personal assurance of hope beyond this life. When we consider the complexity of life, and the necessary chemical interactions that comprise multifaceted living organisms, it is impossible to imagine how thousands of chemical bonds could have occurred all at the same time by random chance alone. Another indispensable force is essential for these systems to function as a viable living entity, and to maintain this complex order. "You alone are the Lord, you have made heaven, the heaven of heavens with all their host, the earth and everything on it, the seas and all that is in them, and you preserve them all." (Nehemiah 9:6)

Believing in the unseen God is not blind faith; it is logical faith, well supported by many known facts and finite details witnessed all around us. While the atheist will invoke his or her own view of science as a reason to deny God, other open-minded scientists relate a much different story. Having been exposed to the compelling evidence of logic and design, as observed through current research findings, many of today's top-notch scientists have initiated a radical shift in their own personal worldview, having finally reached a dead end with evolution theory. The atheist has no way of explaining the existence of an orderly universe. If he or she argues that matter is eternal. without cause, factual science will deny this claim based on the dependable laws of thermodynamics. These vital laws teach us that the universe had a beginning, and is gradually running down. On the other hand, if the atheist should acknowledge the fact that the universe indeed had a beginning, then his or her belief system must adequately account for an initial cause. Even though atheists object to the existence of God, they have failed to offer a logical explanation for human life. Atheist Julian Huxley once said, "We are as much a product of blind forces as is the falling of a stone to the earth, or the ebb and flow of the tides. We have just happened, and man was made flesh by a long series of singularly beneficial accidents."[24] I would submit, instead, that current revealed evidence encourages us to embrace the idea that we are the purposeful plan of a supremely wise and benevolent designer.

Even liberal writers of *Time* magazine were forced to admit that naturalism has failed to explain the emergence of the physical world around us. They wrote: "In the quiet revolution in thought and argument that hardly anyone would have foreseen only two decades ago, God is making a comeback. Most intriguingly, this is happening ... In the crisp intellectual circles of academic philosophers."[25] Buoyed by the powerful influence of Darwinism, scientific thought in the late nineteenth and early twentieth centuries appeared to be leading mankind down the road of naturalism. In the perception of the academic and scientific communities, public opinion alleged a "god of the gaps".

However, in the last seventy years or so experimental science is leading mankind in a new direction—a direction that is clearly pointing, once again, toward the more logical conclusion of design and purpose in the natural world. The current paradigm shift toward 'theistic design' has been gaining momentum because of the revelation of new data, which is beginning to require that reluctant men and women must once again consider an explanation involving intelligence, an admission that science does not deal with very well.

Numerous scientists are finally expressing a genuine belief in a real creator God. Consider a revealing quotation penned by a staff writer from the *Wall Street Journal* in December of 1997. The journalist offered this surprising remark after considering the degree to which science is now pointing its brightest minds in a new and different direction. "I was reminded of this a few months ago when I saw a survey in the journal *Nature*. It revealed that 40 percent of American physicists, biologists, and mathematicians believe in God–and not just some metaphysical abstraction, but a deity, who takes an active interest in our affairs and hears our prayers: ...The God of Abraham, Isaac, and Jacob."[26] This emerging trend toward worldview transformation within the scientific community reveals that the more we discover about life and the universe, the more we are pointed in the direction of a magnificent, loving God. As we follow the evidence wherever it would naturally lead, the truth about the vast material world comes more and more sharply into focus. The remarkable degree to which the major constants of physics must match razor-sharp criteria to explain our physical existence is causing great numbers of previously agnostic scientists to conclude there is some sort of supernatural plan or Agency at work.

"God created everything by number, weight, and measure ... In the absence of any other proof, the thumb alone would convince me of God's existence."[27] This is the conclusion of an iconic figure by the name of Sir Isaac Newton, who is arguably the greatest scientist of all time. How about a more contemporary view: "There is for me powerful evidence that there is something going on behind it all ... It seems as though somebody has fine-tuned nature's numbers to make the universe ... The impression of design is overwhelming."[28] This statement came once again from a present-day world-class scientist. British-born cosmologist Dr. Paul Davies is highly regarded as one of the greatest living physicists of our time; he was the winner of the 2002 Faraday Prize issued by the Royal Society and the 1995 winner of the Templeton Prize.

Consider these compelling statements from two other leading scientists of the twentieth century, both of whom are Nobel Prize winners. "Both religion and science require a belief in God. For believers, God is in the beginning. and for physicists he is at the end of all considerations. To the former he is the foundation, to the latter, the crown of the edifice of every generalized worldview."[29] This statement comes from the mind of German theoretical physicist, Dr. Max Planck, who originated quantum theory, a discovery that won him the Nobel Prize for physics in 1918. 'Quantum theory' has now revolutionized man's understanding of atomic and subatomic processes in the way that Albert Einstein's 'general theory of relativity' greatly enhanced our understanding of space and time. Commonly regarded as one of the outstanding physicists of the last century, Dr. Planck explains that when arriving at the current apex of man's knowledge and understanding, we are forced to conclude that a very powerful unseen intelligence was actively at work in creating, empowering, and sustaining the physical universe. Dr. Antony Hewish is a British scientist and radio astronomer who was awarded the Nobel Prize for physics in 1974. He said: "The ghostly presence of virtual particles defies rational common sense, and is non-intuitive for those unacquainted with physics. Religious belief in God, and Christian belief that God became Man around two thousand years ago, may seem strange to common-sense thinking ... But when the most elementary physical things behave in this way, we should be prepared to accept that the deepest aspects of our existence go beyond our common-sense intuitions."[30] Many honest men of science would express a need to look beyond nature for the answers they seek, if only peer pressure was not holding them back. Consider this refreshing comment of another American physicist, Dr. Tony Rothman: "When confronted with the order and beauty of the universe, and the strange coincidences of nature, it's very tempting to take the leap of faith from science into religion. I am sure many physicists want to. I only wish they would admit it."[31] The deeper we look into the marvelous mysteries of the natural world, the more convinced we become that 'the signature of God' is clearly written in the complexity of living things—a complexity that the greatest minds in history are unable to comprehend.

The discipline of cosmology on an unimaginably colossal scale and the study of elementary particle physics on an inconceivably minute scale have opened our searching minds to the spectacular splendor and astonishing precision of the remarkable universe in which we live. As what would seem to be insignificant specks of dust in a vast unknowable universe, how does one determine the value of a single human soul? By viewing our lives through the lens of creationism and a biblical worldview, we recognize that God sees every living soul as having infinite redeeming value. Scientists, like Dr. Paul Davies, have rejected the idea that we are unplanned and insignificant in the grand scheme of things. Davies wrote: "I cannot believe that our existence in this universe is a mere quirk of fate, an accident of history, an incidental blip in the great cosmic drama. Our involvement is too intimate ... We are truly meant to be here"[32] The rational intelligibility of the universe points toward a transcendent, supernatural being that is responsible for creating and sustaining these uniquely complex physical designs. Those who sincerely seek after knowledge must consider that the goal of empirical science is essentially to discover what we can about God's grand design, and determine, to some degree, how it works.

Observation: Religion and Morality

If evolution is true, how can we account for the universal expression of religious faith? Many careful observers of human behavior have pointed to the fact that, in every human culture the world over, we find religious people. This spectacle demonstrates the innate tendency within mankind to worship and serve a power greater than, or at least more worthy, than himself. This trend is found throughout the world, and only a few people have a real problem acknowledging a divine presence of some kind. What causes man to seek strength and wisdom from a source outside of himself? In cultures the world over, people are passionately religious. The evolution model has no way of explaining this worldwide phenomenon. Just because mortal life is all many people know, does not mean that is all there is. People the world over are still asking fundamental questions: Who am I? Where did I come from? Why am I here? And, where am I going when my life on earth is over? The evolutionary worldview is unable to answer any of these vital questions. What I find very interesting is that atheism is rather uncommon in the underdeveloped nations of the world. Not everyone has the same understanding of who God is, but most people recognize a supreme power greater than themselves. This alone is compelling evidence that no one was ever born an atheist. People must learn to deny God's existence; it is not in human nature to lean this way naturally. It has been the highly developed nations of the world that have promoted the anti-God theory of evolution through education systems, and a public domain that has been unknowingly designed to indoctrinate the masses. When we fully consider the powerful evidence seen in the widespread expression of religious faith, we can easily discern three major commonalities among people who pursue spiritual endeavors.

1. Everyone is inclined to practice some form of religion, worship, or to demonstrate some level of faith in an intelligence or life force that they consider to be greater than themselves, or at least most worthy of their devotion and allegiance. (Remember, evolution is a religion)

2. People have a strong sense of morality, which helps to govern the way they act and the way they treat others. This universal

system of ethical values comes from a divine download of moral principles that are governed by conscience. Although people are often unable to live up to God's standard, (The ten commandments) the conviction within man to live rightly before God reflects the fact that man was created in His image to represent His glory on the earth.

3. Many of us are to some degree psychologically unsatisfied. The real unhappiness that some people experience is, perhaps, because they have hinged their sense of fulfillment on the wrong set of objectives. Aside from and in addition to their obvious emotional, physical, and material needs, humans have a compelling desire for a genuine spiritual connection that will give lasting purpose and meaning to their lives. They want to know and, in some way, relate to the spiritual power they believe to be guiding their uncertain steps through an often-difficult existence.

The reality of worldwide religious expression is strong evidence for the existence of the supernatural. And, now, there is a movement among the so-called 'new atheists' to label all religions as dangerous, evil, and immoral. I suspect this is a knee-jerk reaction to the tragic events of 9-11. Many have sought to universally condemn the exceptional worth of religious practice based upon the extremist views and actions of a few terribly deranged fanatics. To make such a broad negative judgment about all religions is a classic example of the desperation seen in today's determined atheists, who have lost their footing in the great debate, and are left with only this sad overreach. To use an old cliché, these new atheists have "thrown the baby out with the bathwater." Despite terrible mistakes in times past, when men of violence mistakenly believed they were fulfilling the will of God, an honest observer of history would have to admit that Christianity, in its true sense, is the very bedrock upon which Western civilization rests. It has become the foundation for the genuine cause of freedom and peace in a hostile world that is forever plagued with religious persecution, extreme hatred, and appalling violence.

Another problem for evolution is morality, which is difficult to make sense of in a world without God. Science has a difficult time explaining why morality exists at all. For example, science can tell you that if you laced someone's food with cyanide, it would kill them. However, science cannot tell you if it is morally right or wrong to poison your spouse in order to collect the life insurance money. This constitutes an ethical question that is beyond the reach of science. This is not to say that secular society is absent of moral values, which would not be true. However, you cannot measure morality in a test tube. Atheist philosopher, Bertrand Russell, once wrote: "Whatever knowledge is attainable, must be attained by scientific methods; and what science cannot discover, man cannot know."[33] I was appalled by this statement, which is far from true. There are many crucial questions that scientific inquiry cannot answer, or even begin to explore, and therefore, an outside source of information is required—information such as revelation from God. Somebody had to set the standard for acceptable, and unacceptable human behavior. Because of his self-serving goals, man is clearly not qualified to do that. This ethical insight must be provided by someone who is moral in nature and not self-serving, but is completely devoted to the happiness and welfare of others. God is the only one qualified to define the proper ethical guidelines for humanity. He represents the perfect standard of moral justice, and selfless generosity. If the evolution theory were true, where did morality come from? Is morality something found in humanity because of the mindless work of natural selection? If all men were created in the image of the same eternal God, and all men were given a conscience to help govern life with respect to right and wrong, then this could explain why ethical values are largely universal within the human family, even among those who deny that God exists. How is man able to consistently distinguish between what is right and what is wrong? With evolution, there is no one to set the ethical standards that define objective moral values. Atheism cannot provide mankind with a sound basis for determining good from evil, and, therefore, the 'theory' cannot account for the ability all humans have to distinguish between good and bad behaviors. Apart from the existence of God, (the moral lawgiver and perfect standard of absolute good) atheism can offer no rational explanation for why moral values exist at all. Therefore, without God, there is no sure standard for judging something to be good or evil. And there is little doubt that certain well-defined moral precepts and principles exist in all human cultures.

Researchers in anthropology have found a deep and consistent understanding that certain selfish behaviors are deemed immoral and damaging to human lives. They find that hatred, lying, cheating, stealing, murder, rape, adultery, incest, and other very selfish behaviors are consistently seen as morally wrong in human cultures the world over. From where did this universal sense of right and wrong come, if not from God? If mankind has evolved from the apes as evolutionists claim, morality within the heart of man makes no sense; therefore, evolution is powerless to explain why moral standards are universal. The only explanation that makes sense is that moral laws were given to mankind by a moral lawgiver. Guilt, shame, and remorse are the feelings we experience when we violate one of these moral laws. Why is this? This is because everyone has been given a moral conscience to help govern their lives with respect to right and wrong. When we choose to disregard the clear warnings of conscience, by ignoring that inner voice and acting upon selfish impulses, the result is feelings of guilt and condemnation. These feelings are a reminder that we have violated a moral law established by the 'lawgiver' to whom we are ultimately accountable. Evolution has no answer as to why we possess a moral conscience at all. Why do the people long for justice, fairness, and decency? Could it be because they were created in the image of a just, fair, and decent God, who is moral in nature, and has created mankind in His own likeness? This is the only sensible way to explain the strong reluctance of man to recklessly engage in certain universally condemned behaviors.

Most people would surely agree that having sexual relations with another man's wife is morally reprehensible. Why? Could it be because the Holy Bible says, "Thou shall not commit adultery"? We feel violated when someone wrongs us in a way that is emotionally or physically painful, because a moral law has been breached. If evolution is true, and humankind is simply a higher form of animal life, human nature would be driven by survival instinct (survival of the fittest), and there would be nothing to prevent you and me from acts of assault, theft, and even murder, if it improved our chance of survival. The well-known Oxford professor, John Lennox, who has traveled and lectured extensively throughout the former Soviet Union, reports that the famous Russian writer and intellectual Fyodor Dostoyevsky is known to have said, "If God does not exist, everything is permissible!"[34]

Evolution can provide no moral reason to restrain from rape, robbery, or incest. I would submit that objective moral values would be completely foreign to a nontheistic universe. To say that as highly evolved primates, we have developed a strong sense of morality through an evolutionary process is a desperate stretch of imagination. I remember recently watching a nature program on television that featured the nesting habits of a certain breed of penguins. The male penguins were observed gathering rocks to prepare a nesting area for the female, much like you would gather rocks to prepare a fire pit. While some penguins were searching for additional rocks to complete the nesting area, a few tuxedoed rascals would be stealing rocks that were previously gathered by others instead of finding their own. We know that penguins do not understand this behavior to be morally wrong. They are simply acting out their inborn survival instinct, and whatever it takes to survive is fair play to them. Not so with humans who must answer to each other and to a higher power for the moral choices they make.

When animals reproduce, the strong instinct for survival of a species can be seen in the fact that nesting mothers will sometimes abandon or kill one of their own offspring if it allows more robust offspring to survive. In this way, the hardiest newborns, which would have the greatest chance for survival, are preserved. This is not the case with humans, who possess a moral obligation to care for and preserve every life and will often go to great lengths and great expense to do so. (Tragically, however, many people have decided that it is somehow okay to destroy life prior to birth if they should determine the unborn to be inconvenient, unwanted, or defective.) Human morality, to a large extent, cannot be determined by the individual, or by the culture, because certain behaviors are objectively wrong in all cultures, at all times. In the moral mind of Oxford professor, John Lennox, "Everyone would agree that torturing babies is ethically and morally wrong." According to Professor Lennox, "Ethics involves top-down causation, and cannot be deduced from a bottom-up system of arguments."[35] In other words, having a clear sense of moral responsibility requires input from an outside influence greater than man, and cannot logically be generated from the accidental manipulation of simple chemical elements in a closed universe. The sovereign, eternal God, who created mankind, has also provided the moral laws that were intended to govern our ethical behavior in a fallen world, and to protect us. The human conscience is the active restraining force in the life of every individual. The atheist certainly can be moral, and do good, without question. At the same time, certain religious people can and sometimes are very much immoral and corrupt. However, the actions of some, who are defiant, rebellious, or perhaps fanatical, certainly do not negate the value of religious faith in those who are consistently faithful and extremely devout. Everyone has the capacity for good, because we were all created in the image of the same God, and we all possess a truly moral consciousness within. Richard Dawkins agrees that "it is pretty hard to define absolute morals on anything other than religious grounds." However, he is quick to add that "DNA is neither good nor bad; it just is, and we dance to its music."[36] If we are dancing to the direction of our DNA, when we do harm to others, how then, can we be considered evil? I submit that our DNA has no control over our moral choices when we act by imposing our selfish will upon others. We must take responsibility for our actions, and not attempt to blame them on something like "natural selection."

If we repeatedly violate or disregard the voice of conscience, we will eventually lose that healthy God-given conviction that certain behaviors offend Him. Because of God's convicting moral laws, man is forced to face the problem of sin and to acknowledge that only God is able to do something about it. The crucifixion of Jesus is a definite fact of history that raises a profound question– a question that each of us must eventually come to grips with: 'What was God doing on a Roman cross'? The famous Russian dissident Alexander Solzhenitsyn explained it like this: "The line dividing good and evil cuts through every human heart including me. The cross of Jesus Christ not only gives me hope that evil will one day be conquered, it shows me that God loves me enough to deal radically with the guilt of my own heart, to forgive me, and give me a living relationship with him."[37] You and I will forever be challenged in our thinking by the starry heavens above, and compelled to ponder the moral law within. Our moral conviction is neither subjective nor is it relative; moral principle far transcends the individual and his culture. This conviction within every human soul has a common source--a transcendent, moral lawgiver! There are liberal professors teaching at prestigious universities throughout America, who are instructing students that there are no moral absolutes. This flawed thinking is reckless, and a dangerous pattern of intimidation! The idea that morality is subjective (open to one's personal interpretation) and should be custom fit to meet the selfish desire of each individual, has led to disastrous consequences. When a culture discards the need for moral absolutes, that society is on a path of ethical decline leading to moral chaos. If the trend goes unchecked, it could result in a loss of cultural identity. This has been the case with many human civilizations in the past. The existence of conscience, morality, and universally held values represents strong evidence for God, who created man to live under the conviction of moral law. Without God, men become a law unto themselves, and who can say which person's set of values are the right ones

Through the terribly misguided teaching of evolution, man has created a god in his own image, a god that does not require moral accountability. This explains why evolution has been widely embraced by recent generations. Many choose to serve a man-made god, so their personal desires and private impulses are not subjected to moral restraint. We have a clearly defined legal system to govern society, but our entire legal structure is based on the moral standards that were given to us by the God of the Bible. If atheism is true, no victim of moral terrorism will ever have the satisfaction of seeing true justice prevail. Adolph Hitler, Joseph Stalin, Pol Pot, and Mao Tse-tung will have gotten away with their heinous acts of pure carnage against humanity! On the other hand, the Judeo-Christian heritage, which, in the course of time, gave birth to civilized Western society, teaches us that justice will prevail at the end of days, when humankind faces his eternal Maker, and the scales of divine justice are ultimately balanced. I am convinced that without God-given values and sound, morally based principles there could never have become any sophisticated, civilized society.

In the early part of his distinguished career, Oxford mathematician, Dr. John Lennox became compelled to learn and speak both Russian and German, and ultimately, he would travel and lecture extensively in both of these countries. Dr. Lennox has reported some striking thoughts from a very prominent German philosopher by the name of Jurgen Habermas, whom he has quoted as referencing Western culture to say: "Christianity, and nothing else, is the 'ultimate foundation' of liberty, conscience, human rights, and democracy, the benchmarks of Western civilization."[38] What a remarkable confirmation of the influence the world has received through the historical footprint of Christian faith, in which one solitary figure in history provided the foundational influence upon which all of Western civilization rests.

When considering the question of morality, the atheist will often point at the existence of evil as a just reason to deny God. After all, how could a moral God allow suffering and evil to exist in a world that he supposedly cares about? Those who deny the very existence of God, and yet fail to explain how humans can consistently make moral judgments, demonstrates the illogical nature of the atheist's belief system. Consider the following statement from business learning specialist Rolaant L. McKenzie, who explains, "It is impossible to tell the difference between good and evil in the absence of some reference point that is absolutely good. One cannot determine 3 inches from 4 inches without using a ruler as an absolute reference. One cannot determine actions to be good or evil, without an absolute moral standard. The very presence of evil actually demonstrates the existence of God rather than disproves it."[39] We can surmise that it is the very presence of evil that fully illuminates the presence of goodness. The atheist may also argue: if God exists, God is good, and He is all-powerful, why does He not purge the world of evil? Rejecting God is a poor way of dealing with the disturbing problem of evil, pain, and suffering, because it removes all hope, and still leaves the pain and suffering. With atheism, death is the end. It removes the hope of a world without pain and suffering, and denies that the scales of divine justice will, ultimately, be balanced in the life to come.

All of us have experienced terrible pain, sorrow, suffering, and injustice at some time. It has touched every life, and will impact our world until the end of this age. Instead of blaming God for all that is wrong in the world, we should stop to consider just why pain, death, and suffering are a part of the human story. Consider this quotation from the late C.S. Lewis, the British novelist, literary critic, and Christian apologist, who became famous for his fictional series of novels called the *Chronicles of Narnia.* "God created things with a free will--that means creatures, which can go either wrong or right. Some people think they can imagine a creature, which was free, but had no possibility of going wrong; I cannot! If a thing is free to be good, it is also free to be bad, and free will, though it makes evil possible, is also the only thing that makes possible any love or goodness or joy worth having."[40] When we have considered all that God has done in creating mankind, at the emotional and moral levels, we must all agree that you could never have genuine love in a robotic universe. Authentic love requires freedom of choice! God chose to create us in His image, and to empower us as free moral agents. He did this because he wanted a relationship with man that was based on mutual love and respect. Yes, life is chocked full of jagged edges, with raw, inexplicable pain and suffering. This world is filled with broken bodies, broken hearts, and broken dreams, and unfortunately, that is the reality of the earthly human experience. How could I ever explain to my adult son why his beautiful thirty-three-year-old wife, Amy, was tragically taken from him very suddenly by a brain aneurysm, leaving him with their infant son to raise alone? Nobody can possibly offer an adequate explanation to soften the blow from such an unexpected and devastating loss.

Some things in life we cannot possibly explain, and we will never fully understand many of the 'why' questions on this side of eternity; we can, however, know for certain that our Heavenly Father is still in control, and his abiding love for all of us is so intense that it is not even possible to explain it in human language. We can and must, therefore, give an answer in response to the argument presented by the question of pain and evil in this world. If God were to abolish all evil in a single act, a major part of humanity would be instantly wiped out. Why is this? The answer can be found in our fallen nature. We all possess within a compelling inclination to sometimes engage in behaviors that are morally wrong and offensive to God; these acts include many self-centered behaviors that are also proven offensive to one another. Therefore, 'man is and always has been the cause of evil in this world'. To blame God is a clear miscarriage of justice.

By contrast, evolutionists view humans as basically good and suggest that the inclinations of the human heart should not be stifled by moral constraint. The Bible clearly teaches otherwise—revealing that man was born with a sinful nature that often gravitates toward evil. In Ephesians 2:3, the Bible explains that by nature, we humans are the "children of wrath." Isaiah 64:6 even compares the righteousness of man to "filthy rags" when compared to the perfect standard of a Holy God. Also, Jesus himself acknowledged the sinful nature of man when he said, "If you then, '*being evil*', know how to give good gifts to your children, how much more will your Father who is in heaven give good things to those who ask Him! (Matthew 7:11).

At first glance, most people would appear to be basically good, especially when their personal lives are measured in broad human standards. However, when weighed against the absolute standard of a morally righteous God, they are not! And yes, God has dealt with the problem of evil, pain, and suffering, but He did it in a loving and merciful way that did not require the instant annihilation of unredeemed humanity. Why are people so quick to blame God when things go terribly wrong? I suppose it is because they believe he has the power to change everything for the better, and they are right, he does! However, we seem never to be satisfied without instant gratification in response to our most urgent and pressing needs. But this is not usually how God works. The difficulties of life can sometimes result in bitterness toward Him by some people, but they can also lead us toward dependence on God, when we face difficult trials over which we have no control. It is all a matter of how we choose to respond to the troubling challenges life will inevitably throw our way. Perhaps God can use our troubles to work His will in our lives. The burdensome obstacles we often face on life's journey should draw us closer to God as we make much of Him and less of ourselves.

God's original creation was perfect in every way! Sin (man's selfish rebellion against God's authority) destroyed that original perfect world designed to be our eternal home. However, because of God's unspeakable love and compassion for mankind, he has faithfully prepared and thoroughly carried out a flawless plan by which to restore to mankind everything that was lost and forfeited through sin. By virtue of his sacrificial death, Jesus has made the way for a complete restoration culminating in the fulfillment of God's plan and purpose for the ages.

We have observed that some form of religion is found everywhere in the world. The universal belief in God, or some divine, unseen force serves to demonstrate man's instinctive awareness of a supreme and majestic being that is able to elevate him above his troubling circumstances. We see this belief in young children, who naturally display an innate perception of God, when they express words and actions of simple faith, thus demonstrating a unique awareness of his undeniable presence. The amazing ease with which small children are able grasp the concept of God suggests that the reality of his existence has been instilled in their innocent young lives from birth. "But God shows his anger from heaven against all sinful, evil men who push away the truth from them. For the truth about God is known to them instinctively, God has put this knowledge in their hearts" (Romans 1:18, 19 TLB).

We have discussed the universal existence of the moral laws, values, and principles that serve to govern the ethical behavior of people in all cultures. In a well-attended debate with Antony Flew in 1998, William Lane Craig, an American philosopher, and a very capable Christian apologist, calmly proposed five powerful arguments for the existence of God. When he had eloquently presented the obvious fact of the moral nature of man, he offered this conclusion as related to objective moral values: "If God does not exist, then objective moral values do not exist."[41] As his argument continued, he explained that observations the world over clearly reveal that objective moral values are apparent in all human cultures: "Therefore, a moral God exists!" The conclusion we can and must draw is that the religious nature and basic moral character of man is universally confirmed and clearly serves to represent powerful evidence for the handiwork of a personal, unseen, creator God.

Observation of the physical world and the nature of man demonstrates the reality of the unseen God, who created the universe and governs it with his sovereign power and divine authority. However, His creative handiwork has given us only a partial revelation of the super intelligence that has authored life in the physical dimension. In the following chapter, we will take a close look at the way in which a powerful creator God has personally revealed himself to his creation in more specific ways that cannot be denied.

Chapter Eight

* * *

The Revelation of God

Revelation: Israel-Miracle Nation

Anyone who is aware of the unique history of the Jewish people knows that God intervened in miraculous ways to establish, protect, and preserve this peculiar nation, whom he chose for a very special purpose. The story of the Hebrew nation is unlike the story of any other people. Most of us living today cannot remember a time when there wasn't a recognized state of Israel. However, before 1948 for nearly 1,900 years there was no such country. A brief review of this nation will explain how God revealed himself to the world through this tiny miracle nation.

Through the patriarchs Abraham, Isaac, and Jacob, God established a divine covenant that would bring about great blessing to the whole earth and provided a plan by which he would restore fallen humanity to a right relationship with himself. The seed of the Jewish nation was planted in the heart of Abraham whom God chose, and through whom God promised to fulfill covenant blessing, allowing his seed to become as the stars of heaven in number, and the land to which God was taking him would be given to his direct descendants through divine promise. This irrevocable covenant was passed to Abraham's miracle son Isaac, who in turn bore two sons, Jacob, and Esau. The covenant blessing followed Jacob whose family included twelve sons. Jacob's sons would become the twelve tribes of Israel. Jacob dwelt in the land of Canaan, which God had promised to his descendants through Abraham and Isaac. By providence a severe famine came into the land and forced the chosen ones to flee into Egypt where food provisions were still plentiful, thanks to God's divine intervention in the life of Jacob's son Joseph, who was sold into slavery in a cruel act of jealousy by his brothers and became an Egyptian captive. What was meant for harm in the life of Joseph, God meant for good to preserve many lives, who would have otherwise faced sure starvation. In Egypt, Jacob's tribe became a vibrant nation of ambitious, energetic people. However, after many generations had passed, successive ruling pharaohs found themselves threatened by the strength and number of this emerging nationality within their borders. This fear resulted in the harsh captivity and oppression of God's people, as Israel became enslaved under the heavy hand of Egyptian rule. After centuries of bondage, God sent Moses to deliver Israel from their oppression. They were able to escape the cruel hand of pharaoh by miraculous events God displayed through Moses.

By these miraculous signs and wonders, God profoundly demonstrated his power to preserve and protect his chosen people. The decisive blow was the death of the first born of Egypt, releasing the cruel grip of pharaoh upon the Hebrew captives. With Moses leading them they escaped from bondage, and a miraculous parting of the Red Sea broke the power of Egypt over the fleeing refugees. Israel moved as one nation toward the promised land of Canaan.

In their difficult migration across the hot desert, this Hebrew band was miraculously led by a cloud during the daytime and a pillar of fire at night, as they made their way through the harsh, dry, empty land. The powerful hand of God upon their lives, by divine providence, afforded one miracle after another on their long, arduous journey through the barren wilderness. Miraculous signs offered safety, protection, and provision for the wandering Hebrew refugees. God was preparing this tiny nation to be the instrument through which he would reveal himself to the whole earth. They ate manna from heaven, drank water that sprang from the rocks, and their clothes and sandals did not wear out during the long, grueling trek across the empty wasteland. God's law, delivered to them through his faithful servant Moses, would shape the character of this people, making them into his messenger of hope to a fallen world. After forty years of wandering in the desert while God was preparing his chosen people, Hebrew descendants finally entered the Promised Land where God gave them great victories over the heathen nations who opposed them. These adversaries, whom they defeated by reliance on a miracle-working God, comprised the tribes that occupied Israel's inheritance during their prolonged and difficult captivity in the land of Egypt. In the centuries that followed, God gave his people many prophets, judges, and kings. He blessed them when they obeyed him, and he punished them when they refused. And unto them alone were entrusted the oracles of God (the OT biblical revelation), as this tiny Hebrew nation began to faithfully receive, record, and meticulously preserve God's written revelation— his eternal message of hope and promise to mankind. The scriptures were received by divine impartation as a gift of knowledge and understanding to be shared abroad throughout the entire world for the beneficence of all mankind.

In fulfillment of God's promise to Abraham, the Messiah was miraculously born into the world to deliver the people from their sins. Man's sin brought the terrible curse of death, sorrow, grief, and despair upon the whole earth, and alienated man from his relationship with his creator. For the most part, the Jewish people rejected Jesus as their promised Messiah, and for this they paid a dreadful price. God brought severe judgment on them at the hand of their enemy when the brutal Roman legions proceeded to destroy their sacred temple, the Jewish state, and a vibrant civilization that had endured since before the days of Moses. They were all scattered throughout the world in the Great Diaspora that lasted almost two millennia. However, on May 14, 1948, a devoted remnant of Hebrew nationals would once again declare their sovereign independence after being restored to their beloved homeland. (The land promised to their forefathers by God himself) The Jews finally became a self-governing and independent nation once again. Never in the history of planet earth, has any single nationality of people been so thoroughly demoralized, devastated, and utterly forsaken as Israel was. Yet somehow, through all the turmoil, which culminated in the dreadful Holocaust, they miraculously maintained their national identity as a unique people who longed to return to their promised inheritance. Their miraculous survival as a unique people was surely due, in part, to the holiday feasts God had instructed them to observe wherever they were living. The Hebrew tongue became a discarded language, but astonishingly, it has been revived once again and restored to the Jewish people against insurmountable odds. The nation Israel, whom the Almighty had divinely chosen to bring the whole earth back to the worship of the one true God, was and still is a miracle nation. The unique and fascinating history of Israel (the Jewish people) was orchestrated and sustained by the miraculous power of the true and living God and demonstrates unmistakable, and very compelling evidence for the existence of a powerful supreme being, a divine intelligence who has unpredictably, but quite thoroughly, revealed himself to a skeptical and unbelieving world.

Revelation: What about the Bible?

Man cannot understand a supreme being who dwells outside of time and space, unless that intelligence chooses to reveal himself. Because of man's exalted position in God's eternal plan, the Creator has gone to great lengths to make himself personally known. We understand God only in a general sense by observing his amazing creative works. The basic knowledge of God, which we glean from observation tells us nothing about the depths of his love, grace, mercy, compassion, and forgiveness. Because of love, God has chosen to disclose a much greater revelation of himself through a personal written record that He has made widely available. For a transcendent God to reveal himself to feeble mankind, he would surely have to say and do things that are completely foreign to the normal, mundane experience of human life. This he most certainly accomplished when he openly performed many miracles, fulfilled countless ancient prophecies written centuries before his miraculous incarnation, and publicly unveiled to first-century witnesses the greatest miracle of all. Jesus of Nazareth became the central figure in human history when He rose from the grave a few days after the crucifixion.

The Holy Bible, which was delivered to mankind through the miracle nation Israel, has provided humanity with a very profound understanding of who the God of creation really is, and what he has done through his love without limits for mankind. In these inspired writings, the Holy Spirit has revealed to us much understanding of the character and nature of the Creator. Among other things, the biblical record serves as God's autobiography. God's transcending nature as the eternal sovereign of the universe is revealed— the one has set the physical laws of nature into motion, but has never been limited by those laws. The Bible clearly credits the God of Israel with many divine attributes that are known to no other being in existence.

1. God is omnipotent (all powerful) (Revelation 19:6).

2. God is omniscient (all knowing) (Hebrews 4:13).

3. God is omnipresent (everywhere present) (Proverbs 15:3).

4. God is infinite (without beginning or ending) (Psalm 90:2).

5. God is perfect (without character flaw) (Matthew 5:48).

6. God is immutable (not subject to change) (Malachi 3:6).

7. God is holy (unable to look upon sin with approval) (Leviticus 19:2).

8. God is sovereign (without equal and supreme in authority) (Exodus 8:10).

God is life, energy, and personality, having unlimited authority over his vast creative works. He is the ultimate judge to whom all of mankind is accountable. Men and women would be wise to humbly reverence and fear him, instead of trying to explain away the reality of his existence with some irrational theory. Mortal man will never fully comprehend the magnificence of the Creator. The remarkable ingenuity, creativity, and imagination of God is far beyond the comprehension of the most brilliant minds. The message of the Bible is often found to be counterintuitive to the way man is inclined to think and act. For example, we are admonished to "love our enemy," to "turn the other cheek," to "walk by faith, not by sight," We are encouraged to "give and it will be given unto you." Man is inclined to believe an all-powerful God should simply rid the world of evil and everything will be fine. God is more interested in ridding our self-centered hearts of sin than in taking away our free will (the human characteristic that led to a terrible curse on this world). As C. S. Lewis famously suggested– choice allows for both good and evil, but without it there could never be the kind of love worth having. "For my thoughts are not your thoughts, neither are your ways my ways, says the Lord. For as the heavens are higher than the earth, so are my ways higher than your ways, and my thoughts than your thoughts" (Isaiah 55:8,9).

"God's ways are as mysterious as the pathway of the wind, and as the manner in which a human spirit is infused into the little body of a baby, while it is yet in its mother's womb." (Ecclesiastes 11:5 TLB) "The secret things belong unto the lord our God; but those things which are revealed belong to us and to our children forever ..." (Deuteronomy 29:29) Of course, we cannot understand the great miracles of God that have defied the laws of nature! However, if the God of the Bible is the author of the physical laws that govern our world, then why would it seem so strange if he should choose to transcend those laws to accomplish His purpose?

In addition to his unique attributes of deity, the creator God also possesses many of the noble characteristics that are found in mankind as well. These character qualities of God were born in humankind when Adam was created in the image and likeness of his Maker. Characteristics that are found in both God and man include many interpersonal attributes and emotional responses like love, compassion, mercy, forgiveness, grace, joy, generosity, sorrow, sympathy, kindness, hope, and trust. These and many other distinctively human character traits were given by divine authority, but are found to be drastically limited in the instinct-driven world of the animal kingdom, to which evolutionists have mistakenly assigned mankind.

Some worldviews suggest there are many gods inhabiting realms of authority in the universe. The God of the Bible would flatly deny this erroneous claim! "I am the Lord, and there is no other, there is no God besides me: I will gird you, though you have not known me." (Isaiah 45:5) "Remember the former things of old, for I am God and there is no other; I am God, and there is none like me." (Isaiah 46:9) "You are my witnesses, says the Lord, and my servant whom I have chosen, that you may know and believe me, and understand that I am He. Before me there was no God formed, nor shall there be after me." (Isaiah 43:10) The God of the Bible is a jealous God who will not share his glory with any imagined, superficial, or false deity.

Most of the world's religions present a form of written guidelines to direct their followers, but the Holy Bible is uniquely different from all others, standing alone as the complete, inspired, inerrant revelation of the Almighty. Only the Bible offers abundant evidence that sets it apart from all other holy books as God's eternal Word, and is able to back this claim with many infallible proofs. The bible was God-breathed and thoroughly infused with divine power and authority. The Holy Spirit communicated thoughts and intentions to human writers who have faithfully recorded and preserved them. The powerful evidence that supports the Bible alone as God's Word will be more thoroughly discussed in the epilogue section at the end of the book.

The Holy Bible has presented man a unified, internally consistent message proclaiming one grand central theme: Redemption! Redemption reveals God's love for mankind and the amazing lengths to which He was willing to go to demonstrate that love! The Holy Scriptures present to mankind unique and compelling proof of our Creator's existence! No other written material ever penned offers such remarkable prophetic details written hundreds of years before the events they describe are so clearly and decisively fulfilled. Only the omnipotent, omniscient God of heaven could make that happen. How else can we possibly explain thousands of detailed facts of history recorded far in advance of the events they unmistakably describe?

As opposed to evolutionary assumption, which relies on blind faith in unguided processes, the Bible teaches that the physical elements came into existence out of nothing but at the command of 'intelligent cause'. We must be careful not to misinterpret the intended meaning of scripture. The biblical narrative was meant to be interpreted just like any other written document; in which case, the basic rules of literary interpretation and grammar require us to accept the biblical framework as literal and factual, unless figurative language is clearly intended. For example, Jesus said, "I am the door." This short phrase is a metaphor, but it represents something very real regarding his role as it relates to man. In the Psalms we find these words: "the earth shall not be moved." This does not mean that the earth is stationary in space, but that the earth is stable and consistent in its rotation and movement through the solar system. We dare not accept millions of years of death and struggle, and add it to the Bible without undermining the message of the cross. The biblical record clearly states that death has come upon this planet because of man's sin. "Therefore, just as through one man sin entered the world and death through sin, and thus death spread to all men, because all (have) sinned." (Romans 5:12) The message of the cross is that the ravages of death and decay will ultimately be reversed, and not only for those who have placed their hope in the atoning Christ, but for all of nature and the whole of creation as well. "For the creation was subjected to futility … because [of Jesus Christ] the creation itself also will be delivered from the *bondage of corruption* into the glorious liberty of the children of God" (Romans 8:20, 21). We must return to accepting the literal authority of scripture as the primary basis of our thinking if we hope to understand "the mystery of life".

Many perplexing questions have been considered in man's effort to fully comprehend the mysterious nature of God. Certain aspects of God's divine character and purpose will no doubt run contrary to human logic and reasoning. Consequently, the strange paradox of God's mysterious nature is often used as a reason to deny him, when it is actually strong evidence affirming the reality of his existence. We must ask ourselves again, how could God be superior to men if mankind was able to fully comprehend every aspect of his divine nature? If the Bible reveals something about the Creator we cannot grasp with our intellect or perhaps something that we object to on moral grounds, we should accept it by faith, rather than attempting to explain it away or to deny God altogether, based on some weak-minded human reasoning. For example, we hear many people complain that the Old Testament records human carnage on an unprecedented scale and that God condoned the violence. Many people find it hard to get past this, so they judge God as evil and hateful. I believe these people fail to understand the big picture of what God was attempting to accomplish through the nation Israel. We must keep in mind that God's goal for mankind is to reveal his love to them by providing a lasting means of redemption. Above all, He wants fallen humanity to return to a right-standing relationship with himself and become a part of his eternal family. Because of man's volition, he has the capacity to choose either good or evil. All men are morally culpable before their Maker. Because real love and relationship requires free choice, God knew that evil was not only possible but also inevitable as a consequence of bad choices. In fact, the choices of mankind became so evil in the Old Testament that on one occasion God destroyed everything he had made save Noah's family. "The Lord saw that the wickedness of man was great in the earth, and that every intent of the thoughts of his heart was only evil continually. And the Lord was sorry that He had made man on the earth, and He was grieved in His heart." (Genesis 6: 5,6)

To establish a chosen nation through whom God could reveal himself to the whole world, it was critical that Israel become a separated people unto God who would not become polluted by the heathen practices of the other ancient tribes around them. The wickedness and idolatrous practices in the land of Canaan had become so depraved that God could not allow this terrible evil to infect the children of Israel or His plan to redeem humanity would be thwarted. The gross depravity and sexual immorality in Sodom and Gomorrah, for example, had become so corrupt that God chose to destroy those cities with fire from heaven because the inhabitants refused to repent of their evil deeds. The Canaanite tribes, who occupied Israel's inheritance during their long captivity in Egypt, had become so depraved that they commonly practiced bestiality, incest, and even child sacrifice to their gods of stone. Deviant sexual behaviors became the norm in heathen cultures that had become so defiled that God said: "The land vomits out its inhabitants" (Leviticus 18:25). These vile abominations had approached the level of depravity that caused God to send the universal flood to destroy mankind in the days of Noah. If you go to a doctor who discovers that you have cancer, the doctor will say we must remove it or you could die. Are you going to have the doctor remove all evidence of the disease or just the main tumor? There was a cancer in the land of Canaan when the children of Israel returned from captivity and God instructed them to remove it lest they should become like the heathen nations around them, which would negate their national divine calling to represent the creator before all nations. The Canaanite tribes were given over four hundred years to repent of their wickedness, and they refused. Therefore, God used the ensuing conquest of Canaan to generate the historical context and religious climate that would lead to the introduction of the promised Messiah.

God our Creator has revealed to us everything that our feeble minds can possibly comprehend, and it is more than enough to find fulfillment in this life and the next through a personal, ongoing relationship with him. The Holy Spirit has revealed to our inquisitive minds that "we can see and understand only a little about God now, as if we were peering at his reflection in a poor mirror; but one day we are going to see him in his completeness, face to face. Now, all that I know is hazy and blurred, but then, I will see everything clearly, just as clearly as God sees into my heart right now." (1 Corinthians 13:12 TLB) The supernatural aspects of the biblical text make it obvious that a supremely intelligent and benevolent Creator has thoroughly revealed himself to fallen humanity. You and I can be absolutely certain that the powerful evidence revealed in God's written revelation of scripture confirms and validates the reality of his existence. There is no other way to explain the remarkable supernatural material in the biblical text, which is far beyond the intellectual capacity of man to knowingly record in and of his own willful intent. The remarkable nature of scripture must be considered overwhelming evidence for the existence of the personal, and loving God who created us!

Revelation: The Incarnation

God has instilled in man an awareness of himself. He has made himself known by the beauty and splendor of nature and the extremely complex design of the universe. He provided mankind with a healthy conscience to govern his response to established moral laws intended to protect him from choices that would do harm to himself and those around him. The sovereign Lord has made himself known throughout the earth through the impressive historical record of his chosen people, Israel. God has also revealed himself in great detail through the divinely inspired pages of the Holy Bible. And finally, and most amazing of all, God has completed the revelation of himself through the person of his Son, Jesus Christ.

To many people, the Word of God remains systematically vague. It is only when we look at the person of Jesus Christ that we find God's Word coming into clear focus and making perfect sense. The most astonishing revelation God has ever given to us of himself was the day he stepped out of eternity, passing through the veil between heaven and earth, to clothe himself in human flesh, becoming one of us. In so doing, he subjected himself to the same pain, sorrow, and suffering common to humanity. By this world-altering event, God has demonstrated a love without limits for his 'crowning creation'. He withheld nothing in subjecting his only Son to public humiliation, and the most horrible death imaginable on a cruel Roman cross– proving for all time that the 'redemption of mankind' is the compelling motivation that drives the heart of God.

Jesus Christ was much more than a great moral teacher or prophet. He said things that would be considered totally outrageous if they were not true, and then he backed up his words with miraculous signs, proving his divine nature to all those willing to follow him. He told his few followers, "If you do not want to believe what I'm telling you (that I am the way, that I and the Father are one), then you should at least believe me based on the many miracles you have seen." (John 14:11, paraphrase) When God took the form of man, faced the hardships, trials, and temptations we all face, and accepted, by his humiliating death on the cross, the terrible punishment that was due for our transgressions, he proved beyond doubt his unspeakable love for mankind. This passionate love for image-bearing mankind is something that we can barely comprehend! And of all the religions throughout world history, only in Christianity, through the life and work of Jesus Christ, do we see God reaching out to man, providing a sure way for us to be reconciled to a right relationship with himself. God will never force us to believe in him, but it should be understood that he has provided us with more than enough proof to validate his existence. He has also given us an understanding of the unselfish motivation that led to his voluntary and sacrificial death. The hope of God Almighty is that those, whom he has lovingly created in his own image, would ultimately recognize the unspeakable sacrifice (his atonement for sin) that was once and for all time offered on their behalf, and that they would willingly accept and gladly embrace this overwhelming demonstration of unconditional love.

Richard Dawkins, Christopher Hitchens, and others have said that historical evidence for the life of Jesus Christ is still in dispute. This unwarranted attack on the very core doctrine of Christian theology has been decisively and unequivocally refuted. The eminent British archaeologist and ancient-literature specialist Sir Grahame Clark, professor at the University of Cambridge, wrote a very definitive statement: "Frankly, I know of no ancient historian, who would have a twinge of doubt about the existence of Jesus Christ–the documentary evidence is simply overwhelming."[1] According to Dr. Gary Habermas, author of *The Historical Jesus*, there are no less than twenty different non-Christian written sources that document important details of the life of Christ. When these details are all put together, they provide enough information to give us a reasonable outline of the life of Jesus without ever touching the New Testament.

The Bible claims Jesus to be God incarnate on numerous occasions. "In the beginning was the Word, and the Word was with God, and the Word was God ... and the Word became flesh and dwelt among us ... (John 1:1, 14) Jesus proclaimed, "He who has seen me has seen the Father" (John 14:9). "I and My Father are one" (John 10:30). "Before Abraham was, I Am" (John 8:58). The simple fact that many of his unbelieving listeners wanted to stone him was proof that they understood he was claiming to be God. By fulfilling hundreds of prophecies concerning the coming messiah, and backing his identity with miraculous signs and wonders, he offered us the perfect credentials to support his claim of deity. Jesus was God in human flesh. There is no question about it! Everything he said and did supports this fact! He provided everything we need to know about God and withheld nothing, including his own life. Jesus Christ was and is the ultimate proof of God's existence!

In addition to the powerful evidence I have already cited, we are compelled to consider the grand finale in God's historical revelation of himself. This occurred when Jesus Christ rose from the grave triumphant over death as he promised he would. How can we possibly explain a death-defying miracle like the resurrection using human logic and reasoning? One thing we know for sure is the God of heaven is not bound by the physical laws of nature. The miracles recorded in Scripture are not a matter of science, but a matter of the ancient historical record that was faithfully documented and passed down to each generation. Since an all-powerful creator God is never limited by the natural laws he has ordained to govern the physical realm, he can choose to supersede these laws and to interject divine power into the human experience whenever he chooses to do so. The resurrection miracle required a colossal input of divine power characterized by a deliberate overriding of natural law. This single event in human history was the God-given guarantee offered to validate Christian faith. The bodily resurrection of Jesus is the final victory, and the most compelling proof that could ever be cited to confirm God's existence! This indisputable fact of history provides the greatest confirmation, yet that we live in a natural world that is subject to supernatural events at God's discretion in response to human need. Without this miraculous, well-documented historical event, there would never have been a faith-filled Christian community as we observe throughout the world. Christian faith has forever won its place in history by virtue of the sheer weight of evidence confirming that Jesus had risen from the dead in a world-altering and life-changing event. Scientific investigation cannot tell us if the resurrection of Christ was possible, but revealed evidence complied over the centuries can and has established the fact that this event has taken place in recorded human history. The resurrection of the Lord Jesus Christ is a well-established event in the historical record that has been proven far beyond reasonable doubt, resulting in widespread religious zeal the world over.

When the apostle Paul wrote to the Corinthian church regarding the resurrection, he said, "If Christ is not risen, your faith is in vain; you are still in your sins" (1 Corinthians 15:17). This astonishing miracle, which offers the world the central evidence to validate Christianity, has subsequently led to the committed and uncompromising faith of many millions worldwide. When Jesus entered the time-space-material dimension with a physical body, he showed us exactly who God is and he opened wide for us the doorway to heaven, so that you and I would have the opportunity to live in fellowship with our Creator forever in the kind of mutual love relationship God had originally intended for mankind.

If a person came into the lives of ordinary people like us claiming to be the Son of God and the Savior of the world, how could this person convince mankind of his identity? What could he do to set himself apart from every other self-proclaimed messiah that the world has ever known?

1. Perhaps he could say incredible things, such as the words spoken in the Sermon on the Mount, which would generate a large public following.

2. Perhaps he could perform breath-taking miracles in the presence of eyewitnesses— incredibly mystifying supernatural acts such as healing the sick, casting out demons, walking on water, and the most fascinating feat of all, raising the dead.

3. Perhaps he could accurately fulfill hundreds of ancient prophecies that were recorded centuries before his appearance— detailed predictions that would identify the true messiah spoken of and well documented by the great prophets of the Old Testament.

4. Perhaps he could read the hearts and minds of those whom he was attempting to reach in his effort to reveal the will of God.

5. Perhaps he could predict future events as a genuine prophet in his own right.

6. Perhaps he could demonstrate an unconditional love and compassion for mankind that would be unprecedented in all recorded human history.

7. Perhaps he could reveal many astounding things about the true nature and character of his Father in heaven with whom he communed daily.

8. Perhaps he could set aside his own personal interests and choose instead to go the distance in fulfilling the will of his Father by laying down his innocent life for the salvation of others.

9. Perhaps he could rise from the dead and appear before his followers and many other eye witnesses on multiple occasions, demonstrating that he was from beyond this world, that what he said was true, and that death could not hold back his life-giving power.

10. Perhaps he could ascend into heaven in the presence of his disciples, offering the final confirmation of who he was, and establishing for all time the reality of his divine nature.

Every one of these credentials was demonstrated in the life of one historical figure. Jesus Christ drastically set himself apart from all other men when he lived and moved among the occupants of ancient Israel in the first century. Without modern communication, Jesus left an overwhelming impression on mankind that has literally transformed the course of human history. What this single man said and did, in three short years of public ministry, served to alter forever the direction and course of humanity by drastically transforming the civilized world.

I ask you to consider the following question very carefully: what else could a personal, loving creator God possibly do for humanity to reveal his existence to a mundane, three-dimensional world and, while doing so, fully demonstrate his great compassion and unconditional love for fallen humanity? Jesus was and is God!

He loves every member of the sinful human race. And he proved it by his miraculous birth. He proved it by his history-altering life, and he proved it by a voluntary, sacrificial death, followed by an astonishing and world-transforming resurrection.

Leaders of major religious movements throughout human history— like Buddha, Mohammed, and Joseph Smith have all identified themselves as teachers or prophets. Only Jesus claimed to be God! When we seriously examine his history-altering life, take a close look at his precise fulfillment of messianic prophecy, consider his life-transforming miracles, and honestly examine his earth-shaking resurrection from the dead, it becomes obvious that he had the credentials to back that claim. Jesus never told his audience they must follow certain doctrines or guidelines to find the truth. He said, "Follow me" (Matthew 9:9). "I am the light of the world, he who follows me will not walk in darkness" (John 8:12). "I am the way, the truth, and the life. No one comes to the father, except through me" (John 14:6). By these statements he inferred that in following him one would receive the power to obey the laws of God, like the Ten Commandments. Jesus was the total embodiment of deity, who voluntarily left the spirit world behind to fully reveal the heart of Father God to fallen humanity. "For in Him dwells all the fullness of the Godhead bodily" (Colossians 2:9). "No one has seen God at any time. The only begotten Son, who is in the bosom of the Father, He has declared Him" (John 1:18).

Humanity has been inclined to worship this Jesus of Nazareth. Why is this? Because, in him, we find a complete revelation of the God of Israel, the sovereign Creator of the universe, the purpose-driven Savior of the world, and the One who promised that some future day he would return in glory to forever be with those who love him. When God the Father sent his only Son into the world, he revealed himself to humanity in the most personal way possible. By doing so, he left no doubt whatsoever that the intent of his heart was to forever 'redeem mankind'. God's deliberate plan to become intimately known to his 'crowning creation' was the driving force that led to the crucifixion, demonstrating his everlasting love for every single human soul.

Revelation: The Church of God

The worldwide Christian community is powerful evidence that God did not remain distant from human suffering, but became part of it to the extent that we can never fully understand. The message of the church is that God loves us beyond human comprehension. He demonstrated that love by providing a means (the atonement) by which the penalty for sin could be removed from each person who ultimately places his or her faith in the completed work of Christ on the cross. Those who trust in what God has done for humanity to address the problem of sin have entered into a personal relationship with their Creator, a relationship that enables them to spend eternity with the sovereign Lord of the universe, and not because they deserve it, but because of God's grace and unfailing love. It is that personal experience with God, through a heartfelt relationship, that has given millions of people the world over hope for a new and unimaginable life beyond this one, far transcending any optimism of carnal human expectation. Therefore, another overpowering evidence for the existence of God can be found in the simple lives of ordinary people who have responded in believing faith to the words of Jesus, when he declared, "I say to you, he who hears My word and believes in Him who sent Me has everlasting life, and shall not come into judgment, but has passed from death into life" (John 5:24). This wide world is full of people who have believed in their hearts that Jesus was eternal God in human flesh, and what he said was and is divine truth. They have believed in the reliable promises of God as revealed in his eternal Word, have repented of their sins, and have received the forgiveness that was made available to them by the death of Jesus Christ on a cruel Roman cross. After the final atonement for sin, humanity received sure confirmation of his divine rescue mission when Christ rose up triumphantly from the grave three days later in a well-documented, supernatural event that has never been refuted. In believing this truth, many millions of lives have been remarkably transformed from the inside out, changing them into the very people God created them to be.

World-famous Reverend Billy Graham served his generation as a world-renowned author and preacher who will surely be remembered as the greatest evangelist of the twentieth century. His monumental effort in building and strengthening the church of God in America and around the world is recognized everywhere by both Christian and secular sources. After more than seventy years in public ministry, at the age of ninety-five he recorded these profound words in his most recent book, *The Reason for My Hope*. "If I did not believe that Christ overcame death on the cross and bodily rose from the grave, I would have quit preaching years ago. I am absolutely convinced that Jesus is living at this moment at the right hand of God the Father and reigns in my heart. I believe it by faith, and I believe it by evidence found in the Scriptures."[2]

Perhaps you knew some Christian men or women before they made the decision to declare their faith in Jesus Christ as their personal Lord and Savior. If so, then you have no doubt witnessed a remarkable transformation that comes over someone when surrendered to the gospel message. The transformed lives of God's people, who have been supernaturally *born again* (having entered into redemption by faith alone), is powerful evidence of a personal creator God who works in the lives of ordinary people. Those who have proclaimed the name of Jesus and have willingly submitted to his eternal truth and divine authority, have become members of the worldwide church of God. The transformed lives of believing men and women have provided incontrovertible evidence to confirm the powerful work of a merciful Savior, who actively seeks to restore unto himself and to fully redeem each and every member of the fallen human race.

Purpose-Driven God

In the words of the well-known creation scientist Dr. Henry Morris, we are reminded that "the evidences of intelligent design and special creation abound everywhere in nature, and there is no real excuse for anyone to believe in naturalistic evolution, for which there is no legitimate scientific or historical evidence whatever."[3] When we have examine the compelling evidence for the existence of God, it becomes obvious that people who are set on denying Him will refuse to believe no matter what convincing proof is offered. These devoted 'skeptics' have created for themselves a very comfortable worldview— one that requires no personal accountability, and opens the door to every kind of carnal pleasure through self-gratification, at the expense of God's commands. Many people stubbornly refuse to allow their self-serving lifestyle to become disrupted by even the most compelling evidence. These misguided souls may find themselves beyond the reach of a loving, purpose-driven God by their own choosing. There are certainly many who are open to the possibility of a merciful God reaching out to them in unconditional love. The God of heaven helps all who seek him to begin a life governed by saving faith. "Come unto me all you who labor and are heavy laden (by sin) and I will give you rest" (Matthew 11:28).

God has revealed volumes to us about who he is and what he has done. "But without faith it is impossible to please Him" (Hebrews 11:6). The Lord will never submit to anyone's stubborn insistence to make himself completely known to the absolute satisfaction of the human mind. He has already revealed to us everything we need to know about him; anything more, we will have to accept by faith. The Bible tells us in Romans 1:17 that "the just shall live by faith." It also teaches that "he who comes to God must believe that He is (that He exists) and that He is a rewarder of those who diligently seek Him" (Hebrews 11:6). "By faith we understand that the worlds were framed by the Word of God (spoken forth by His command), so that the things which are seen were not made of things which are visible" (Hebrews 11:3).

Without divine revelation unveiled to the finite human mind by supernatural means, the God of heaven would remain a mystery to all. He would be essentially unknowable apart from divine revelation. The mystery and wonder of the creator will continue to amaze us, and for finite man it will always be that way on this side of eternity. Because of our limited understanding of divine providence, and the absolute requirement to exercise faith in the matter of our relationship with God, we are left with only two choices. We can accept God on the basis of what we observe and what he has faithfully revealed to us, or we can reject him outright because of human pride by insisting that we live life on our own terms, stubbornly refusing to submit to divine authority. If, by a personal act of our free will, we choose the latter, we will forfeit everything when we pass on from this life, and face the consequences of our choices and actions in the next.

The compelling evidence that verifies the reality of God leaves modern man in a very serious predicament. As the certainty of death approaches for each of us, upon what possible grounds does any man have to stand if he ignores the evidence, rejects God's truth, and repeatedly denies that the atonement of Christ was required to address his sinful nature? All men and women who seek after revealed truth must understand that to acknowledge God and believe he is our Creator, is to recognize divine authority to whom we are each accountable. We must fear the living God, but at the same time we must understand that his purpose-driven love for humanity is intended to address the fallen condition of every human heart.

Is God really waiting— longing to hear from us? In the final analysis, all of us will have to answer that question for ourselves. I cannot begin to prove the existence of God, in the scientific sense, any more than I can prove that my wife loves me. However, even though we cannot detect the Creator with our physical senses, the existence and power of God is actually something that few people are willing to deny. Have you ever heard anyone question the existence of the wind, because they cannot see it? I sure have not! Who among us could possibly deny the reality of this invisible force, when its profound effects are so clearly seen and felt? In the same way, we must ask ourselves: does the natural world, and the human story played out on earth, exhibit the creative work of an unseen force? Do we see credible evidence to verify a world that was carefully planned and purposefully designed? Has 'observation' and 'revelation' given us sufficient confirmation to acknowledge the reality of our Creator? If men and women weigh these 'many infallible proofs' in the balance, while assessing *the mystery of life*, the conclusion should read, "God cannot be denied!" "Woe to the man who fights with his Creator. Does the pot argue with its maker? Does the clay dispute with him who forms it?" (Isaiah 45:9 TLB)

Although the evidence for a transcendent superior mind is overwhelming, people choose to deny the existence of God because they cannot deal emotionally with the accountability a just God would require for sin. As the wisdom of man has dominated the primary cultural institutions within Western societies– education, politics, science, communication, and the entertainment industry have made it difficult to make a rational judgment about the meaning of life. Listen to these words of British columnist Peter Hitchens, former atheist, and brother of Christopher Hitchens (who was known for public debate opposing God). As a prize-winning journalist, Peter Hitchens has publicly challenged the famed British Broadcasting Corporation (BBC) about their trite, mocking approach toward Christian faith. He wrote, "The BBC teased religious leaders by asking them if they believe in the literal truth of the great Bible stories ... I would like to ask BBC chiefs, and the rest of our secular establishment if they believe in the literal truth of evolution. Evolution is an unproven theory. If what its fundamentalist supporters believe is true, fishes decided to grow lungs and legs and walk up on the beach. The idea is so comically daft that only one thing explains its survival-that lonely, frightened people wanted to expel God from the universe, because they found the idea that He exists profoundly uncomfortable."[4]

The Lord God is pursuing us, and asking each of us to seriously consider *the mystery of life*. He wants to be known by each one of us in a personal way, and to be in right relationship with him when we ultimately cross over into life after death. A lasting relationship of mutual love with his image-bearing sons and daughters was God's plan and purpose for mankind from the very beginning. The great God of heaven's glory has always passionately desired to have a family— children that he could love and have fellowship with, who would also love him in return.

Allow me to paraphrase some inspiring words of senior pastor Rick Warren of the world-renowned Saddleback Church: "You were planned and created for God's own pleasure; you were formed and designed with the intention of becoming a part of His eternal family."[5] Pastor Rick wrote a powerful book some years ago entitled *The Purpose-Driven Life*. Today, we are told this book has sold more copies worldwide than any book other than the Bible. How is this possible? It was possible because people desperately want to believe there is some actual purpose for their lives; they want to know they are here for a reason! The evolutionary theory has left mankind without any reason for existing. The crucial issue of God's existence remains before us, and we will never be rid of it on this side of eternity. God wants us to believe in him, trust in him, and know that he created us with a very special purpose in mind.

C. S. Lewis, Josh McDowell, and Lee Strobel are a few of the many who initially set out to disprove and discredit Christianity and the Bible. After being confronted with the evidence, however, they were all forced to seriously reconsider their atheist view. Ultimately, all of them abandoned a self-directed lifestyle that consistently denied divine purpose. At the end of their search for truth, these men opened their hearts and submitted their will to a new found faith in the Christian God, whom they determined to be very real. Each of them began a personal relationship with the sovereign Lord they had stubbornly denied for much of their adult lives. I believe the only reason to maintain that human life is merely a freakish product of blind chance, with no intended purpose, is because of personal choice. The evidence does not point there. In fact, the bible says that there would come a time when, despite the evidence, men will choose to be willingly ignorant—scoffing at the promises of God and his so-called love for humanity. "... There shall come in the last days scoffers, walking after their own lusts... And saying, where is the promise of his coming? ... For this they willingly are ignorant of ... (2 Peter 3: 3-5) KJV.

Many refuse to examine the evidence, and to understand how God has revealed himself to a fallen world. The message of humanistic evolution is that you and I are completely insignificant in the grand scheme of things. No wonder so many people live like there is no tomorrow. No wonder sexual perversion and woke ideology is so pervasive in our culture. If man has no real purpose, then why not take whatever you want in this life at the expense of others. On the other hand, the message of the bible is that you and I are 'so treasured' that the God of creation was willing to die in order to purchase and secure a relationship with each one of us. If there really is a purpose-driven God who loves sinful man, and desires fellowship with him, then it is surely his responsibility to reveal himself to those he loves. Man needs to understand God's heart, and the plan he has ordained for the infinitely valued human lives that he has created in his own image.

What is this plan? What is God's will for mankind? The bottom line is 'God wants each one of us to become part of his family'. Because God sees man as sinful and separated from him, this is why Christ had to die— to make a way for us to get back into a right relationship with our Maker. He is "not willing that any should perish but that all should come to repentance" (2 Peter 3:9).

How would a transcendent God go about the task of relating, in a personal way, to a fallen, material world? Perhaps he could design exceptional beauty and complexity in nature to an extent that it would demand the acknowledgment of a designer--a chief architect of great intellectual capacity and ingenious creativity. Maybe he could instill within mankind an innate awareness of himself. Perhaps he could give humanity a moral conscience bringing conviction over wrong thoughts and actions. Possibly, he could use a nation of flawed people, such as Israel, to record his very thoughts and actions in a divinely inspired book of remembrance— a book containing the vast information and knowledge that could only have been known to a sovereign, omniscient God. Perhaps the most astonishing action he could take would be to set aside his divine majesty, choosing to leave the spirit world behind to clothe himself in human flesh becoming one of us. In humility he became poor in order to personally walk among his creation and demonstrate for us the true nature of God. "For you know the grace of our Lord Jesus Christ, that though He was rich, yet for your sakes 'He became poor', that you, through His poverty, might become rich" (2 Corinthians 7:9). Then, perhaps, in his final act of love and devotion, he could humble himself and take the place of sinners by dying the most agonizing death a criminal could ever face. By paying that terrible price to redeem mankind, he would balance, for all time, the scales of divine justice. Then, in one final action of triumph and victory, he could rise from the dead, as he promised he would, proving to all mankind his divine nature. Surely, this would ultimately demonstrate God's endless love for humanity, whom he has always considered to have great redeeming value and unspeakable eternal worth. I will ask once again, what else could He have done?

The Reality of God

When each one of my four children were born, I had to come to grips with the possibility that, in exercising their personal free will they could grow up one day and choose to hate and reject me. Because of man's moral agency, God does not have control over the choices people make any more than I had control over the decisions my growing children would one day make. Years ago in Olympia, Washington, I remember speaking to a homeless young man on the street outside of a grocery store. I asked him why he found himself in such dire circumstances at a very young age. He told me that his father was quite well off financially, but he had come to disregard his father's authority over him after a family dispute. As a result of these unfortunate circumstances, this young man was no longer willing to submit himself to the earthly authority God had placed over his life. Guess what was keeping him from returning home, and repenting of his foolish rebellion. Pride! Isn't this exactly what all of us have done to our heavenly Father? We have each despised his authority, gone our own way, and have often allowed our foolish pride to keep us from turning back to him. The Bible says that each one of us have sinned against our Creator, and, in some way, we have all rejected his commandments, and have broken his laws. We have not wanted to admit that our sin and rebellion separates us from him, and that foolish pride has kept us from acknowledging our need to repent. As pride was keeping the homeless man in poverty, pride is what keeps many people today in the hopelessness of spiritual poverty.

I chose to have children, because it was well worth the risk in order to have a relationship with them based on mutual love and respect. God also knew that man could choose to rebel against his authority, disregard his Word, and even attempt to explain away his very existence in order to justify a selfish lifestyle that often refuses to submit to moral restraint. God was taking a big risk when he created mankind with personal volition, but it was worth the risk, so that he might have an opportunity to enjoy each one of us as part of his own family.

It was not God's fault that Hitler, Stalin, Pol Pot, and Mao made the awful choice to exterminate millions of their own people. It is so easy to blame God for the cruelty and hatred that people direct toward each other, when the truth is that mankind has exercised free will in committing horrendous acts of carnage against his fellow man. And none of the blame rests with God! Yes, there is a terrible curse on this world that, sadly, resulted in horrifying disease, pain, suffering, and death. God warned us in Genesis 2 that this would happen if Adam chose to rebel against his authority. We have no one to blame for this fallen world except ourselves! God's original creation was perfect in every way, but man chose to rebel against God's commandant, and now we are suffering from the awful consequences of sin. Keep in mind that we still have free will, which means we can choose to repent of our selfish acts and return, in faith, to a merciful, loving God, who offers us much more than we deserve. One thousand years before Christ, the psalmist King David asked God a question with these searching words: "When I consider your heavens, the work of your fingers, the moon and the stars, which you have ordained, what is man that you are mindful of him?" (Psalm 8:3,4). This question is still resounding in the minds of many today— why does God care so much about humankind? It seems rather strange to think of ourselves as special considering that we are merely a smidgeon of protoplasm, inhabiting a tiny speck of a planet, orbiting a very undistinguished and ordinary star, on the outer arm of a spiral galaxy that is populated by billions of similar stars, most of which are far greater in splendor and more magnificent than our sun. Mankind's home, which is in a rather small galaxy called the Milky Way, would seem to be lost in space considering that our galaxy is only one of multiple billions of galaxies that God distributed throughout the unimaginable vastness of the cosmos.

As strange as it sounds, God knows exactly where we are and he even knows the number of hairs on our heads. He created us in His image to become a vital part of his eternal family. He did not create us because he needed us! He created us only because he wanted us! And that's why I love him and want to be his friend, because out of unconditional love, and the desire for mutual relationship, he first chose me, and delighted in possibility of my personal association with him. Without God, it is very difficult to understand the unselfish concept of unconditional love. I do not place conditions on the God I serve, such as "I will love you and serve you if you will do this or that for me." No! I love him because he has first loved me, and I worship him because he is worthy! This brings me to a consideration of the core issue of Christian faith. *The heart of Christianity is the cross*! Our eternal Maker, who found himself in a broken relationship with mankind because of humanity's rebellious use of free will, wanted to set things right so badly that he was more than willing to place himself under the curse of fallen humanity. He walked in our shoes, has suffered the many trials of human experience, and withheld nothing from us by dying in our place, taking upon himself the penalty due for our selfish transgressions.

The most astonishing and incomprehensible thing about God is his tremendous personal sacrifice on our behalf, because of his passionate love for humanity. Our Creator, ultimately, became our Redeemer! There is certainly no greater love than this! In the 1990s, a poll commissioned by the Pew Research Center reported that 71 percent of respondents say they never doubt the existence of God. In this same poll, 61 percent of Americans said they believe miracles come from the power of God, and 53 percent said prayer was important in their daily lives. Some people argue that you could never prove God is real. Apparently, these people are confused about what is meant by *evidential proof.* There are two basic ways to prove something is factual. The first method is a scientific one based on observation. If, for example, I wanted to prove to you that a bar of Ivory soap floats, all I would have to do is to prepare a basin of water and toss the bar of soap into it, in your presence. The fact of the matter would be confirmed by your personal observation. The second method of proof is called judicial or evidential; this is also referred to as the legal-historical method. The legal-historical method of proof is used in courtrooms throughout America every day. Judicial proof must rely upon a collection of all available physical evidence, eyewitness accounts, and expert testimony in order to reach a verdict that is considered "beyond reasonable doubt." This, of course, is the method we must consider as we render an opinion about the existence of God, and the validity of the Christian faith. I cannot prove the existence of God scientifically. And for that matter, I cannot prove that my wife loves me, but I would stake my life on it based on the overwhelming weight of evidence that would confirm this fact to anyone who is at all interested to know.

Now, suppose I were to tell you that there is a tall man with glasses and a mustache standing on the other side of my front door at this very moment. Perhaps you would question me as to how I could possibly know this to be true. This prediction would certainly not be outside the remote realm of possibility. However, I would be a fool to stake my life on it. Why? Because I have absolutely no reliable evidence to confirm this is in fact true. The atheist likewise is foolish. Why? Because he has no reliable evidence, and yet he appears to be wagering his eternal soul on the belief that God does not exist. Blind faith in an unproven theory that places all hope in 'pure chance' and the imagination of man is a poor way to invest in the future. We must always ask ourselves, what does the evidence teach us, and where is it leading? If we just follow the evidence, we will never be misled! Open-minded scientists are considering the current volume and reliability of scientific data, and are beginning to follow where it would naturally lead them.

Consider the bold remarks of Dr. Frank Tipler, a professor of physics and cosmology at Tulane University: "When I began my career as a cosmologist some 20 years ago, I was a convinced atheist. I never in my wildest dreams imagined that one day I would be writing a book purporting to show that the central claims of Judeo-Christian theology are in fact true, that these claims are straightforward deductions of the laws of physics, as we now understand them. I have been forced into these conclusions by the inexorable logic of my own special branch of physics."[6] In his book entitled *The Physics of Christianity*, Dr. Frank Tipler makes this remarkable assertion: "From the perspective of the latest physical theories, Christianity is not a mere religion, but an experimentally testable science."[7] Another complementary conclusion is offered by Dr. Arthur L. Schawlow, a former professor of physics at Stanford University, and the esteemed winner of the 1981 Nobel Prize for his remarkable work in developing laser spectroscopy. "It seems to me that when confronted with the marvels of life and the universe, one must ask why and not just how. The only possible answers are religious ... I find a need for God in the universe and in my own life."[8] He also added that "we are fortunate to have the Bible and especially the New Testament, which tells us so much about God in widely accessible human terms."[9]

Christian faith is by no means blind faith! Confirming evidence that validates our faith in God is overwhelming. Those who deny God's existence simply do not want him to be real. In general, it can be stated that genuine faith in anything is only as valid as the evidence that can be offered to back it up! The basic truth about the Christian gospel remains valid because of a massive weight of physical and historical evidence, reliable testimony, and multiple eyewitness accounts that validate the historical claims of Jesus Christ. We learn of his remarkable life as a Jewish carpenter and an itinerant preacher, prophet, and healer of the first century. We would all be foolish to believe in a living Messiah, if there was not any valid evidence to back his claim of deity. Faith in historical Christian theology is a serious commitment that is based on volumes of reliable evidence and well-documented eyewitness accounts. The New Testament record is full of historically accurate truth upon which a life of evidence-based faith can begin. Thousands of men and women have studied and documented the compelling evidence pointing to the reality of a Judeo-Christian God and his redemptive work on behalf of fallen humanity.

Consider the words of internationally known, Indian-born Christian apologist and evangelist Dr. Ravi Zacharias, who has personally concluded that "a man rejects God, neither because of intellectual demands nor because of scarcity of evidence. A man rejects God because of a moral resistance that refuses to admit his need of God."[10] Dr. Zacharias adds: "God has put enough into the world to make faith in him a most reasonable thing, and he is left enough out to make it impossible to live by sheer reason and observation alone."[11]

World famous twentieth century philosopher Antony Flew was arguably Mr. Richard Dawkins predecessor as the world's most notorious atheist. Near the end of his life, however, Dr. Flew explained to Oxford Professor, John Lennox, exactly why he chose to abandon his decades long commitment to atheism. He explained his revised worldview by pointing out that the study of DNA had shown an almost 'unbelievable complexity' of the arrangements that were needed to produce life. Flew, therefore, concluded that an intelligent agent must have been involved. Many people, who find themselves on the cutting edge of advancing scientific research, are concluding, as Antony Flew did, that the complexity of the universe points to the existence of a benevolent Creator. When abundant evidence is carefully weighed, atheism can be seen as little more than an 'escape mechanism' by retreating from responsibility for one's choices and actions in a self-centered world, where people stubbornly reject moral absolutes.

Declaring a personal worldview should never become a choice between God and science. We must have both! God is the active agent who designed and created the universe. He is the brainpower behind the incomprehensible mechanisms that spells out life. God is to life and the physical world what Henry Ford was to the mechanized motor vehicle. The new atheists have confused the issue between mechanism and agency. They say that life can be explained by mechanism alone (natural selection) without agency (a designer). This is like saying that people can enjoy the modern motor vehicle only because of the principle of internal combustion (mechanism), and, therefore, we have no need of Henry Ford (agency). We need both the causing agent and the science to explain the diversity of life in this world! God is the innovator, but he is also the brilliant mind behind the established mechanisms (laws) by which everything operates with such amazing regularity in the physical world.

The pursuit of science can comprise man's compelling ambition to seek the knowledge leading to a better understanding of what a sovereign God has created. The new atheists have fabricated the idea of a war between science and God, but this attempt is not working, because the more we learn about the physical world, through empirical science, the more we recognize that a design agent was required to conceive of and initiate the complexities of a remarkable world. Consider these remarks from Dr. Albert Einstein, winner of the Nobel Prize for physics in 1921 for his groundbreaking 'general theory of relativity'. Some believe that Einstein possessed the most brilliant analytical mind of the twentieth century. Some scholars have even suggested that Albert Einstein was actually an agnostic, if not a full-fledged atheist. I beg to differ. This brilliant scientist had understood that a powerful presence far beyond nature is required to explain the physical universe, and the remarkable complexities of organic life on earth. Albert Einstein once wrote that: "Everyone who is seriously engaged in the pursuit of science becomes convinced that the laws of nature manifest the existence of a spirit vastly superior to that of men, and one in the face of which we, with our modest powers, must feel humble."[12] He then continued with: "My religiosity consists of a humble admiration of the 'infinitely superior spirit' who reveals himself in the slight details we are able to perceive with our frail and feeble minds. That deeply emotional conviction of the presence of a superior reasoning power, which is revealed in the incomprehensible universe, forms my idea of God."[13] "Human beings, vegetables, or cosmic dust--we all dance to a mysterious tune, intoned in the distance by an invisible piper."[14] By now we should know who that mysterious, invisible piper really is!

God designed our world with extraordinary beauty and complexity because he foreknew that, (human nature being what it is) the search for knowledge would have a tempting tendency to lead man into believing that there is no need for 'divine intelligence' to explain the physical universe and the magnificent diversity of life on Earth. In fact, he made it so complex that the more we learn, the more inclined we are to acknowledge that, in a material world operating according to the mathematical laws of physics, it would be unimaginable for planets and proteins to generate themselves by chance. If some people are determined to wager their eternal soul on an infinitely impossible biological accident, they certainly have a right to do so. However, knowing what I have discovered in a lifelong search for truth and understanding as revealed through science, history, and logic, I am compelled to do everything in my power to persuade them otherwise. The consequence of that choice will prove to be catastrophic! When presented with so many verifiable facts and offered a compelling weight of evidence, it would seem foolish to conclude that there is no God! These are not my words, but words that should ring loud and clear in the minds of those who refuse to acknowledge their Creator. These words were recorded twice by the psalmist David nearly three thousand years ago, and they are still reverberating today: "The fool has said in his heart, 'there is no God'" (Psalm 14:1, 53:1).

This presents a question that demands very serious consideration. Why is it so foolish to deny God? Because Christian faith is not blind faith unlike evolutionism. Persuasive and compelling evidence pointing to God's existence is rock solid. The terrifying reality that should be keeping the atheists and agnostics awake is the fact that they will not escape the divine judgment of their Maker– a judgment that will surely come upon their lives, and their posterity some future day whether they have believed in a theistic universe or not. "Why does the wicked man revile God? Why does he say to himself, He will not call me to account (for my sins)?" (Psalm 10:13 NIV) Don't be a fool! Believe in Him! Acknowledge and trust in Him– the sovereign God of creation who set the universe into motion by his creative power. We have everything to gain and nothing to lose, by believing, we are well on our way to finding the whole truth about 'the mystery of life'.

Many are beginning to recognize that evolutionary thought is like a ball and chain around the neck of modern civilization.

This theory threatens to rob us of our creative purpose and the hope of a glorious future in the presence of our Maker. Listen to the words of American engineer and mathematician, Dr. I. L. Cohen. "It is not the duty of science to defend the theory of evolution, and to stick by it to the bitter end ... no matter which illogical and unsupported conclusions it offers. On the contrary, it is expected that scientists recognize the patently obvious impossibility of Darwin's pronouncements and predictions ... let's cut the umbilical cord that tied us down to Darwin for such a long time. It is choking us and holding us back."[15]

We need to expose evolution for what it really is to protect the minds of young people and point men and women toward the hope that is found in Christ alone. Dr. Pierre Paul Grasse, former chair of evolution at Sorbonne University, concurs. A world-class zoologist, Dr. Grasse is very blunt about the problem: "Today, our duty is to destroy the myth of evolution, considered as a simple, understood and explained phenomenon which keeps rapidly unfolding before us."[16]

"The heavens declare the glory of God; and the firmament shows His handiwork. Day unto day utters speech, and night unto night reveals knowledge. There is no speech nor language where their voice is not heard. Their line has gone out through all the earth and their words to the end of the world" (Psalm 19:1,4). "Since earliest times, men have seen the earth and sky and all God made, and have known of his existence and great eternal power. So they will have no excuse [when they stand before God in judgment]" (Romans 1:20 TLB). "For thus saith the Lord that created the heavens; God himself that formed the earth and made it; he hath established it, he created it not in vain, he formed it to be inhabited: I am the Lord; and there is none else" (Isaiah 45:18 KJV). "Thus says the Lord, the King of Israel, and his Redeemer, the Lord of hosts: I am the First and I am the Last; Besides me there is no God" (Isaiah 44:6). "By faith we understand that the worlds were framed by the Word of God, so that the things which are seen were not made of things which are visible" (Hebrews 11:3). "Remember the former things of old, for I am God and there is no other; I am God, and there is none like Me. Declaring the end from the beginning, and from ancient time things that are not yet done ... (Isaiah 46:9,10).

God Reigns: Concluding Remarks

A lack of understanding about the reality of God and widespread absence of the fear of God, along with a stubborn refusal to obey His Word, lies at the root of every weighty problem that besets us as a human family. "All the ills from which America suffers can be traced to the teaching of evolution."[17] So said William Jennings Bryan, a leading politician and presidential candidate in the late 1800s. "The god of this age has 'blinded the minds of unbelievers', so that they cannot see the light of the Gospel of the glory of Christ, who is the image of God" (2 Corinthians 4:4 NIV). "You will ever be hearing but never understanding. You will be ever seeing, but never perceiving. (The hearts of these people have become calloused.) They hardly hear with their ears, and they have closed their eyes" (Matthew 13:14-15 NIV, paraphrase). They are "always learning and never able to come to the knowledge of the truth" (2 Timothy 3:7). Hear these powerful words from famed civil rights reformer, Dr. Martin Luther King Jr., who was clearly disturbed by the way in which science has threatened to replace faith in our culture: "Modern man has brought this whole world to an awe-inspiring threshold of the future. He has reached new and astonishing peaks of scientific success ... yet, in spite of these spectacular strides in science and technology, something basic is missing. There is a sort of poverty of the spirit, which stands in glaring contrast to our scientific and technological abundance. The richer we have become materially, the poorer we have become morally and spiritually ... We have allowed the means by which we live to outdistance the ends for which we live ... Our scientific power has outrun our spiritual power. We have guided missiles and misguided men ... If we are to survive today, our moral and spiritual 'lag' must be eliminated."[18]

A firm rebuttal of naturalist theology comes from British physicist John Houghton: "The fact that we understand some of the mechanisms of the working of the universe or of living systems does not preclude the existence of a designer, any more than the possession of insight into the processes by which a watch has been put together, however automatic these processes may appear, implies there can be no watchmaker."[19] In 1948, while presenting his Armistice Day address at the close of World War II, famous American general Omar Bradley concluded his remarks with this monumental statement: "We have too many men of science, too few men of God. We have grasped the mystery of the atom and rejected the Sermon on the Mount. Ours is a world of nuclear giants and ethical infants. We know more about war than we know about peace, more about killing than we know about living."[20] These exceptional words ring so true today more than ever before. When all the evidence is weighed, and after every last argument is systematically analyzed, it brings us to the obvious question: what is keeping us from getting back to the worship of the one true God, who has faithfully revealed himself to humankind.

What are we teaching our children? Will we remain silent, while allowing them to be robbed of faith in a magnificent Creator? We must do whatever we can to push back this terrible cloud of deception that hangs over our beloved nation. William Jennings Bryan is best known for his defense of creationism in the famous Scopes monkey trial at Dayton, Tennessee, in July of 1925. In this well-known court case, he gave the following remarks: "Parents have a right to say that no teacher paid by their money shall rob their children of faith in God and send them back to their homes skeptical, or infidels, or agnostics, or atheists."[21] It is disturbing to acknowledge that this is exactly what has happened in America over the last eighty years. We have witnessed a steady decline of 'faith in God' leading many families into the post-Christian era of America.

"And He has made from one blood every nation of men to dwell on all the face of the earth and has determined their pre-appointed times and the boundaries of their dwellings, so that they should seek the Lord, in the hope that they might grope for Him and find Him, though He is not far from each one of us; for in Him we live and move and have our being" (Acts 17:26-28). Have you noticed how the devil loves to divide people based on race, and many people are falling for this nonsense. There is only one race! It's called the 'human race'! Do not make too much of skin color, this is only genetic variation programmed into human DNA (variation within the human kind). It is protection the creator has given us for adaptation to various climates and environments. Hatred toward those of another skin shade is nothing but plain ignorance.

You and I are no accident; God planned us and appointed us to dwell on the earth for such a time as it pleased to him. Trust in God, not the ever-changing opinions of man. "There is a way that seems right to a man, but its end is the way of death" (Proverbs 14:12). "Look to me, and be saved, all you ends of the earth! For I am God, and there is no other" (Isaiah 45:22). Even the Old Testament promises that God can be found, and God can be known. "And you will seek me and find me when you search for me *with all your heart*!" (Jeremiah 29:13). Please note that Jeremiah did not say 'with all your mind'! The search for God through intellect alone has led to great frustration, characterized by a terrible lack of fulfillment. Jesus said, "I tell you the truth, anyone who will not receive the kingdom of God like a little child will never enter it" (Mark 10:15 NIV). The simple faith of an innocent child demonstrates the humility God requires of mankind.

Who can prove the non-existence of God? Upon what evidence will he or she stand? Consider the words of German-born rocket scientist, Werner von Braun, who became a major architect of the American space program at the height of the Cold War. Regarding the powerful evidence he discovered in his special branch of rocket propulsion engineering, Dr. von Braun was compelled to publicly declare the reality of a super-intelligence governing the universe. He penned these words in a letter to the California Board of Education: "As we learn more and more about nature, we become more deeply impressed and humbled by its orderliness and unerring perfection ... One cannot be exposed to the law and order of the universe without concluding that there must be design and purpose behind it all ... Speaking for myself, I can only say that the grandeur of the cosmos serves to confirm my belief in the certainty of a Creator .... the creation of the fundamental laws of nature ... they challenge science to prove the existence of God. But must we really light a candle to see the sun?"[22]

We are accountable to the creator. We have broken his laws and stand guilty before a Holy God. "For the word of the Lord is living and powerful and sharper than any two-edged sword, piercing even to the division of soul and spirit and of joints and marrow, and is a discerner of the thoughts and intents of the heart; and there is no creature hidden from his sight, but all things are naked and open to the eyes of Him to whom we must give account (Hebrews 4:12, 13).

"But God shows his anger from heaven against all sinful, evil men who push away the truth from them. For the truth about God is known to them instinctively; God has put this knowledge in their hearts..... Since earliest times, men have seen the earth and sky and all God made and have known of his existence and great eternal power. So they will have no excuse [when they stand before God at judgment day]. Yes, they knew about him all right, but they wouldn't admit it or worship him or even thank him for all his daily care. And after a while, they began to think up silly ideas of what God was like and what he wanted them to do. The result was that their foolish minds became dark and confused. Claiming themselves to be wise without God, they became utter fools instead" (Romans 1:18-22 TLB).

I will end this chapter with a quotation from the Greek philosopher Plato: "We can easily forgive a child who is afraid of the dark, the real tragedy of life is when men are afraid of the light."[23]

There is a magnificent God in heaven— a benevolent Creator of the vast physical reality that you and I have come to know. This Creator God loves us beyond measure, and we are certainly accountable to him! The evidence for the existence of the Almighty goes far beyond reasonable doubt and constitutes a much more intellectually sound position than that of atheism. If people do not believe in God, it is not because of a lack of evidence pointing unmistakably toward a benevolent Creator. People do not believe because they do not want to believe. In re-quoting the American Neurobiological scientist Dr. George Wald, we can see clearly why many people do not believe in God. Even though he knew scientifically that 'spontaneous generation' was a fallacy, Wald commented: "that leaves us with the only possible conclusion— that life arose as a supernatural creative act of God. I will not accept that philosophically, because I do not want to believe in God. Therefore, I choose to believe in that which I know is scientifically impossible: spontaneous generation arising to evolution."[24]

Now that we know God is real, we must make every effort to find out who he is. Then we will need to understand exactly what

God expects from us. And finally, and most importantly, we should be compelled to do what he says! That, my friend, is what we call "fearing God," and our entire pursuit of true happiness, grounded in meaning, purpose, and fulfillment depends on it.

CHAPTER NINE

* * *

THE MERGING OF SCIENCE AND FAITH

THE FINAL ACT—NO CURTAIN CALL

There are two issues facing humanity today that must be reconciled for man to experience true peace. The first requires every individual to face up to his or her impending 'mortality'. We must become conscious to the fact of our approaching death, and to fully embrace the ramifications of this uncontested reality. From the moment of birth, we have been set on a path toward our eventual demise. It is easy to push this thought to the back of our minds, or to chalk it up as a matter for future consideration. People generally set out to live as though their tomorrows will never end. By ignoring or postponing the thought of his own mortality, man has often been able to ignore the second critical issue facing humanity: 'immortality.' Will everyone continue to exist in some state forever as the Bible clearly teaches? This menacing question has been circling around in the back of every human consciousness; is it possible that man will face an endless eternity when this mortal life comes to its inevitable conclusion? Like a dramatic theatrical play, the curtain of mortality will come down upon the final act— perhaps at a time least expected. Can we afford to ignore this ever-persistent question? We are certainly all aging and the sands of time are running out, forcing each of us to face up to the fact of our impending departure. We need answers to these life-and-death questions, and we dare not settle for the wrong answers. There is so much at stake when men and women are facing eternity every day. If we will only consider the evidence, this will undoubtedly lead to the understanding we so desperately need in order to satisfy our deep longing for truth, meaning, and hope beyond this world.

Atheism, as a growing ideology, is a dangerous position to embrace. It places people in the situation of having to face certain death without knowing what lies beyond. This view of life forces men and women to believe that— if they must face an eternity beyond the grave, the merits of good personal behavior in this life should warrant God's favor in the next. This is a philosophical position the Word of God flatly denies. Millions of people, who have persisted in believing it would be better served if they understood the biblical concept of 'redemption'.

There has been too much hate, war, and brutality on this earth since the beginning of man's self-indulgent occupation. Many ideas and ideals have been fought over throughout the passing centuries and, even today, mankind cannot agree on how to live in peace and friendship with one another. It seems that as soon as one war quiets down, another major conflict heats up. We so often fail to respect and appreciate our differences. Of course, black lives matter, but so does every other human created in God's image, regardless of ethnic background. To our shame we have made far too much of skin color and have allowed it to divide us instead of embracing our differences for what they are— a gift from God. The greatest single issue that divides mankind is that we cannot agree on the reality of a sovereign God who created us and who will hold us accountable for our rebellion against his Word and his authority. Only the truth will prevail in the end, so we must ask ourselves the question, 'how then shall we live'?

Science: A Powerful Witness

Albert Einstein once mused, "Science without religion is lame; religion without science is blind."[1]

Examine the following short list of names. I am sure you will recognize some of them right away, especially if you have an interest in science, as I do. Other names may seem unfamiliar to you. What you see here is just a random list, in no particular order, of some of the most impressive scientists of past generations, as well as many contemporary names whose scientific research has carried the bright torch of man's curiosity well into the twenty-first century.

Albert Einstein	1921	Robert Boyle
Max Planck	1918	Willian Kelvin

Joseph J. Thomson	1906	Sir Isaac Newton
Arthur Compton	1927	James Maxwell
Max Born	1954	Michael Faraday
Nevill Mott	1977	Johannes Kepler
Werner Heisenberg	1932	Nicolaus Copernicus
Antony Hewish	1974	James Joule
Arno Penzias	1978	Louis Pasteur
Ernst Chain	1945	Gallileo Galilei
Arthur Schawlow	1981	Ernst Mayr
Christian Anfinsen	1972	Fred Hoyle
Richard Smalley	1996	Werner von Braun
Robert Millikan	1923	Michael Denton
Charles Townes	1964	Michael Behe
William D. Phillips	1997	Paul Davies
William H. Bragg	1915	Tony Rothman
Guglielmo Marconi	1909	Freeman Dyson
Isidor Rabi	1944	Frank Tippler
Erwin Schrodinger	1933	Robert Jastrow
Alexis Carrel	1912	Allan Sandage
John Eccles	1963	James Tour
Joseph Murray	1990	Henry Schaefer
Derek Barton	1969	Gregor Mendel
Walter Kohn	1998	Blaise Pascal
Ronald Ross	1902	Frances Bacon
Alexander Fleming	1945	Frances Collins
Albert Schweitzer	1952	Wolfgang Smith
Charles Sherrington	1932	Rene Descartes

What do all the men listed above have in common? The list includes a growing number of men whose dedicated study and research has shaped the world of historical and contemporary science. The men who I list in the left column have all been awarded the coveted Nobel Prize (date received following each name) for major contributions to the advancement of science and a better understanding of our world. The names listed to the right are inclusive of many great scientists of the past as well as the present whose contributions have led to major advancement in our current knowledge of the basic laws of nature, organic life, and the physical universe. All these early and modern-day scientific pioneers listed have one other thing in common. Each one of them believes in a creator God, and would affirm that they have gained this recognition, not by the influence of any church or religious community necessarily, but through a search for understanding in their dedicated pursuit of science. Their great love for knowledge about the unknown having led to a remarkable, undeniable accumulation of reliable evidence pointing unmistakably toward this bold conclusion. These devoted men became compelled to accept a controversial worldview in spite of the many time-honored traditions of science that have been grudgingly cast aside in acquiring this knowledge. In fact, it is reported that 60 percent of the Nobel Prize winners over the last century have confessed belief in a supreme God or authority who is responsible for the vast physical cosmos that caused them such irresistible curiosity. Why does the ongoing pursuit of empirical scientific achievement ultimately lead us back to God? In creating the universe and conscious carbon-based life on this earth, God calculated everything with such razor-sharp precision and complexity that man's successful discovery of even the slightest bit of knowledge would inevitably point back to Him.

The hope of future generations lies in the fact that modern science, the most powerful force in contemporary thought, is beginning to get it right. As you can see from the impressive list of names cited here, an ever-growing number of top research scientists are becoming more determined than ever before to 'discover the truth' and reveal the facts, even if it means having to admit that over one hundred years of biased speculation has surely led humanity down the wrong path. These brave and increasingly open-minded researchers are no longer willing to hide newly discovered scientific data for the purpose of propping up an outdated theory that has been refuted by an avalanche of new information acquired over the last century or more., This new generation of investigators is becoming more willing than ever to pursue the evidence wherever it leads them, as more scientists are putting aside their initial bias and seeking only verifiable facts in what has become a life and death struggle hanging over the civilized world. These rational-thinking scientists are questioning the merits of such an obsolete theory as Darwinism. Consider the words from James Gorman, a science reporter for the *New York Times*: "Evolution is not only under attack by fundamentalist Christians, but is also being questioned by reputable scientists. Among paleontologists, scientists who study the fossil record, there is growing dissent from the prevailing view of Darwinism."[2]

Werner von Braun was a Prussian-born scientist, who fled Nazi Germany at the end of World War II. He soon began use his talents and experience to insure a more peaceful world. In America, he was to become a world-class rocket propulsion engineer who led the way in the development of the American space program at the height of the Cold War. Because of his outstanding leadership in structuring the space program, he became known by NASA as the "father of rocket science." This American icon, who loved science with great passion, was also a devoted Christian believer who viewed science and faith as comrades, not enemies. Consider the implication of his bold remarks declaring faith as an important virtue to any serious scientist who is willing to admit that the grandeur of the cosmos clearly points to a superior intelligence. "Whereas all other living beings seem to find their places in the natural order and fulfill their role in life with a kind of calm acceptance, man clearly exhibits confusion. Why the anxiety? Why the storm and stress? Man really seems to be the only living thing uncertain of his role in the universe; and in his uncertainty, he has been calling since time immemorial upon the stars and the heavens for salvation and for answers to his eternal questions: Who am I? Why am I here?" He continues: "The main spring of science is curiosity. But there would not be a single great accomplishment in the history of mankind without faith. Any man who strives to accomplish something needs a degree of faith ... But many people find the churches, those old ramparts of faith, badly battered by the onslaught of three hundred years of scientific skepticism. This has led many to believe that science and religion are not compatible, that 'knowing' and 'believing' cannot live side-by-side. Nothing could be further from the truth! .... The two most powerful forces shaping our civilization today are science and religion ... While science tries to learn more about the creation, religion tries to better understand the Creator ... One cannot be exposed to the law and order of the universe without concluding that there must be a divine intent behind it all .... as we learn more and more about 'nature', we become more deeply impressed and humbled by its orderliness and unerring perfection ... Speaking for myself, I can only say that the grandeur of the cosmos serves to confirm my belief in the certainty of a Creator."

Dr. von Braun reveals his passion for both the spiritual and the scientific pursuits of life when he further explains the close and essential connection between the two. "The knowledge that man chose between good and evil should draw him closer to his Creator. Next, the realization should dawn that his survival here and hereafter depends on his adherence to the spiritual rather than the scientific ... the ethical guidelines of religion are bonds that can hold our civilization together. ......Without them, man can never attain the cherished goal of 'lasting peace' with himself, his God, and his fellow man. We should not be dismayed by the relative insignificance of our planet in the vast universe as modern science now sees it ... In fact, God deliberately reduced Himself to the stature of humanity in order to visit the earth in person, because the 'cumulative effect' over the centuries of millions of individuals choosing to please themselves rather than God had infected the whole planet... When God became a man himself, the experience proved to be nothing short of pure agony. The stage was set for a situation without parallel in the history of the earth. God would visit creatures, and they would nail him to a cross! I am certain that, were He among us today, Christ would encourage scientific research as modern man's most noble striving to comprehend and admire His Father's handiwork."[3]

While science attempts to answer for us the *how* questions, religion admonishes us to consider the *why* questions. If we can only understand a fraction of the how and the why questions of life, we will surely experience a small measure of the fulfillment we were meant to enjoy. Where did we come from? How did we all get here? Why are we here? Where are we going when we die?

Follow the Evidence

Men and women today, who are determined to *follow the evidence* wherever it leads, are now finding themselves up against enormous resistance and antagonism. It certainly takes a strong person to buck the tide and risk being discredited, denounced, and even humiliated by his or her peers; but this is what happened long ago to men like Kepler and Galileo who bravely stood up to the establishment. Their determination and persistent pursuit of knowledge ultimately shaped a new scientific paradigm for the next generation. This important principle first originated with Plato, and was adopted by Dr. Antony Flew, who decided that an open-minded approach to the question of origins must be engaged in by anyone sincerely seeking after the truth. In Plato's *Republic*, Socrates is scripted to insist that "we must follow the argument wherever it leads."

In contrast, a firm position held by the new atheists is clearly represented in the written works of Richard Dawkins and others who maintain a firmly fixed but narrow-minded view of the concept of God and the origin of life. This rigid position, with respect to the modern-day conflict of ideas, firmly maintains that, "either you are with us all the way, or one with the enemy ... even eminent thinkers who express some sympathy for the other side [of the argument] are denounced as traitors."[4] Calling your philosophical opponent enemy or trader is uncalled for by the dogmatic proponents of atheism and demonstrates the extreme measures they will often use to demean anyone who denies Darwinism. Of course, the argument rages on, but these Darwinian zealots have consistently failed to address the most critical issue at stake in this hotly contested cultural debate over the origin of life. No rational argument is proposed as to how conscious life may have emerged from the interaction of lifeless chemical elements. Dawkins offers only a weak stab at the initial origin of life and consciousness with a simple remark that it was based on events triggered by "an initial stroke of luck."[5]

There was once a public perception that modern science buried God, and as undertaker, each modern advancement provided another nail in his coffin. This bizarre idea received support in the 60s and 70s, but now has become obsolete, because the more man discovers, the more scientific achievements are pointing unmistakably toward 'intelligence'. The contemporary Christian worldview continues to boldly assert that biblical faith and science are inseparable.

The creation view of life continues to assert that biblical faith and scientific achievement are completely compatible. As it turns out, faith is the natural human response initiated by reliable evidence. In his world-famous biography of the life of Jesus, the first-century apostle St. John wrote the following words: "These (things) are written that you may believe" (John 20:31). John fully understood that his extraordinary eyewitness account of the life of Christ would become powerful evidence upon which many would discover a life of saving faith. It has never been a part of the biblical worldview that major spiritual ideals should become earnestly believed and embraced in the absence of sufficient evidence. Just as we observe in science, faith, coupled with logic, reason, and evidence are constant intellectual companions that cannot be separated. In rejecting and condemning biblical faith as blind faith, the extreme actions of the militant new atheists have seriously undermined any scientific credibility they may have once had.

Neo-Darwinian thinkers, like Richard Dawkins and others, hold firmly to the erroneous and misguided notion that faith in God is blind, with no reliable supporting evidence to back it up. This is far from true! They are completely wrong in making this assertion. Faith is inseparable from any attempt at a scientific inquiry to understand the physical world around us. World-renowned physicist, Paul Davies, says it like this, "Just because the sun has risen every day of your life, there is no guarantee it will rise tomorrow. The belief that it will--that there are indeed dependable regularities of nature--is an act of faith, but one which is indispensable to the progress of science."[6] Ill-fated attempts by new atheists to cleanse the human consciousness of religious faith are destined to fail. Humanists have brought great pressure upon society by insisting that faith in God must be restricted to the private domain, while theoretical science should have unrestricted access to public consciousness in every area of human society.

British philosopher Bertrand Russell, a winner of the Nobel Prize for literature in 1950, was an inspirational figure in the life of Richard Dawkins and greatly influenced his effort to carry the torch of atheism well into the twenty-first century. In a revealing book about her famous father, Bertrand Russell, British essayist Katherine Tait shared that her devoutly atheist dad was "never open to any serious discussion about the existence of God." She wrote: "I could not even talk to him about religion."[7] Nevertheless, Ms. Tait, who would marry a Christian minister, believed that her father's "whole life was a search for God ... somewhere at the back of her father's mind, at the bottom of his heart, in the depths of his soul there was an empty space that had once been filled by God, and he never found anything else to put in it."[8] She said he had the "ghost-like feeling of not belonging, of having no home in this world."[9] However, the persistent inquiry and soul-searching questions of his only daughter must have had some effect on Russell, for in a rather poignant passage he once wrote, "Nothing can penetrate the loneliness of the human heart, except the highest intensity of the sort of love the religious teachers have preached."[10]

You will find nothing remotely resembling this remark in the writings of Richard Dawkins. He seems determined to champion an unrelenting theme common among today's atheists, who make the contemptible claim that no scientist worth his or her salt believes in God! This very provocative and outrageous claim has now been swept aside by the current tide of world-view transformation among today's top-notch scientists. Dawkins is wrong! He has made the classic mistake that nearly all evolutionists are making. They have always believed that an extremely old age for earth makes anything seem possible, even if it is not. Those who have placed uncompromising faith in neo-Darwinian theory, leading to the conclusion of no creative Agency in the universe, have made a life-altering decision based on a poor theory and flimsy evidence.

With an ever-increasing amount of contemporary knowledge strongly supporting the time-tested creation worldview, many naturalists are becoming anxious about the current widespread use of terminology implying intelligent design. Evolutionary thinkers view any claim of intelligence as a dangerous revival of the creationist movement, characterized by an unscientific approach to the problem of identifying the meaning of life. The concept of a super-intelligent causation behind a remarkably complex universe is certainly not a recent phenomenon. This popular idea is as old as the universally prevalent notion of philosophy and religion themselves. The commonplace belief in a theistically designed universe has stood the test of time over many millennia. Only in the last couple of centuries, due to its widespread dissemination in Western culture, has blind, unguided naturalism become a popular worldview. Consider this remark: "It is absolutely safe to say that if you meet somebody who claims not to believe in evolution, that person is ignorant, stupid, or insane (or wicked, but I'd rather not consider that)."[11] Dogmatic humanists like Richard Dawkins, who spue this unwarranted, fallacious rhetoric seem to have gone off the deep end in their absurd disdain for "theism." However, I am convinced they will gain very few followers in their misguided war against God and religious faith.

At least to some degree, the motivation behind current efforts to disseminate neo-Darwinian dogma worldwide is an attempt to banish the idea of God from human consciousness and public discourse. Misguided nonbelievers must furnish arguments in support of their strictly anti-God worldview, but they have failed to do so. Instead, all they can do is train their sights on many well-known abuses in the historical past involving the world's major religions, by arguing that religion, including Christianity, has failed to deliver on the theme of peace and prosperity for all. Many people agree that atrocities have occurred in our world under the guise of religion. This, however, should not be a stumbling block to those who would earnestly seek after knowledge and understanding, in order to find meaning and purpose for their lives. Even Christians often fail to live up the commandment of God to 'love thy neighbor'.

It is equally true that many benevolent, world-altering changes have come forth through the dedicated efforts of those who have faithfully lived out their Christian values. For example, after years of unwavering perseverance and personal sacrifice, William Wilberforce was finally able to prevail upon the English Parliament to end the dreadful practice of a human slave trading. This bold act of Christian charity and courage literally transformed the English-speaking world. Abraham Lincoln waged a very long and bitter fight to end the uncivilized practice of slavery in America. He was completely dedicated to the biblical ideal that all men were created equal, and are endowed by their Creator with certain inalienable rights, including the right to life, liberty, and the pursuit of happiness. Because of his unwavering commitment to this Christian virtue, Lincoln would go on to pay the ultimate price. The Reverend Martin Luther King Jr. was able to successfully lead a civil-rights movement that helped to revolutionize the American way of life. This effort placed the Western world on a course toward respecting human rights and human dignity far and wide. America's decision to enter World War II was inspired by our Christian duty to champion the cause of freedom in a world where evil tyrants were imposing their heinous ambition on innocent lives. This action would lead to the liberation of the death camps of Nazi Germany, and ended the Japanese reign of terror in the Pacific. After the founding of America, the Christian community established the first hospitals, schools, colleges, and universities. The virtue of Christian charity has become widespread throughout the world as American history will attest. Countless charitable organizations have emerged by way of the Christian command to *love thy neighbor as thyself.* These efforts have provided food, clean water, clothing, shelter, medicine, and medical services to the underprivileged worldwide. Examples of such charitable organizations would include Feed the Hungry, the Salvation Army. Samaritan's Purse, St. Jude hospital, Habitat for Humanity, and Never Thirst, not to mention Catholic hospitals, homeless shelters, crisis pregnancy centers, and countless of other humanitarian organizations that have addressed human need and suffering in America and many other nations the world over.

With free will man is at liberty to choose either good or evil. Consider the words of writer Mr. Roy Varghese, "The excesses and atrocities of organized religion have no bearing whatsoever on the existence of God, just as the threat of nuclear proliferation has no bearing on the question of whether $E=mc^2$."[12] Evil is and always has been born in the heart of man who has the capacity to impose his selfish will on the less fortunate. The suffering of humanity has often been imposed by what humans have done to one another. Do not forget that it was man, not God, who invented the implements of war, torture, and slavery. The mindless, selfish actions of those who disregard God's moral laws will often impact innocent lives— like the child out for a ride on his new birthday bicycle only to be struck down by a drunken driver, or the schoolgirl walking home after classes, who is cut down by a stray bullet in a gang-infested and crime-ridden neighborhood. The Lord would have to revoke free will, and thereby impose a robotic state upon mankind in order to prevent tragedies like these from happening to innocent victims.

British philosopher Antony Flew was a former Oxford atheist and professor, the late son of a Methodist minister, who became a decorated, internationally known scholar. Early in his storied career, Dr. Flew had claimed to be a left-wing socialist and professing communist, who led the charge of atheism for nearly fifty years. The respectable thing about Dr. Flew is that he became open-minded, and determined that he would live by the Socratic principle, which suggests that "we should follow the argument wherever it leads." This ultimately led him to make a life-altering decision. Antony Flew's initial reason for denying God's existence was his inability to reconcile the problem of evil and suffering in a world with an all-loving, all-powerful God. He reasoned that if there is a God of love, then the existence of pain and suffering would be inconsistent with his divine nature. This argument, which has become all too common among atheists, was covered in significant detail in chapter 7. Even so, people who consistently deny God will say, "Only statements that can be verified using the direct methods of science are considered to be meaningful." Therefore, by their own admission, evolution is meaningless, because it has not and cannot be verified by the empirical scientific method.

As a world-class philosopher, Antony Flew would privately investigate where the evidence was leading, while he maintained a public view of atheism. However, as the trend toward unbiased scientific research in the late twentieth century began to reveal evidence that was inconsistent with evolutionary thought, Flew became determined to follow the data wherever it would lead. As it turns out, the tipping point for Dr. Flew was DNA research. The amazing discoveries produced by these scientific investigations was the final straw that broke the camel's back regarding Dr. Antony Flew's worldview. Soon after, Flew wrote: "what I think DNA material has done is that it has shown, by the almost unbelievable complexity of the arrangements which are needed to produce (life), that intelligence must have been involved in getting these extraordinarily diverse elements to work together ... It's the enormous complexity of the number of elements and the enormous subtlety of the way they work together. The meeting of these two parts at the right time, by chance, is simply minute. It is all a matter of the enormous complexity by which the results were achieved, which looked to me like the work of intelligence."[13]

This bold statement marked a major shift from his life-long philosophical point of view. After this radical worldview transformation, Antony Flew subsequently explained that "science spotlights three dimensions of nature that point to God. The first is the fact that nature obeys laws. The second is the dimension of life, of intelligently organized and purpose driven beings, which arose from matter. The third is the very existence of nature ... the origin of life cannot be explained if you start with matter alone."[14] In the mid-to-late twentieth century, no single individual embodied the argument for atheism more so than Antony Flew. However, after explaining the unexpected conversion to 'theism' in the later years of his life, Dr. Flew has offered these parting remarks: "I have followed the argument where it led me, and it has led me to accept the existence of a self-existent, immutable, immaterial, omnipotent and omniscient being."[15] "We have all the evidence we need in our immediate experience ... only a deliberate refusal to look (at the evidence) is responsible for atheism of any variety."[16] And now, this single man's storied career has become highly representative of thousands of scientists, philosophers and educators who have shaken off their initial bias, and have come to embrace the honest, knowledge-seeking principle compelling them to follow the argument wherever it leads.

A good man, especially a scientist should focus primarily on the facts, and be wise enough to follow wherever they lead, even if they lead in a direction that opposes his preferred worldview. Because of this important principle, we see that worldview transformations like that of Antony Flew are becoming commonplace. The discovery of the laws of physics, the fine-tuning of a vast and remarkable universe, and the transcendent complexity in organic design has led unbiased, knowledge-seeking scientists, educators, and philosophers, to embrace the certainty of an infinitely 'superior mind'. If we will only examine the evidence, the truth will become crystal clear.

The Evidence Stands Tall

I have followed a clear trail of evidence where it would naturally lead through many years of study and research. I have examined this unmistakable path through the many disciplines of science, over the clear and well-preserved footprints of history, and represented by a wide variety of expert testimony gathered from scientific literature over thirty years. What you have read throughout these pages is the focused accumulation of many thousands of hours of study. research, writing, and editing. I am determined to make it my personal mission and lifelong aspiration to know and to understand, to some degree, the observed and revealed truth represented by this accumulation of knowledge and revelation about life's most pressing questions. This diligent search has directed me toward a powerful creator God, who alone is responsible for the vast physical reality we are privileged to enjoy. We observe a material world that is filled with mystery, life, and possibility for those who were created in God's image. This all-powerful creator God has made himself known to his creation in a variety of unmistakable ways. We should open our hearts and our minds to the one who said, "Let there be light! -and there was light." For many years now, there has been a major effort on the part of so-called 'enlightened individuals' to circumvent the Word of God by contrasting scripture with secular scientific knowledge—to the end that secular opinions have dominated the cultural divide between church and state. No single issue has caused greater harm to man's perception of biblical truth than the ongoing creation-versus-evolution controversy. Two hundred years of scientific skepticism has left the message of the God's Word in doubt in the minds of millions worldwide. Before we abandon the Bible, on the basis of misguided public opinion, we must first consider that the Bible has revealed scientific knowledge about the physical world, and the nature of living things, long before mankind documented so many of these fascinating details.

We must consider the false claim of many that the Bible is in direct conflict with science and, therefore, cannot be trusted. Evolutionary thought is a man-made philosophy, contrived by people who wanted to make their own rules. This bankrupt theory has fostered disdain for authority, has indirectly generated widespread immorality, and instills the false hope of people avoiding all accountability to a higher power. How can people afford to risk their future on the mathematically impossible prospect that lifeless, impersonal matter, subject to random chance, can explain a remarkable universe? The God of creation has documented an account of what happened at the dawn of history in the book of *Genesis*, and we would be foolish to ignore it. Why does God care so much about feeble mankind? God's love for humanity (his crowning creation) is so boundless, that he could not bear the thought of spending eternity without us. Isn't it about time to put unproven, unfounded skepticism behind us and to seriously consider what really happened when life began?

The following is a partial list of scientific knowledge that was documented in the biblical text thousands of years before mankind had any firsthand understanding of these facts and details. Also included are the general dates when mankind's discovery confirmed these scientific facts. This is nothing less than scientific proof that the bible was inspired by God.

Known scientific fact	**Biblical Reference**	**Date of man's discovery**
The Law of Entropy	Psalm 102:25-27	After the fall
Northern sky void of stars	Job 26:7	17th century

Earth suspended on nothing	Job 26:7	1650
The earth is circular in shape	Isaiah 40:22	15th century
Animals carry diseases	Leviticus 11	16th century
Quarantine for disease	Leviticus 13	17th century
Animal blood carries disease	Leviticus 17	17th century
Blood is essential for life	Leviticus 17:11	19th century
Oceans have natural paths	Psalm 8:8	1854
Earth in nebular form initially	Genesis 1:2	1911
Most seaworthy design ratio	Genesis 6	1860
Light is a particle with mass	Job 38:16	1932
Radio astronomy signals	Job 38:7	1945
Oceans contain fresh springs	Job 38:16	1920
Snow has material value	Job 38:22	1905, 1966
Innumerable number of stars	Genesis 15:5	1940
Dust is important to survival	Isaiah 40:12	1935
Gas particles have weight	Job 28:25	16th century
Light splits up into colors	Job 38:24	1650
Matter is invisible particles	Romans 1:20	18th century
Plants + sunlight = food	Job 8:16	1920
Stars move through space	Job 38:32	19th century
Description of water cycle	Eccles. 1:6,7; Job 36	17th century
Lightning and thunder related	Job 38:25	19th century
The universe had a beginning	Genesis 1:1	20th century
Universe governed by laws	Job 38:33, Jer. 33:25	16th-19th centuries

Having studied and worked in the healthcare industry for fifty years, my favorite Bible-science revelation is God's command to

Moses that all newborn male children from the tribes of Israel were to be circumcised on the eighth day (Genesis 17:12, Leviticus 12:3, Luke 1:59). Why was it so important to wait that long before performing the act of male circumcision in Israel? As it turns out, the intestines of a newborn are sterile. It requires a number of days of breathing and feeding before an infant is able to lay down a healthy flora of natural bacteria often referred to today as "probiotic organisms." As these highly beneficial organisms accumulate in the gut of a newborn, they begin to produce vitamin K, which of course is essential for blood clotting; also, the level of an important blood-clotting protein prothrombin peaks in the newborn at this time. For these reasons, the eighth day was the safest time to circumcise a child before the advent of modern medicine. This critical knowledge would have prevented newborn male infants from bleeding to death at the time this minor surgical procedure was performed in ancient Israel.

How could Moses have known it was critical to wait until the eighth day to perform the rite of circumcision? That knowledge came from God (the Creator) who alone knew the unique details he had designed in the human body. God revealed many other important details about health, nature, and science to mankind long before they were verified by science. The Bible anticipated, as well, the eventual rise of pseudo-science, such as the evolutionary hypothesis. As we read in 1 Timothy 6:20, the Bible warned us, in advance, of a strong deception would arise in our world. "Guard what was committed to your trust, avoiding the profane and idle babble (irreverent talk) and contradictions (traditions of men that conflict with God's Word) of what is falsely called knowledge (science); by professing it some have strayed concerning the faith."

And continuing in 2 Timothy we read, "In the last days perilous times will come (like unprovoked acts of pure terrorism motivated by hatred). For men will be lovers of themselves, lovers of money, boasters, proud, blasphemers, disobedient to parents, unthankful, unholy, despisers of good, lovers of pleasure rather than lovers of God ... always learning and never able to come to the knowledge of the truth. But evil men and imposters (teachers of false worldviews) will grow worse and worse, deceiving and being deceived. For the time will come when they will not endure sound doctrine (biblical truth), but according to their desires, because they have itching ears, they will heap up for themselves teachers (men who reject God's eternal truth) ... and they will turn their ears away from the truth, and be turned aside to fables". (2 Timothy 3:1-4,7,13; chap 4:3,4)

What is information but the passage of knowledge from one intelligent being to another? Information, conveyed by letters, words, and language, can only be generated by an intelligent mind. It has nothing to do with the accidental interaction of mass and energy in a mindless universe. Information cannot exist without intelligence! Suppose you were walking along the seashore, and you came upon some large letters scratched in the sand that spelled out "David loves Linda." Can you imagine anyone coming to the conclusion that these words appeared magically on the seashore because of some mindless, unguided, random process to produce an intelligible message by pure chance? It is absurd to imagine that any educated person could draw such a conclusion. Human DNA includes around seven billion bits of information contained within about 3.5 billion chemical letters of detailed genetic code. Life is built on information and information comes from intelligence. Therefore, the information contained in DNA had to come from an intelligent mind. This implies a creator! Now, which came first, DNA or protein? Since each requires the other, the most logical answer is they both originated at the same time. This clearly infers supreme intelligence and a creation worldview.

I find it strange that many of our scientists are searching for meaningful signs of electronic communication far beyond our planet that would indicate the presence of intelligent life well beyond our world. Yet these same scientists would most likely refuse to acknowledge that the extremely complex language and information (biological communication) appearing in organic DNA points toward design by a "living intelligence." According to contemporary, world-renowned physicist, professor, and writer Dr. Paul Davies, "Making a protein simply by injecting energy is rather like exploding a stick of dynamite under a pile of bricks and expecting it to form a house ...You may liberate enough energy to raise the bricks, but without coupling energy to the bricks in a controlled and ordered way, there is little hope of producing anything other than a chaotic mess."[17]

The Demise of Darwinism

Today, after the accumulation of several hundred years of scientific investigation and much historical research, it has become more obvious than ever that evolution theory is grossly inadequate to explain how organic life could have emerged from nonliving matter to form the complexity of living organisms we see today. How should we respond to the scientific and historical evidence that has cast a dim view on Darwinism and has made so many revered scientists of the twentieth century look foolish for their outlandish claims? It is presumptuous of man, with such limited knowledge and intelligence, to think that he could possibly explain events that occurred in the ancient past when Earth's history began. Evolution theory is simply a widely accepted hypothesis that attempts to explain the origin and diversity of life, not because it can be proven through a logical examination of the evidence, but because the only alternative, 'special creation', required the work of supernatural intelligence. The unproven evolution model is much more difficult to believe than creation, and requires a great deal more faith because any presumed evidence continues to vanish in light of today's advancing scientific knowledge.

Consider a more complete quotation by I. L. Cohen, a member of the New York Academy of Sciences. "Any suppression, which undermines and destroys that very foundation on which scientific methodology and research was erected, evolutionist or otherwise, cannot and must not be allowed to flourish ... it is a confrontation between scientific objectivity and ingrained prejudice–between logic and emotion–between fact and fiction ... in the final analysis, objective scientific logic has to prevail–no matter what the final result is–no matter how many time-honored idols have to be discarded in the process... after all, it is not the duty of science to defend the theory of evolution and stick by it to the bitter end--no matter what the illogical and unsupported conclusions it offers ... If in the process of impartial scientific logic, they find that creation by outside intelligence is the solution to our quandary, then let's cut the umbilical cord that tied us down to Darwin for such a long time. It is choking us and holding us back ... Every single concept advanced by the theory of evolution (and amended thereafter) is imaginary, as it is not supported by the scientifically established probability concepts. Darwin was wrong ... the theory of evolution may be the worst mistake made in science."[18]

Is there, perhaps, a self-gratifying motivation behind a theory that leaves man in control. Listen to the confessions of two prominent atheists, each of whom has had a noteworthy influence on this debate in years past. One of them was the grandson of Thomas Huxley, often referred to as Darwin's right-hand man ("Darwin's bulldog"). A writer and philosopher, Aldous Huxley stated: "I had motives for not wanting the world to have meaning: consequently, assumed that it had none ... and was able, without any difficulty, to find satisfying reasons for this assumption ... For myself, as no doubt, for most of my contemporaries, the philosophy of meaninglessness was essentially an instrument of liberation. The liberation we desired was simultaneous liberation from a certain political and economic system, and liberation from a certain system of morality. We objected to the morality because it interfered with our sexual freedom."[19]

Thomas Nagel is a former professor of philosophy and law at New York University. While honestly considering his view of life as an atheist, he recorded these remarks: "I speak from experience, being strongly subject to this fear myself. I want atheism to be true and am 'made uneasy by the fact that some of the most intelligent and well-informed people I know are religious believers. It isn't just that I don't believe in God and, naturally, hope that I'm right in my belief. It's that I hope there is no God! I don't want there to be a God; I don't want the universe to be like that ... My guess is that this cosmic authority problem is not a rare condition, and that it is responsible for much of this scientism and reductionism of our time. One of the tendencies it supports is the 'ludicrous overuse' of evolutionary biology to explain everything about human life, including everything about the human mind ... this is a somewhat ridiculous situation ... it is just as irrational to be influenced in one's beliefs by the hope that God does not exist as by the hope that God does exist."[20]

Debunking evolution is not difficult if we are willing to focus our attention on a convincing trail of evidence currently available. It's only those who flatly refuse to consider the evidence that are missing the truth about such a vital question as the origin of life. The teaching of evolution theory leads to a sense of purposelessness, which can have disastrous consequences. In a speech delivered at America's prestigious Harvard University some years ago, it was the world-famous Russian dissident Alexander Solzhenitsyn who was harshly booed by many of those in attendance when he uttered the following words, which aptly describe the general direction and trend of nearly every culture in the Western world today. "The greatest burdens will come upon any society when they forget God!"[21]

Chinese-born scientist, Dr. Kenneth Hsu, speaks out boldly concerning his view on Darwinism: "We have all heard of The Origin of Species, although few of us have had time to read it ... a casual perusal of the classic made me understand the rage of Paul Feyerabend... I agree with him that Darwinism contains wicked lies; it is not a 'natural law' formulated on the basis of factual evidence, but a 'dogma, reflecting the dominating social philosophy of the last century."[22] According to the late Dr. Henry M. Morris, the unholy mantra adamantly maintained by modern evolutionists remains the same as it always has been in the past. After absorbing many heavy blows and powerful contradictions that true objective science has dealt 'the theory', through computer-age scientific advances, the basic refrain of naturalism simply will not go away. This litany proclaims: "We know evolution is true, even though we don't know how it works, and have never seen it happen."[23] To the ardent evolutionist, because intelligent design is not an option, and because the fossil record has left traditional evolution without its much-needed support, the only conclusion they can possibly accept is that the mechanism driving evolution must be revised to fit the currently revealed facts. This has been the typical strategy of evolutionary humanists all along— start with the conclusions that you know to be true, then methodically fill in the blanks by whatever means necessary to support your conclusions. Steadfastly follow this approach, even if it means you should ignore, deny, or perhaps minimize a growing body of evidence that increasingly lends its support to intelligent causation and a creation worldview.

The Power of Worldview Thinking

I want to strongly emphasize the critical importance of embracing an accurate worldview. Our personal view of life will hold a very powerful influence over the way we live each day. It will dominate our beliefs, actions, and attitudes in every area of human experience. It will control our todays and determine the outcome of our tomorrows, but most of all it will serve to shape our eternal destiny. The origin of conscious organic life is vital to our overall understanding. How we view this crucial question will have profound implications on how we are likely to view the most important issues of life. The dangerous doctrine of evolution is based upon the self-serving and prideful wisdom of man, and its foundation rests on the shifting sands of time. This man-made ideology is beginning to collapse before our eyes, and the vacuum left must be filled with truth, giving lasting purpose and meaning to our lives well beyond the few short years of our mortality.

Hundreds of philosophies and scores of religions have been created by the wisdom and will of man in his unrelenting effort to circumvent the Word of God. The greatest danger of evolutionary doctrine is that it ultimately leads to a universal breakdown of personal, as well as governmental morality, resulting in a rejection of the authority of the Bible, and a failure to appropriate its many promises leading to salvation and eternal life. If human lives are left grounded in evolutionary humanism, life soon becomes subjective, in which case everyone can choose their own set of values, and can impose their will upon whomever they wish. Adolph Hitler believed he was advancing human evolution by ridding the world of an inferior race along with abnormal people, whom he believed to be unfit contributors to a superior race of men. As it turns out, he was a deranged psychopath whose sense of moral superiority became a crime against humanity rivaled by only a few other men in world history. Belief in the nonexistence of God can be seen as an escape mechanism to avoid moral responsibility for one's behavior in a self-centered world where many people simply do not like being told how they should live. Because of this, many people who desire to explain the origin of life without any reference to a supernatural authority have eagerly embraced the theory of evolution. In becoming self-proclaimed gods, who believe they can set the boundaries of morality, humanists have sought to repel any sense of accountability to an intelligent power greater than themselves.

Evolution is the most devastating social philosophy the modern world has ever known, and it continues to be a major factor in the moral decline of contemporary cultures across the globe, including the United States of America. Evolution provides the pseudoscientific rationale for all kinds of evil influences throughout the world, including everything from abortion to communism to ethnic cleansing. Most people are oblivious to the monumental absurdities presented by the bizarre tenets of Darwinism. They find themselves living in the moment and completely unaware that the ramifications of evolution theory have control of their lives. They may regard evolution as merely an abstract biological theory, with little or no consequence to themselves. However, this religious ideology is more like a giant, insidious octopus with tentacles that have reached into every area of modern culture, every field of study, every level of education, and every practicing discipline, with very destructive consequences. The twentieth century was by far the deadliest century in human history in terms of the bloody carnage mankind inflicted upon his fellow man (with an estimated 150 million lives lost through human conflict alone); not to mention the unspeakable genocide perpetrated by deranged world leaders who held to an evolutionary worldview, and used it to justify their actions. I do not think it coincidental that the twentieth century also saw the greatest degree of evolutionary indoctrination the world has ever seen. Evolution is not merely a crude, ambiguous biological theory but a dangerous and very destructive philosophical worldview that is in direct opposition to the Judeo-Christian ethics and values that gave rise to Western civilization. Every single life that it touches is, at least to some degree, pulled away from a sovereign and benevolent God who lovingly fashioned mankind for a noble purpose and a magnificent future.

Scientists Are Speaking Out

If your personal quest for truth and meaning excludes any mention of God, perhaps it is time to consider where the evidence is leading so many of today's top-notch scientists, who have had enough with presumption and speculation that does not measure up to the facts. More than ever before, scientists are speaking out their protest toward the unfounded assumptions of evolution theory, which most of Western civilization has been dragging around like a ball and chain for the last century or more. Consider what is suggested by these courageous, outspoken experts. French zoologist, Pierre-Paul Grasse, author of over three hundred scientific publications, has drawn the following conclusion: "Through use and abuse of hidden postulates, of bold, often ill-founded extrapolations, a pseudo-science has been created. It is taking root in the very heart of biology, and is leading astray many biochemists and biologists who sincerely believe that the accuracy of fundamental concepts has been demonstrated, which is not the case."[24]

Loren Eiseley, an American anthropologist, educator, and philosopher, also became a natural science writer, and subsequently received no less than thirty-six honorary degrees during his outstanding career. His written works include an evaluation of the failed effort on the part of twentieth-century science to fully establish credibility for the evolution model. Regarding a clear failure to painstakingly legitimize evolution theory, he made this assessment: "With the failure of these many efforts, science was left in a somewhat embarrassing position of having to postulate theories of living organisms which it could not demonstrate. After having chided the theologian for his reliance on myth and miracle, science found itself in the inevitable position of having to create a 'mythology' of its own: namely, the assumption that what, after long effort it could not prove to take place today had, in truth, taken place in the primeval past."[25]

Luther Sunderland served as an American aerospace engineer working with General Electric for many years. He became an avid writer who thoroughly researched the question of origins and frequently published his findings. Consider these remarks from his book entitled *Darwin's Enigma*: "Now, after 120 years of the most extensive and painstaking geological exploration of every continent and ocean bottom, the picture is infinitely more vivid and complete than it was in 1859. Formations have been discovered containing hundreds of billions of fossils and our museums are filled with over 100 million fossils of 250,000 different species ... The availability of this 'profusion of hard scientific data' should permit objective investigators to determine if Darwin was on the right track. What is the picture which the fossils have given us? ... The gaps between major groups of organisms have been growing even wider and more undeniable. They can no longer be ignored or rationalized away with appeals to imperfection in the fossil record."[26]

Louis Agassiz, a Swiss-born, European-educated biologist, and geologist was recognized as an innovative scholar in the study of earth's natural history. He eventually came to America and became a professor at Harvard University, where he taught zoology and geology. After a career in the natural sciences, he made a bold declaration that has driven a stake into the very heart of the evolutionary model: "The theory [of evolution] is a scientific mistake."[27] He backed that statement with another observation. "The origin of all diversity among living beings remains a mystery as totally unexplained as if the 'book of Mr. Darwin' had never been written, for no theory unsupported by facts, however plausible it may appear, can be admitted in science."[28]

Nils Herbert-Nilsson was honored as a Swedish botanist and research geneticist for his work in extensive breeding experiments as a professor of botany at Lund University. He was elected a member of the Royal Swedish Academy of Sciences in 1943. After a lifetime of research in attempting to demonstrate the process of evolutionary change in organic species, he published this conclusion. "My attempts to demonstrate evolution by an experiment carried on for more than 40 years has completely failed. At least I should hardly be accused of having started from any preconceived anti-evolutionary standpoint."[29]

Chinese scientist, Dr. Kenneth Hsu, has studied geology, paleoclimatology, and oceanography, among other disciplines.

After his immigration to the United States in 1948, he completed his formal studies at Ohio State University, and then received a PhD from UCLA in 1953. Dr. Hsu taught applied sciences throughout the United States for many years, while working for private industry (The Shell Oil Company) as a petroleum geologist and an environmental engineer. He became a member of the U.S. National Academy of Science in 1986. Dr. Kenneth Hsu has always been a straight shooter, and has never been afraid to report what he knew to be true. Here is an example of the bold comments he has made, which are driving home the idea that Darwinism was a colossal scientific mistake: "George Bernard Shaw wisecracked once that Darwin had the luck to please everyone who had an ax to grind. Well, I also have an ax to grind, but I am not pleased. We have suffered through two wars and are threatened by an Armageddon. We have had enough of Darwinian fallacy."[30]

James Perloff was an avid writer who, among other things, had a special interest in investigative science for many years. The subject matter found in his publications include topics such as Darwinism, U.S. politics, and world history. In his popular 1999 book, *Tornado in a Junkyard*, he made a bold assertion about evolution theory: "The evolutionary establishment fears creation science, because evolution itself crumbles when challenged by evidence. In the 1970s and 1980s hundreds of public debates were arranged between evolutionary scientists and creation scientists. The latter scored resounding victories, with the result that, today, few evolutionists will debate. Isaac Asimov, Stephen Jay Gould, and the late Carl Sagan, while highly critical of creationism, all declined to debate."[31]

Dr. David Raup, the Harvard-educated, University of Chicago paleontologist studied the fossil record extensively for many years during a long and fruitful academic career. He observed first-hand how weak the argument for evolution had become after the onslaught of recent evidence that has strengthened the creationist position. He commented: "I doubt if there is a single individual within the scientific community, who could cope with the full range of [creationist's] arguments without the help of an army of consultants in special fields."[32]

It may be appropriate here to clarify, somewhat, how the role for God has been presented in the long-standing debate on origins. The support for God in 'the great debate' has generally appeared in two different but related formats: 'Creationism' and 'Intelligent Design'. The term Creationism constitutes a belief that the universe and life in it have originated from specific acts of divine creation by a transcendent God as described in the biblical genesis account. The conclusion is that evolution theory is terribly inadequate and cannot possibly account for the history, diversity, and complexity of life on Earth. This belief is scientifically supported and promoted by several well-funded organizations based in the United States, including ICR, Answers in Genesis, and the Creation Research Society, among others. Intelligent Design, as closely related origin model, constitutes the broad argument for the existence of God and is promoted as an evidence-based scientific theory about life's origin. The claim is that certain features of the universe and of living things are best explained by intelligent causation rather than by a mindless, undirected process. The Intelligent Design movement was championed by the Discovery Institute, a politically conservative think tank, based in Seattle, Washington.

Dr. Phillip E. Johnson is a retired UC Berkeley law professor, whose work in the ID movement has earned him the unofficial title, 'Father of Intelligent Design'. He became a co-founder of the Discovery Institute's Center for Science and Culture, an organization which seeks to promote the Intelligent Design hypothesis in scientific circles internationally. As a true champion of the intelligent cause argument, he sought to convince the scientific community to allow a role for God in scientific theory. In these remarks, he explains how the false premise of evolution has sustained itself for so long in the light of new evidence that flatly denies Darwinian claims: "It [evolution] is sustained largely by a propaganda campaign that relies on all the usual tricks of rhetorical persuasion: ... hidden assumptions, question-begging statements of what is at issue, terms that are vaguely defined and change their meaning in mid-argument, attacks of straw men, selective citation of evidence, and so on. The theory is also protected by its cultural importance. It is the officially sanctioned 'creation story' of modern society, and publicly funded educational authorities spare no effort to persuade people to believe it."[33]

Dr. Randy L. Wysong is an American physician who served as an instructor of human anatomy and physiology for many years. He may be best known for a book he had published in 1976, entitled *The Creation-Evolution Controversy.* In the book, he had this to say about modern evolutionary ideology: "Evolution can be thought of as sort of a magical religion. Magic is simply an effect without a cause, or at least a competent cause, 'Chance,' time, and 'nature,' are the small gods enshrined at evolutionary temples. Yet these gods cannot explain the origin of life. These gods are impotent. Thus, evolution is left without competent cause and is, therefore, only a magical explanation for the existence of life."[34] Dr. Harold Urey was known for his work as an American physical chemist. He pioneered a major work on isotopes leading to the discovery of deuterium, which earned him the Nobel Prize for chemistry in 1934. Dr. Urey went on to play a significant role in the eventual development of the atomic bomb, but he may be remembered most for his contribution to various theories relating to the development of organic life from nonliving matter. Although a man of his stature should be confident in a theory of origins, the following statement from Dr. Urey reveals considerable doubt. "All of us who study the origin of life find that the more we looked into it, the more we feel that it is too complex to have evolved anywhere ... We believe as an article of faith that life evolved from dead matter on this planet. It is just that its complexity is so great, it is hard for us to imagine that it did."[35]

Michael Denton is a British-Australian author and biochemist, receiving his PhD in biochemistry from King's College in London. His examination of the evidence through a lifetime of work in biochemistry led him to write a book in 1985 entitled *Evolution: A Theory in Crisis*. This timely book has proven to be a valuable resource in the creationist movement. It offers a systematic critique of neo-Darwinian theory that covers a whole gamut of topics from paleontology to molecular biology. In this publication Dr. Denton concludes his argument by explaining how a false and misleading hypothesis could have had such a fashionable influence on Western society. In a complete citation of a previous quote he said: "The influence of evolutionary theory on fields far removed from biology is one of the most spectacular examples in history of how a highly speculative idea for which there is no really hard scientific evidence can come to fashion the thinking of a whole society and dominate the outlook of an age ... Ultimately, the Darwinian theory of evolution is no more nor less than the great cosmogenic myth of the 20th century."[36]

Raising the Bar on Living

In this book, I have documented the reasons for and the evidence favoring the embracing of a biblical worldview. These many reasons (derived primarily from secular sources) have taken the reader well beyond what he or she 'should' believe by providing compelling evidence for 'why' they should believe it. Nothing in life is worth believing unless one can demonstrate a rational and reasonable argument to support it. 'Naturalism' or 'Theism'? There is so much at stake in getting this answer correct. We certainly cannot afford to wait until our life ends to find out which worldview was true. In these few chapters, I have laid out as best I can the background and *the* true nature of this ongoing debate. Now it is up to the reader to decide what to believe about *the mystery of life*, and to choose whether God deserves a place in that decision.

With this ongoing creation-evolution argument, we find ourselves in a life-and-death struggle for the heart and soul of America and the civilized world. The outcome of this struggle will, no doubt, bear eternal consequences for those who have engaged in the debate, and for those who remain oblivious to its exceptionally powerful influence over their lives. Without knowing it, millions have become indoctrinated. Consequently, many have come to a fateful decision regarding which hypothesis will capture their allegiance, shape their worldview, and control their destiny. It is the goal of this book to refute the theory of evolution using sound reason, logic, objective science, and historical truth, all of which being confirmed by the expert testimony of many well-qualified authorities with a wide variety of philosophical and scientific backgrounds. Only the truth will provide the lasting sense of contentment, satisfaction, and peace that we all desire. It is time to raise the bar, and set a higher standard for ourselves and our families when it comes to shaping a personal worldview that will manifest itself in every area of our lives. True faith is not a leap in the dark. It is a commitment based on evidence! We would be foolish to believe in something for which there exists no valid supporting verification. This would be like believing in a living Santa Claus or the tooth fairy. The powerful, faith-filled conviction of several billion Christian believers worldwide cannot be taken lightly; this life-changing experience, based on abundant evidence, is commanding proof for the existence and influence of a Creator God.

You will never experience a greater sense of deep satisfaction than when you can say, 'All is well with my soul'! The mystery of life will never be decided, explained, or understood through our best efforts using human intellect alone, no matter how many more colleges and universities are founded in pursuit of this ideal. We were created in the image of a powerful, personal God who wants to know each of us in a personal way. When it comes to deciding which view of life is right for you, 'follow the evidence' and then ask yourself: What is my heart telling me? The only way I know to overcome the worry in living and the fear of dying is to make peace with God.

CHAPTER TEN

* * *

How Then Shall We Live?

Creator, Redeemer, and Judge

Most people think of Jesus (God's eternal Son) as the Savior of the world; but did you know that, according to the bible, He is the transcendent creator of the universe, and the righteous judge to whom all of humankind will ultimately answer? Our devotion to and worship of him is greatly magnified when we fully understand what the bible teaches about the incarnate God entering our world as a lowly carpenter's son some two thousand years ago. The Bible has declared, "For by Him (Jesus Christ) all things were created, things in heaven and on earth, visible and invisible, whether thrones or powers or rulers or authorities – all things were created by Him and for Him. He is before all things, and in Him all things hold together." (Colossians 1:16,17 NIV) "In the beginning was the Word, and the Word was with God, and the Word was God ... He (Jesus Christ) was in the beginning with God. All things were made through Him, and without Him nothing was made ... He was in the world, and the world was made through Him, and the world did not know Him, ... and the Word (Jesus) became flesh and dwelt among us." (John 1:1-3, 10, 14) "God, who at various times and in various ways spoke in time past to the fathers by the prophets, has in these last days spoken to us by His Son, whom he has appointed heir of all things, through whom also he made the worlds." (Hebrews 1:1,2)

That the Creator of the universe would love us enough to die for our sins, which provides the potential for you and I to be restored to a right-standing relationship with God, is the best news we could possibly ever hope for. How could any person reject such a powerful demonstration of unconditional love? And yet this is exactly what millions of people are doing, while they continue to live their lives as strangers to the very God who created them. It was for the sake of love and relationship that Jesus became a man and identified himself with sinful humanity— experiencing our pain, sorrow and suffering. No one can accuse the living Christ of not understanding the trials they face in this mortal life. He faced more trial, sorrow, and suffering than any of us could ever imagine, when he willingly took our sin upon himself (dying in our place) so that we could know redemption, and the hope of spending eternity in his presence. *How then shall we live*? Many will live for the next big Hollywood blockbuster, the latest and greatest video game, or perhaps the next victory for their favorite team. Do these frivolous pursuits really matter in light of eternity? We would do well to focus some serious attention on the things that matter most!

Nearly everyone understands the significance of the world-changing event we observe every year on December 25th. On this day, we celebrate the miraculous incarnation on the earth of the creator God of the universe, who reduced himself to the physical dimension in order to prove, once and for all, his undeniable love for humanity. God has done everything he possibly can to prove how much he cares, even to the point that he provided us with a sure way out of our miserable circumstances (the trials of living in a cursed world). God has made the gospel truth so simple that even a small child can understand it. It goes something like this: God has created us; God loves us. He gave us free will, which we used to separate ourselves from Him. And finally, God wants to save us from the penalty, power, and presence of sin that is so characteristic of our fallen world. However, He cannot do that if we continue to deny that he exists, and go about the business of inventing our own truth, while completely disregarding His written Word, and His sovereign authority.

Jesus himself offers a stern warning to the ungodly philosophers of our time who assuredly caused the greatest measure of offense– an offense that has been cunningly perpetrated against the innocent young lives of children in America and around the world. The word 'offend' means to lay a snare for, to set a trap for, to cause to stumble and fall. This is what evolutionary humanists have done in our world as the result of a massive campaign to control public opinion and to shape public policy. Children believe what they are told not only by parents but also by other important figures of authority in their lives.

Jesus saw this deception coming. Evolutionary teaching, and secular humanism, has now led millions toward disbelief in God and a blatant disregard for His eternal Word. This dark cloud of deception has been very instrumental in leading us into the most immoral and corrupt time of our nation's history! When considering the many children growing up in each generation, this is what Jesus had to say, "Whoever receives one little child like this in my name receives Me. But whoever causes one of these little ones who believe in Me to sin, it would be better for him if a millstone were hung around his neck, and he were drowned in the depth of the sea. Woe to the world because of offenses! For offenses must come, but woe to that man (men and women) by whom the offense comes!" (Matthew 18:5-7) The Bible teaches that Christ has been given the authority to execute divine judgment upon those for whom he paid the ultimate price. "I charge you therefore before God and the Lord Jesus Christ, 'who will judge the living and the dead' at His appearing and His kingdom" (2 Timothy 4:1).

Dr. Henry Morris was a long-time student of physical science and earth history. As a devoted scholar of biblical truth, as well, he worked tirelessly to connect science and the Bible as the natural outcome of rational understanding. Regarding the spiritual matters, he once wrote: "Of all the world's religions and philosophies, only orthodox biblical Christianity teaches both the supernatural creation of all things by the transcendent, yet personal Creator God, and also the substitutionary death and bodily resurrection of the Creator, the Lord Jesus Christ."[1]

A. E. Wilder-Smith was a British-born organic chemist who earned three doctoral degrees (organic chemistry, pharmacology, and physical chemistry). As a devoted Fellow of the Royal Society of Chemistry and a former Oxford atheist, he became challenged by several of his peers, along with the influence of C.S. Lewis, to examine the weight of physical evidence with an open mind. Wilder-Smith had concluded like Antony Flew that if a loving God really existed, he would surely not have tolerated all the terrible injustice, suffering and heartache that is evident in the world we know. Never-the-less Dr. Wilder-Smith sought to discover the truth as revealed to him through his dedicated love for scientific endeavor. His search for understanding brought him to the realization that a powerful, personal God was the only way to explain a world as remarkable and complex as ours. Upon his discovery of this knowledge, he was transformed into a devoted creationist and wrote several important books expounding on how the depth of scientific knowledge points unmistakably to a creator. He went on to become a leading advocate and strong proponent of creation science worldwide. Although he never used the term himself, his influence had an instrumental effect by inspiring others who would eventually initiate the 'Intelligent Design' movement that has become so popular today. The life-transforming power originated by his devoted love for scientific understanding is revealed in the following personal declaration of Dr. A. E. Wilder Smith. "I believe myself in the living God who did it. I believe that this God, who supplied the information revealed himself in the form of a man so that man could understand him. We are made to understand. I want to understand God. But I can only do this if he comes down to my wavelength, the wavelength of man. I believe that God revealed himself in the form of Christ, and that we can serve him and know him in our hearts as the source of the 'logos'– all the information that is necessary to make the universe and make life itself. Look at the beauty of nature around us! When you consider that it all grew out of matter injected with information of a type I have been describing, you can only be filled with wonder at the wisdom of the Creator, who, first of all had the sense of beauty to do it, and then the technical ability. I am filled with wonder as I look at nature, to see how God technically did it ... The Scripture teaches perfectly plainly, and it fits in with my science perfectly well, that the one who did, called Himself the Logos. That Logos was Jesus. Jesus called Himself the Creator who made everything— 'for Him and by Him'. Now, if that is the case, then I am very happy and filled with joy that He made the Creation so beautiful, and that He also valued me enough to die for me, to become my Redeemer as well."[2]

Why Did He Visit Earth?

People really want to know and understand the truth. It is just that so many of them do not know where to find it. Jesus said, "I am the way, *the truth*, and the life. No one comes to the Father except through me." (John 14:6) If what He said is true, then none of us will ever know real contentment if we ignore his words and deny his power to save us. Jesus came into our world to cancel the debt of sin, and to balance for all time the scales of divine justice. The King of glory–the sovereign Lord of all creation–chose to take my punishment and yours upon himself. How can anyone ever comprehend such great love as this?

As we continue to deliberate on the evidence, there is one

other important question that we should ask ourselves. Why is Jesus so popular throughout the world today, and what does this acknowledgment imply? He never wrote a book. He never went to college. He never led an army or held public office. He barely traveled as far as two hundred miles from the place of his birth, yet his powerful words, humble life, and undeniable influence have impacted the entire world for over two thousand years. More books have been written about him than any other person who ever lived. More songs have been sung to honor his great name. And this one solitary life has inspired more movies, and works of art, than any other person in the history. Two millennia of time have passed from the days Jesus walked on the earth as a wandering itinerant, and yet his popularity and monumental influence continues to flourish worldwide. All the world's greatest leaders, military commanders, scientists, and educators combined in the history of the earth have not commanded the sheer magnetism of this single solitary life. There is only one possible way to explain this unique and commanding level of undeniable worldwide influence and popularity. This phenomenon can only be understood by the fact that Jesus truly was God in human flesh. What he said and did in three short years of public ministry has literally impacted every life on planet Earth in some way for two millennia. And he did it all as a lowly, isolated figure, who grew up in a rather primitive culture, coming from an obscure village, on the barren hillside of a tiny country. This humble servant left a trail of colossal consequence, and achieved his worldwide renown and prominence without the advantage of any superior technology, while basically traveling everywhere he went on foot.

The Christian faith is based on undeniable historical evidence that can and has been verified as accurate and trustworthy. If a person were to believe in something for which there was no truly reliable evidence to justify a rational argument, this would make him or her a fool! Christianity is not foolish speculation. If a person wanted to discredit Christianity, there is only one thing they would have to do. They would have to disprove one single event in history. They would have to debunk the resurrection of Jesus Christ. By doing so. the entire Christian faith would crumble and fall. Many have attempted to do just that, but all have failed. The resurrection is the bedrock upon which the Christian faith rests worldwide, and it has been verified by many eyewitness accounts, and historical facts, making this event the most easily verified miracle of antiquity.

After the resurrection, Jesus appeared multiple times to public witnesses over a period of forty days. No less than six key appearances are recorded in Scripture, and have been corroborated by eyewitnesses, including an appearance to some five hundred people all at one time in the region of Galilee. Most atheists are afraid to challenge the resurrection, because the evidence supporting it is rock solid. Consequently, since atheists cannot discredit the most captivating and miraculous event ever recorded, skeptics are inclined to attack the only Christian doctrine that historical facts alone cannot ultimately prove—the existence of God, and the accuracy of early events recorded in the venerated book of Genesis. The doctrine of creation has come under greater assault and scrutiny than any other Christian pillar of faith. By waging this war, atheists have appealed to the intellectual pride and emotional inclination of those who want to deny the existence of God. It sits easier on the conscience to believe that no accountability awaits those who live self-centered lives and have continually denied moral absolutes.

This protracted warfare against God, fueled by decades of evolutionary propaganda, has generated a false sense of security for many people who wish to justify their sin. This false confidence will only last for a short time. Suddenly, at the moment of death, the reality of God's existence will become apparent to all, as many deceived and unbelieving souls face an eternity of separation from their Creator, whom they have personally rejected and have denied during a lifetime of self-directed living. The Bible defines death (the second death) as separation of the human soul from the Creator forever (see Revelation 21:8).

The humanist worldview claims that man is basically good. God's view is that mankind, although desperately loved, is sinful and separated from him through an inborn rebellious nature. The sin of our forefather [Adam] led to a fallen world that has affected the entire human race. God's Word explains it like this: "Therefore, just as through one man [Adam] sin entered the world and death through sin, and thus death spread to all men, because all (have) sinned." (Romans 5:12) "Through one man's offense [original sin] judgment came to all men, resulting in condemnation, even so, through one man's righteous act [the sacrificial death of Christ], the free gift [salvation by faith alone] came to all men, resulting in justification [as if I had never sinned] ... For by one man's disobedience [Adam's fall) many were made sinners, so also by one man's obedience (Christ's sacrifice] many will be made righteous" (Romans 5: 18-19). The Bible states that "for all have sinned and fall short of the glory of God" (Romans 3:23). It further explains that "the wages of sin is death, but the gift of God is eternal life through Jesus Christ our Lord" (Romans 6:23).

Redemption is the central theme of the Bible. This divine act is the culmination of God's great love for humanity and it demonstrates how far he was willing to go to prove that love. Everything the Bible teaches points us toward our desperate need for reconciliation with God. When man fell into sin, his relationship with God was broken, and he now carries a debt (a sin debt) that separates him from his Maker. Living with guilt, and being unable to atone for it himself, man became completely dependent upon God's grace and mercy. Only God could pay the costly debt that mankind was burdened with. To reconcile that broken relationship, God determined to make the payment himself, a payment that could potentially free all men of their guilt and shame thus restoring that original relationship. Because man was created in the image of God, and carries free will, it was now up to man to accept, by choice, God's offer of salvation. God has offered us a gift of eternal value that cost mankind nothing. (Romans 6:23) By definition, a gift cannot be earned. The sacrifice is completely with the giver. The recipient can only choose to accept or reject the gift by an act of his or her personal will. Only one person can save you and me from the wrath of God. Jesus, who was without sin, was the perfect sacrifice and God's only means of atonement for man's sinful transgressions (Hebrews 4:15, 2 Corinthians 5:21). Each one of us must decide to accept God's gracious offer of salvation through Jesus Christ. Once we understand, making no decision is a clear choice to reject God's plan of redemption.

To understand the gift of God, it is necessary to know what is meant by the term "redemption." Because this concept is central to the biblical narrative, we must thoroughly define the term so this essential doctrine will be fresh in our minds. The word 'redeem' means to buy back, to pay off, to set free upon full payment of ransom, to atone for guilt, to reclaim after recompense of a debt, 'to fully recover by payment and restore a person or thing to its original state or condition'. If, for example, my wife and I had a desperate need for immediate cash and we decided to take her valuable wedding ring to a pawn shop in order to obtain the necessary funds, the store's shopkeeper would gladly loan a certain amount of money against the ring based on the value of the diamond. I would then have so many months to pay back the money, plus interest, in order to redeem my property and return it to my wife in which case, it would be fully restored to its original state or condition. If, however, I failed to pay back the required amount by the due date according to the contractual agreement, in order to fully redeem and repossess my diamond, the ring would then become the property of the shopkeeper and I would have a very unhappy wife. Now, because of sin, you and I became subject to Satan (see 1 John 3:8). However, once we have accepted God's offer, and have thus become fully redeemed through faith in the blood of Jesus Christ that was shed on our behalf, the devil loses his power over us because we have been bought back and restored to our original state or condition (a right standing relationship with God). This is the very same condition that existed between God and Adam before the fall.

If you and I were never redeemed through a simple act of repentance and faith, by accepting the payment offered for our sin, then we would remain under Satan's dominion, and subject to an eternity with him instead of a state of bliss that could be enjoyed in the very presence of God. *Redemption* of the human soul restores the original state or condition that man enjoyed at the time of creation, when God declared his works to be very good. We are the children of Adam. We have inherited a sin nature from him– 'but God,' apart from any useless effort by mankind, has provided a way to reverse the curse brought on by sin and to restore man and woman to their original state or condition of uninterrupted fellowship with their Maker. The perfect world of Genesis was forfeited because of man's transgression. Mankind became alienated from God, and separately in terms of relationship. The main purpose of the Bible is to explain how God has fulfilled his promise to redeem mankind from the tragic curse of death and separation, restoring him to the original state or condition that existed before the fall. This does not mean that those who come to saving faith in Christ will suddenly experience a perfect world. Salvation is three-fold. The believer is immediately saved from the penalty of sin, then the Holy Spirit, working in the life of the believer through sanctification, is continually saving us from the power of sin as we submit our lives to God's will; and ultimately God promises to remove us from the presence of sin. Think of it as an IOU from God. Our salvation is not complete until we are in his presence in a perfect place as God had originally intended for mankind (his crowning creation). This was all made possible when Jesus bore our sins on the cross. God has already completed everything he is going to do about the problem of sin. Now, it is up to us. Will we accept his plan and the unbelievable price Christ paid to restore that broken relationship, or will we foolishly reject God's plan, and attempt to concoct a salvation plan of our own? Some people say they cannot accept a God who sends people to hell. I understand this objection, but they are terribly misinformed.

God has done everything he possibly can to prevent this tragedy from happening. Of their own free will, people have rejected Christ and choose to remain separated from God by neglecting his love, grace, and forgiveness; thus, they have openly condemned their own eternal souls to a godless eternity. God does not send people to hell— they send themselves by rejecting the only remedy that has ever been offered to prevent it. Hell was not prepared for man; it was prepared for the devil and his angels (see Matthew 25:41). Men and women go there only because they reject God's love and refuse to submit to his authority. The prideful arrogance of many people will say, "I don't need God, I don't need his plan of salvation. I will save myself. I will deserve an eternal reward because of my own righteous acts and exemplary behavior." Many people seem to believe that if the good in their lives outweighs the bad, they will be fine in the end, when God examines their life in judgment. This deception, which remains ever so prevalent in the heart of mankind today, is outright rebellion against a merciful God's plan to save us. There is only one unforgivable sin— the sin of persistent unbelief. "For by grace (undeserved favor] you have been saved through faith (believing in Christ's atonement for sin), and that not of yourselves (not by human effort); it is the gift of God (a gift cannot be earned), not of works (a personal effort at righteous living), lest anyone should boast (making a claim that they are in right standing before God because of their own personal effort at good behavior) (Ephesians 2:8, 9). The Bible says that our goodness is like filthy rags compared to the righteousness of God (see Isaiah 64:6).

The standard that God has set for mankind is 'perfection'. "You shall be perfect, just as your Father in heaven is perfect" (Matthew 5:48). Now, this standard will seem too demanding, so many people will say: 'God is not fair, nobody is perfect' so why try— maybe I will just live out my days the way I want and enjoy my life while I can. This is certainly a standard that none of us could ever hope to achieve on our own. Let us say, for example, the Grand Canyon in northern Arizona represents the wide gulf that exists between sinful man and God. Mankind needs a way to get across that expansive chasm. In 1968, the year I graduated from high school, Mexico City was the site of the most remarkable leap in the history of the Olympic long jump competition. A young American competitor by the name of Bob Beamon found himself down to his last jump in order to qualify for the finals among the world's greatest athletes. On his final preliminary jump, he managed to qualify rather easily after some timely advice from his coach. Young Mr. Bob Beamon, with a boost of confidence, was determined to make his greatest leap in the finals to secure the gold medal. Putting forth his best effort on his first attempt he ran as fast as possible, hit the take-off board, and sailed toward the sand pit. It felt like a good jump, but to his dismay and the shock of a worldwide television audience he had not only made his best jump ever, but had eclipsed the current world record by almost two feet. Disbelieving officials were stunned as they brought out another tape to measure and re-measure a distance that was thought to be impossible. They would call it 'the miracle jump of Mexico City'. Now, just suppose the fastest man alive was attempting to leap across the chasm of the Grand Canyon through human effort alone. Unfortunately, his greatest possible leap would still leave him about ten miles short of that goal. That is how futile it is for people to think that human effort alone (apart from Christ) could somehow earn them a right-standing relationship with God. The only possible way we can be found righteous before a Holy God is by crossing over a bridge from our side to His. *That bridge is Christ*! (see 1 Timothy 2:5). We must keep in mind that this *perfection* the scripture is referring to is in 'the eyes of God', not in the eyes of man. Though this perfection seems to be completely unattainable, 'the good news' is that when we repent of our selfish transgressions, God forgives us completely, imputing to us (credits to our account) the righteousness of Christ (see 2 Corinthians 5:21). When that happens, you and I are judged as perfect in the eyes of God and we will never have to stand before Him to account for our sins. This is the best news anyone could ever hope for and the only way perfection will ever be found in any of us (see Philippians 3:9). In the Old Testament the high priest would offer an animal blood sacrifice, on behalf of the people of Israel, each year, but this routine religious practice would only provide a covering for sin for one year; it could never remove sin permanently. God's final judgment for sin was completed once, permanently, and for all time as the New Testament explains: "But when this priest (Christ) had offered for all time one sacrifice for sins, (his death on the cross) he sat down at the right hand of God– because by one sacrifice he has made **perfect forever** those who are being made holy" (Hebrews 10:12, 14 NIV).

Many people are uncertain about the future. They are fearful, anxious, and often worry about what may lie ahead. Countless people are terrified of the thought of dying and live in distress about what may lie beyond the grave. How can man possibly conquer this tormenting fear and terrible worry that holds so many hearts and minds captive? Unfortunately, many people will never come to understand what life is all about. And consequently, they will never be freed from these disturbing thoughts that will stalk them all the way to their grave. The Bible says, "It is a fearful thing to fall into the hands of the living God" [without knowing His forgiveness] (Hebrew 10:31). The doctrine of divine creation is the very foundation of Christian faith and theology (see Hebrews 11:3). It is the one Christian doctrine that the enemy of the human soul has attacked with greater vengeance than all the others combined. Anyone who rejects 'special creation', has denied God, and has demeaned the Bible as a collection of myths, legends and fables. These are likely to embrace a false worldview that will lead to devastating eternal consequences.

Offering a Hope and a Future

The most amazing thing about the Bible is that it explains, in easily understood terms, how a sinful person can exchange their guilt and shame for His righteousness by a simple act of faith. Everyone loves a good trade! I can think of no other bargain that compares with the one God has offered mankind, and I cannot understand why anyone would reject such an offer. God is offering to exchange of our guilt, imposed by a life of sin, for the righteousness of Christ, who knew no sin. What an amazing transaction. Nothing compares with God's grace toward the ones He bled and died for. Do not be deceived! Evolution is a colossal lie! Do not let it keep you from accepting the greatest offer that was ever presented to mankind. "Professing to be wise, they became fools, and changed the glory of the incorruptible God into an image made like corruptible man--and birds and four-footed animals and creeping things. Therefore, God also gave them up to uncleanness, in the lusts of their hearts, to dishonor their bodies among themselves, who exchanged the truth of God for the lie (evolutionary humanism) and worshiped and served the creature rather than the Creator, who is blessed forever." (Romans 1:22-25).

The Bible goes on to explain what happens to men and women who stubbornly reject God's mercy and choose instead a false worldview. This leads to a form of personal morality that embraces a very worldly and self-centered approach to living. The average person in today's high-tech, modern world, neither knows about, or cares about the faulty premise of evolution theory, and yet his or her life is continually influenced by it in a multitude of ways. This theory has led to a type of modern-day morality that regards God's Word as outdated, old-fashioned, and with little value in governing contemporary lifestyle choices. Many so-called 'enlightened' individuals have long since cast off the moral constraints they will no longer tolerate. Thus, evolution has provided mankind with a way to hide from God. Consider these thoughts from evolutionist Julian Huxley: "Darwinism removed the whole idea of God as the creator of organisms from the sphere of rational discussion. Darwin pointed out that no 'supernatural designer' was needed, since 'natural selection' could account for any new form of life, there is no room for a supernatural agency in its evolution."[3] Soon after Huxley made this dogmatic statement in 1960 an avalanche of scientific data, and the personal testimony of hundreds of respected scientists, exposed the failures of the world's 20th century naturalists, who had championed the propaganda campaign that sustained evolutionary thought for so long.

Evolution is a methodically crafted scheme by which the enemy of mankind has attempted to discredit Christianity and debunk the Bible. It leaves man without the wisdom of God and with the appearance of controlling his own destiny in the absence of the 'accountability' that was intended to safely govern his affairs. Darwinian evolution is like a very corrosive acid that eats through traditional Judeo-Christian values in all areas of life, and leaves people with a deceiving ideology. This radical worldview leads to a refusal to acknowledge divine authority and submit to moral restraints that were intended to protect us, as defined for us by the Word of God.

The anti-God religion of evolutionary humanism still holds sway with many people today who believe it gives them the justification needed to pursue a life of self-gratification at the expense of the God's commands. The fruits of humanism reflect a blatant neglect of the moral foundation that has governed our Judeo-Christian nation for over two hundred years. Sinful acts against God are the natural outcome in the lives of people who have chosen to become a law unto themselves. Pornography, fornication, adultery, homosexuality, and many other acts of sexual deviancy, along with abortion and euthanasia are the natural outcome of a humanistic ideology that says man decides for himself what is or is not morally acceptable within a modern culture. Now, we face an all-out assault on gender roles, traditional marriage, religious freedom, and worst of all, censorship of thoughts and ideas sanctioned by government action. The recent surge of 'woke ideology' is unlike anything we have ever seen before. When the government and liberal courts begin to legislate morality, this is a sure sign that we are systematically abandoning our Judeo-Christian heritage. The only possible hope for our beloved nation to escape the pending moral collapse that will weaken us from within, is widespread repentance leading to a revival of the values and principles of our forefathers. These were the principles that, no doubt, motivated God to bless America with great success and prosperity unlike no other nation. The enemy of the human soul continues his work of deception, keeping men in darkness, so that they may not perceive the marvelous light of the gospel, nor experience the priceless gift of redemption that was purchased on their behalf by the sacrificial offering of God's own Son on Calvary. Evolutionary theory is a dark cloud of deceitfulness hanging heavily over the human mind, and placing a death grip upon the human soul. Do not be fooled by it!

No man or woman can consider themselves well educated without some practical knowledge of the Bible. Without this, man struggles to interpret life's meaning with very strange and unrealistic notions designed to ease the human conscience within, and to circumvent and cast off the need we all have for personal accountability. There is a growing lack of understanding of biblical truth in America that is leading toward cultural transformation, characterized by a loss of appreciation for the values and principles that have led to our unprecedented prosperity.

Diehard evolutionists long ago predicted that the next logical step after humanism is socialism. After having embraced

Darwinism for such a long time, we are now beginning to see the public consciousness of our beloved nation trending in this direction more so than ever before. Many people fail to comprehend that the teaching of evolution theory does not leave the youth of America or any other nation 'in a neutral position' regarding religious persuasion, but instead, this ideology promotes the blatantly anti-biblical religion of secular humanism leading toward atheism. This deceiving secular ideology can tug at the heart of man, and in many cases will pull him away from his Creator. Atheism always leads mankind toward greater government control by way of unbridled secular authority, accompanied by a sharp decline in man's traditional reliance on and submission to the governing authority of the Word of God.

All biblical truth hinges on Genesis 1:1. "In the beginning God created the heavens and the earth." When man rejects this basic truth, it opens the door to all kinds of deceiving philosophy. After studying the Bible for much of my adult life, I am thoroughly convinced that it is a God-inspired revelation of divine origin. I have come to believe and will defend the Word of God as the foundational basis for all truth. Who can confidently say, "I flatly deny the existence of God and reject the Bible as a divinely inspired record of His eternal truth"? True Christianity is an evidence-based faith! What I have offered to the reader in these few chapters is only a small portion of the abundant evidences that thoroughly support Christian-based theology. When we are considering these 'many infallible proofs', the question we must ask ourselves is, *how then shall we live*? For those who reject the Bible as the divinely inspired, authoritative Word of God, the quest to understand '*the mystery of life*' will never have a satisfactory outcome. This is true because a 'mystery' is 'that which cannot be discovered by the natural, rational searching of the human mind, but must be revealed'. God--the author of life, has revealed powerful truths to us through his Word and his creative works. Now, it is up to us to believe!

If the worldwide flood recorded in Genesis chapters 6-9 was of the magnitude and intensity indicated in the text, then the entire elaborate scheme of an evolutionary worldview collapses in a pile of rubble. Why? Because evolution theory is entirely dependent upon a fossil record that must be interpreted within the framework of vast geological ages of earth history. There are some who say they believe in God, and yet support the unbiblical view that God has used evolution to create living things. This very unfortunate and misguided compromise requires a very loose interpretation of Scripture, and, in fact, denies its literal meaning. This view relegates biblical truth to symbolic allegory, allowing men to pick and choose what is literal and what is not. This is an attempt to divest God's Word of its miraculous nature in order to satisfy the modern intellectual demand for a naturalistic view of life. This dangerous concession will undermine the integrity of the God's Word, and leaves mankind in control. It allows man to interpret the Bible in whatever way he wishes, so it conforms with his preferential view of life. God's Word is always meant to be taken literally, unless obvious figurative speech is used.

We must never allow the misguided theories of men to color our minds when we interpret the Word of God. This is what people are doing when they accept certain portions of Scripture as literal, while rejecting other portions as "allegorical" or perhaps "poetic", when the text is clearly written in a manner intended to be read and understood as literal truth. When man interprets the biblical text the basic rules of literary interpretation must always be applied. Life is a mystery, but the mystery has been unveiled to us by the Word of God. When we receive its inspired truth at face value, it sets us free from our greatest fears, doubts, and unbelief. If there really is a God in the spiritual dimension, who created the physical dimension, then you and I are certainly accountable to him, and rightly so. We will stand before Him in judgment on some future day when our choices, actions, and the motives of our hearts are to be thoroughly examined. "For the Word of God is living and powerful, and sharper than any two-edged sword, piercing even to the division of soul and spirit, and of joints and marrow, and is a discerner of the thoughts and intents of the heart. And there is no creature hidden from His sight, but all things are naked and open to the eyes of Him to whom we must give account." (Hebrews 4:12,13)

Conclusions We Should Draw

The evolution model presents a false worldview that has had devastating consequences in our world. There is powerful evidence to confirm the existence of a sovereign creator God, who has revealed himself to his creation. I have offered evidence to confirm the Bible is God's written revelation to mankind. God's Word will, no doubt, serve as the basis for divine judgment when He ultimately evaluates each of our lives. "He who rejects Me, and does not receive My words, has that which judges him. The word that I have spoken will judge him in the last day" (John

12:48). (For a more detailed review of why we can trust the Bible as a reliable and accurate source of divine inspiration, please refer to the epilogue section at the end of the book)

Listed below are the conclusions that I and many others have made based on the remarkable evidence that is pointing man, unmistakably, toward the Word of God as the primary source of understanding and the final authority for discovering truth and meaning in this world.

1. The one true God is very real, and he has revealed himself in unmistakable ways to those who remain open to the prospect of his existence (see Psalm 19:1; Isaiah 43:10, 44:6, 45:22).

2. God created everything, and is the intelligent designer who is responsible for both the spiritual and physical realm (see Genesis 1:1-30, Isaiah 45:12, Psalm 33:6).

3. Mankind was created in the image of his Maker, and became God's crowning creation— the centerpiece and focal point of his attention and affection (Genesis 1:27).

4. God's original creation was perfect in every respect, and reflected the awesome nature of his divine character and his extraordinary imagination (see Genesis 1:31).

5. In his relentless desire for personal relationship, God was compelled to create man with free will (the unique ability to choose his own path) (Genesis 1:27, 2:17, 3:17).

6. Mankind used this freedom to rebel against the authority of his Maker (Genesis 3:1-19).

7. The tragic fall of mankind corrupted not only his personal relationship with God, costing him full dominion and authority over the earth, but the whole of creation was drastically altered by man's sin (Genesis 3:17-19, Romans 8:19-23).

8. The terrible consequence of sin resulted in the curse of death and decay that has come upon the entire physical dimension (see Genesis 2:17, Romans 8:19-23).

9. God's everlasting love for humanity prompted Him to reach out to mankind by preparing and executing a brilliant plan of redemption (Genesis 3:15, John 3:16).

10. The overall purpose of God's unmistakable love for his image-bearing sons and daughters was to 'redeem them' to become a part of his eternal family and also restore, to as many as received him, that which was lost when mankind originally chose to rebel against his Maker by disregarding divine authority (Revelation 21:1-4, 22:1-5).

11. As a result of this divine plan, the redeemed of God (those who have entered into a God-fearing relationship with the Creator by faith in Christ) have become the rightful heirs to everything that God has created, and will enjoy his presence for eternity (see Romans 8:16-18).

"Happily, ever after" is not just in fairy tales. God's plan is to make this magnificent dream a reality in the life of every human soul he has created, loved, and died for.

People can choose to believe these conclusions about life and reality or not. Every person must decide for themselves and embrace his or her own worldview, and each of us will live with the consequences of our decision. However, I can guarantee, based on thirty years of study and research, that this is the direction the evidence is pointing toward and we would be foolish to ignore it. We all know that death is inevitable for each one of us. If the Word of God is true then heaven and hell are very real places, and each one of us will be judged by a higher power that so many have been determined to deny or ignore. The meaning of life is one question none of us can afford to get wrong. We must get the truth settled in our hearts and minds now, while we still have hope, because after we die, we will all know for certain how very real God is— only then it will be too late to choose to follow Christ and be delivered from the wrath of God, if we have ignored and rejected God's love all of our mortal lives. Those who choose to reject him and his Word in this life are destined to carry that devastating choice into eternity, and it will cost them everything when they become separated from their Creator's presence forever.

Mankind has struggled through the centuries to discover true meaning in material living. And those who are honest with themselves will have to admit that the competitive pursuit of power, prestige, possession, and pleasure leaves the human soul very empty of the satisfaction that would define a gratifying and purpose-filled life. If there is a living God to whom we must all give account, as the evidence strongly suggests, you and I cannot afford to be ignorant of who he is, of what he has done, and of what he expects from us. What possible justification can there be for anyone to reject him, deny him, and refuse to acknowledge his divine power and authority?

If I were to ask the question, what do you long for most in life? I believe an overwhelming majority of people would express a desire to be loved and accepted. We need to feel that our life has significance to other people. We were created with the need to love and to be loved. This reflects the fact that we were created in the image of a personal, loving God, because this is undoubtedly the top priority of our Creator as well— to love and to be loved. He has designed you and me to become a part of his eternal family, and he desires for each of us to have a personal and ongoing relationship with him. We were created by a purpose-driven God who longs for that kind of relationship with every one of his image-bearing sons and daughters. Ultimately, however, the decision to reciprocate God's love is ours alone. God's primary commitment to us was fulfilled when his only Son was nailed to die on a cruel Roman cross. Now, the decision to respond is up to us. What will we do about the man called *Jesus*? In the final analysis, the Son of God becomes the focal point of all truth regarding *the mystery of life*!

There is no logical reason why anyone should deny him. There is no scientific reason to deny him. And there is no historical reason to deny him. Only the prideful human heart and a brash unwillingness to submit to his authority will keep people living in unbelief. You may be thinking: Why should I invest my devotion and allegiance in a worldview I cannot possibly explain? My short answer to that question is, because this mortal life on planet earth is totally and completely unexplainable apart from God! Many people simply give lip service to God, but have never given him the glory that he alone deserves. In referencing Isaiah, Jesus said, "as it is written: this people honors Me with their lips, but their heart is far from Me. And in vain they worship Me, teaching as doctrines the commandments of men. For laying aside the commandment of God you hold to the tradition of men" (Mark 7:6-8).

Ask this question as well: can I explain how evolution theory works and point to an abundance of validating evidence that would explain why this concept should capture my allegiance and dominate my worldview? Before you answer this question, please consider the penetrating words of Dr. Collin Patterson, who held a very distinguished position for many years as a senior paleontologist with the British Museum of Natural History. In an honest moment, he provided us with some very candid remarks concerning the current state of evolution. "For over 20 years, I thought I was working on evolution ... but there was not one thing I knew about it ... So for the last few weeks I've tried putting a simple question to various people, the question is: 'Can you tell me any one thing that is true' (about evolution]? I tried that question on the geology staff at the Field Museum of Natural History and the only answer I got was silence .... I tried it on the members of the Evolutionary Morphology Seminar in the University of Chicago, a very prestigious body of evolutionists, and all I got there was silence for a long time, and eventually one person said, "Yes, I do know one thing, it ought not to be taught in high school... Over the past few years, you have experienced a shift from evolution as knowledge to evolution as faith ... Evolution not only conveys no knowledge, but it seems somehow to convey anti-knowledge."[4]

Is an admittedly confusing, outdated, and unexplainable theory, by many of the world's greatest evolutionary experts, really where we should invest our personal faith and allegiance?

Jesus is the answer to every need. Every breath of air you take is because of Him. Every bite of food, every drink of water, every bird that sings and tree that grows, every drop of rain— it is all because of him. On the other hand, every deadly disease, every sorrow, and every dying breath is because of us! We are the cause of a cursed world that does not reflect the original intent of its Maker. We have rejected his truth and despised his authority, and now we are all living with the terrible consequences of those tragic choices. How can anyone dismiss from their life the only hope of escaping a world that is reeling under the curse of death and decay.

Revelation 3:17 speaks of a man who thought he was totally self-sufficient and could depend completely on his own resourcefulness, needing nothing from anyone, including God. He was blinded by the devil's deception. He was full of pride in his own success, and was prepared to live out his days in leisure and luxury. God responded to an arrogant and reckless way of living. "You do not realize that you are wretched, pitiful, poor, blind, and naked." How many people have slipped into eternity leaving everything they have worked for to others, only to discover that their lives are void of any eternal treasure beyond this world?

God is saying the same thing to all of those who have decided that they can handle life just fine on our own, having no need of a power greater than themselves. How many people are like this today? I fear the number is staggering! They must be reminded of the God to whom they will give account, and that their days on this earth are numbered. There is no reason for anyone to remain in a self-centered state of separation from the Creator of the universe. Three verses later, in Revelation 3:20, God says to all of us, "Behold, I stand at the door and knock. If anyone hears my voice and opens the door, I will come in to him and dine with him and he with me." God is offering us an opportunity to experience a restored relationship with him. This is the only way we will ever find peace and security in a troubled world ravaged by sin and death. When we say yes to God, he gives us the power to live the God-centered life we were meant to have. His saving power will enable us to love him in return, and to love our neighbor as ourselves.

What about the multitudes that do not believe they are in need of God's love? How do we deliver a message of hope to them? After the crucifixion, the disciples of Christ were huddled together, cowering behind closed doors. They were confused, fearful, disillusioned, and still unaware that the crucified Christ had risen from the dead as he promised. Soon after this, the resurrected Jesus presented himself to his disciples on a day when the apostle Thomas was absent. When all the disciples who were present told Thomas that Jesus was alive, and was resurrected from the dead, his immediate response was, "I will not believe until I see the marks in his hands and put my hand in his side (where he was pierced)" (John 20:25, paraphrase). This incident is the origin of the familiar phrase "doubting Thomas." Thomas was waiting for the physical evidence. He surely wanted to believe, but he needed proof to dispel the doubt in his mind. When Jesus appeared again, with his disciple Thomas in attendance, an interesting thing happened. Jesus could have rebuked Thomas for his unbelief but instead he gladly showed him the nail marks in his hands and the scar in his side (John 20:27). Why did he do this for the unbelieving disciple? It was because the evidence is what Thomas needed to bolster his faith? In response to the evidence, Thomas declared: "My Lord and my God!" (John 20:28). Doubting Thomas was thoroughly convinced by the evidence even though he could only marvel at what he saw without knowing how to explain it. This disciple, along with all his companions, was transformed with boldness, courage, and faith. He went on to share the good news far and wide, and ultimately gave his life for the cause of Christ when he was speared to death in the nation of India. God knew that it would be more difficult for some to believe, so he offered much evidence to convince unbelieving humanity to follow him. Christianity is an evidence-based faith. When I think of the multitudes that just need to "see the evidence" to bolster their faith, I am reminded once again of what compels me as a Christian apologist. In response to the evidence presented in these pages, are we prepared to declare as Thomas did long ago, "My Lord and my God!" If we show people valid evidence for the existence of God, his creative genius, and his undeniable love for humanity, genuine faith is sure to follow. The physical world of the vast, seemingly unknown universe reveals far too much evidence for a Creator to believe otherwise.

Not Too Good to Be True

Now that evolution theory has been thoroughly exposed for the monstrous deception that it is, proof for the existence of God has been exhaustively examined, and the true nature of the Bible as a document of divine origin has been adequately discussed, "how then shall we live"?

Many will continue to live their lives as though God does not exist and, ultimately, will face his wrath in the coming judgment by trusting in their own goodness to save them. Others will place their faith and hope in the eternal Word of God, acknowledging that all have sinned against the God of creation, have broken his laws, and have disobeyed his commands. This kind of thinking is the first step toward true repentance– to acknowledge God, to fear his judgments, and to feel a sense of remorse for one's own sinful acts and attitudes that have grieved our Maker. Why would anyone gamble their eternal soul on the constantly changing 'wisdom of man.' We can ask God to reveal himself to us in a personal way. If he is powerful enough to create the world, he can surely reveal himself to us, if only we care enough to ask. God does not expect us to be perfect in and of ourselves because he knows this is impossible. He is, however, expecting us to acknowledge who he is, confess that we have rebelled against him by transgressing his laws, and then choose to repent before him and forsake our sin. We are all God's children through creation, but we must also become his children by redemption, or we will never partake of his priceless promises and eternal rewards. With all that God has done for us, no excuse will be justified when we stand before our Maker to have our earthly lives examined. (see

Romans1:20)

With eternity at stake, you and I cannot afford to be indifferent and to ignore such a life-and-death matter as this. We must respond to the truth once we understand it, because no decision is a choice to reject God and his undeniable love for each of us. There can be no neutral ground in this debate; either we are with Him (submitting to his unconditional love and atoning death), or we deny the need for a savior! Do not miss out on the single greatest blessing afforded to us in a lifetime— the priceless opportunity to know for certain, before death comes, that our sins will never be counted against us. Be assured that you possess, by faith alone, a sure hope and a secure future that is guaranteed far beyond the mundane physical reality that you and I have experienced during our lifetimes. The voluntary, sacrificial death of God's only Son on a Roman cross was generously offered to all of us out of our Creator's overwhelming love for humanity and the compelling desire of a sovereign God to spare humankind of eternal judgment.

If you think your belief system is right on, and you are currently not living for Christ, there is one sure way to test what you believe to determine if it is deserving of your allegiance. Ask yourself this question: believing as I presently do, am I prepared to die today with the belief system I currently hold, or do I have nagging doubts about whether I am ready to face what lies beyond? If you can honestly say that you have serious doubts, then you are not ready to face death while embracing to your present worldview. Much of the world does not want God to be the answer to their need. They do not want Jesus in their life! Many remain unwilling to admit that their sin has grieved the Holy Spirit, nor will they embrace the accountability a Holy God requires. Why would anyone reject God's priceless 'gift of salvation' that only requires an act of believing faith?

If you should die tomorrow and face eternity, why should God welcome you into heaven to remain in his presence forever? There can be only one answer, and it is not because you have been a 'good person'! Our goodness is like filthy rags compared to the righteousness of a Holy God. The only possible answer is because you have not relied upon your own merit, which we know is completely inadequate with respect to God's standard, but you have relied completely on the sacrificial death of Jesus Christ on your behalf. He paid our sin debt and offered us a full pardon when He laid down his sinless life to satisfy God's divine justice in regard to our sin.

We must repent of our sin and trust completely in the atoning death of Christ, knowing that it was never possible for us to earn our way to heaven. There is nothing you and I could ever do to merit salvation! All we can do is trust the promises and say yes to God. The rest is up to Him. In the Old Testament, a sacrifice offered by the high priest would secure a covering for sin for one year. In the New Testament, the blood of Christ was offered for the sin of all mankind, but this was not a temporary covering. The death of Jesus Christ was the final sacrifice that did away with the debt of sin in all who believe. I am reminded of the word "expunge." It means to erase, to delete or blot out and remember no more. The Bible says when we repent, our sins are removed from us as far as the east is from the west (Psalm 103:12 and

Jeremiah 31:34).

People who think they have no need for God's love, but have life all figured out, and Christianity is just a crutch for weak-minded people to lean on, should consider the following two questions. Considering the evidence presented in this book, can I really afford to take a chance knowing I may be wrong about God and the Bible? What if there really is a supreme God who created all living things, and will hold me accountable one day when I face his righteous judgment?

Nobody wants to face the wrath of Jehovah, which the Bible says will be poured out upon all who do not know God and refuse to obey the gospel of Jesus Christ (see 2 Thessalonians 1:6-9 & 1 Peter 4:17). Do I really believe that heaven and hell are not real places and that there is no life after death, when the Bible clearly teaches otherwise? What if the Bible is the truth and, when I die, I will become separated from God and everything he provides, forever, because I stubbornly chose to deny his words and reject his love? Perhaps God really does love me enough to send his only Son into the world to die for my transgressions, making it possible for me to be reconciled to a right relationship with my Maker. Can I really afford to ignore the clear message of scripture and take a chance that trying to be a good person will be enough to save me from the wrath of God, when the Bible clearly denies any such claim? (see Titus 3:5) The Bible says, "It is appointed for men to die once, but after this the judgment" (Hebrews 9:27). There will be no chance for reconciliation with God after death. The most important question to ask is, am I ready to face the judgment of God in the current condition of my eternal soul?

Please, do not stake eternity on personal opinion, or a popular philosophy of life, when you do not have the evidence to back it up. In the end, it is only God's opinion that will count, and it is his opinion that we have all sinned and have fallen short of his eternal glory. (Romans 3:23)

Some people believe that Christian faith is foolishness, and that the great miracles of Scripture are too outlandish to believe, if taken literally. Are miracles a problem for God? "But God has chosen the foolish things of the world to put to shame the wise, and God has chosen the weak things of the world to put to shame the things which are mighty" (1 Corinthians 1:27). "For the message of the cross is foolishness to those who are perishing, but to us who are being saved- -it is the power of God" (1 Corinthians 1:18). "For what will it profit a man if he gains the whole world, and loses his own soul?" (Mark 8:36; see also Matthew 16:26). The Bible says, "If you confess with your mouth the Lord Jesus, and believe in your heart that God has raised Him from the dead, you will be saved. For with the heart one believes unto righteousness, and with the mouth confession is made unto salvation" (Romans 10:9,10).

God's impartial forgiveness can only be found through an act of personal repentance, and demonstrating genuine faith in Jesus Christ as the mediator between God and man (see 1 Timothy 2:5). Through spiritual growth lives are set on a path of submission and obedience to God's precepts and his commands. Instead of accepting God on his terms, however, many people in today's me-first generation expect God to accept them on their terms, by overlooking their sin, and tolerating choices made with little or no regard for his divine judgments.

When someone offers you a gift, it is not yours until you reach out and receive it. Just because Christ died on the cross, paying the penalty for man's sin, does not mean your sins are forgiven. You must accept this gift of God's grace to take ownership of it. Reconciliation with God is a decision that must come from the heart, not just an acknowledgment of the mind. Consider a young lady standing before a male companion who is proposing marriage to her. She must, of her own free will, give a response to the offer presented. Silence means she is either unsure or is declining the offer. Any indecision must be resolved and a choice made, otherwise the answer is no. God understands what people mean by their indecision. To be unresponsive to the gospel makes a person no less accountable when they face judgment. Jesus told his disciples, "If you do not believe that I am He, [the sacrificial lamb] you will die in your sins" (John 8:24).

While driving in Orange County, California, on a very bright afternoon, I was pulled over by a police officer who informed me that I ran a red light. Several excuses crossed my mind as to why I may have failed to stop at the intersection. However, none of them would have held up to the serious scorn of the authority figure that was holding me accountable. The truth is, I was clearly guilty, and no excuse could justify my action; consequently, I had to pay a stiff penalty and go to traffic school for breaking the law. The Bible says that we have all broken God's laws and that every one of us stands guilty before him. Now, what parent would not do anything to protect their children from harm, pain, and suffering? That is how God feels about us. He was acting just like a good parent when he identified certain selfish behaviors as sinful. These are behaviors that cause harm, pain, and suffering, at some level, to the guilty party, and in most cases to someone else at the same time. God has established moral guidelines to protect us! God did this because he loves us, and he wants the best for our lives. In the same sense that we are accountable to obey traffic laws, we are accountable to obey God's moral laws as well.

You and I are responsible to God in the same way that any young child must answer to his or her earthly father. With this said, God should be treated with respect, reverence, and fear as the authority figure to whom we will give account. Evil is conceived in the mind of man, generated through the heart of man, and acted on by the will of man. The Bible teaches that all men have broken God's laws, and stand condemned to be judged for doing so. However, the good news, we can all rejoice in, is that Christ has taken our place and received the penalty that was due for our selfish transgressions, so that we could receive God's forgiveness and be restored to a right and lasting relationship with him. It cannot be overemphasized that no one can earn the favor of God and escape his wrath through the merits of his or her own personal effort at good behavior. (see Ephesians 2:8, 9) The only way we can escape the penalty for our transgressions against a Holy God is for someone else to take our punishment for us. This is exactly what Christ did when he took our sins upon himself at the cross to fully satisfy the righteous justice of God.

If we stubbornly reject or ignore this substitutionary act of saving grace, the wrath of God will remain upon us, and we will be required to pay the penalty for our own sins (see John 3:36). Who would not want someone to voluntarily step in and pay their traffic fine. The good news is that Christ stepped in to pay our 'sin debt' before a Holy God. God sent his only Son, who knew no sin, to become sin for us. All we can do is accept it! Has there ever been a better deal than this? Without the grace of God and His mercy, heaven would be an empty place indeed.

A New Birth Certificate

God is calling men and women everywhere to repent; for he is not willing that any should perish, but that all should come to repentance. (Acts 17:30 and 2 Peter 3:9) How should we respond to the evidence presented in this book and documented in a hundred other books that have recorded many well-documented reasons to embrace Christian faith? What God has offered humanity seems almost too good to be true, but it's not! When someone (like God) longs for a relationship based on 'mutual love and respect', he will do whatever it takes to secure that relationship; and in our case, it required the death of God's Son on a Roman cross. What we should do in response to this undeserved measure of unconditional love are the following:

1. Acknowledge the living God as our creator, provider, and ultimate judge.

2. Confess that we have broken his laws, and stand guilty before him for doing so.

3. Admit that we cannot save ourselves from transgressions committed against a Holy God.

4. Believe that God sent his only Son to do for us what we could never do for ourselves.

5. Acknowledge that the death of Christ and his resurrection were essential to provide salvation.

6. Repent of our sins before a loving and merciful God who offers us more than we deserve.

7. Receive Jesus, the living Christ, into our lives by faith knowing that he has provided the atonement for sin, and is the only mediator between a Holy God and sinful mankind.

8. Decide to allow the Holy Bible to become our instruction manual, which explains in vivid detail how we should live and how we can begin to grow in our personal relationship with Christ.

If you are deciding, for the first time, that God is real and he loves you enough to send his Son into the world to die for your sins, then right now is the time to offer a response to God's work of redemption on your behalf. If you have never experienced a personal relationship with Jesus, by accepting the only provision God has made to eliminate your sin, then please carefully consider the following pages. Jesus said to Nicodemus in John 3:3 that a man will never see the kingdom of God without experiencing a "new birth." This occurs when we repent of our sin. and receive the atonement God has provided, by faith. The crucifixion of Jesus was the price God required to satisfy the sin debt of humanity. When Jesus spoke to Nicodemus about a spiritual birth, he explained this miracle with a penetrating metaphor: "The wind blows where it wishes, and you hear the sound of it, but cannot tell where it comes from or where it goes. So [he said] is everyone who is born of the Spirit." (John 3:8)

Dr. Henry Morris understood early on that science and faith are to be embraced as companions, unlike many misinformed intellectuals are teaching today. Dr. Morris explains the tragic dilemma that is facing humanity and what a loving Creator God was prepared to do about it. "The Creator has imposed the law of decay and death on his whole creation, because of the rebellion of its human stewards. Therefore, only he can defeat death, and this only by paying the redemption price himself, dying for sin, and then rising victoriously from the dead ... Thus, the great miracle of the resurrection requires the great prior miracle of supernatural creation ... the omnipotent, omniscient Creator cannot fail in his creative purposes, even though man's sin has brought the universal reign of death into the world. Therefore, he must conquer death and redeem his creation. This he can only do by his incarnation, his dying for sin, and his resurrection."[5]

Use the following words as a personal confession of believing faith and spiritual commitment:

I believe in the only true God, the Creator of heaven and earth, who has revealed himself to the world through nature, the cosmos, and the Holy Scriptures. I now understand that according to God's Word, I must have a personal relationship with my Maker in order to gain the hope of eternal life when I die. This requires me to experience what the Bible calls "redemption." The Bible reveals that everyone has sinned and has fallen short of God's righteous standard. To have peace with God, I need the assurance that my sins have been forgiven. "God sent his Son into the world, not to condemn the world, but that the world through Him might be saved." (John 3:17) Jesus died in my place, taking my sin and condemnation upon himself, so that I could experience right standing with God in this life and the next. By offering his only Son for my sins, God did for me what I could never do for myself through human effort. God is offering me the gift of eternal life, made possible only by the death of his eternal Son upon the cross. It is my understanding that there is nothing I have done or ever could do to deserve this priceless gift. I also acknowledge that a gift does not belong to me until I have personally received it. The Bible says, "As many as *received* Him, to them gave he power to become the sons (and daughters) of God (John 1:12 KJV). By receiving the risen Christ into my life, I recognize that his death was necessary to satisfy God's justice in regard to my sin. I believe Jesus has risen from the grave, victorious over death, giving me the assurance that he was, indeed, the living God in human flesh. The Bible defines death as eternal separation from God in a horrible place that was 'not prepared for man', but for the devil and his fallen angels. By applying my signature to this document, I am affirming my personal decision to receive Jesus Christ into my life, and I freely acknowledge that what Jesus did on the cross was necessary to save me from the wrath of God. Today, I will experience the forgiveness of sin and a new relationship with God, which the

Bible promises to all who believe and receive Him. If I were to reject God's generous offer and instead depend on my own good works to merit heaven, the Bible clearly teaches that I would have no hope beyond the grave. God's Word plainly states, "Jesus is the way, the truth, and the life and no one comes to God the father, except through Him." (John 14:6) I have decided to accept God's offer of forgiveness by receiving Christ today and, by making this decision, I now join millions of others around the world who have gladly accepted God's amazing grace into their lives. I will closely examine the following scriptures for better understanding as I begin a new life of faith at the direction of God's Word: Romans 3:23, 6:23, 10:9-10; Mark 8:36; John 1:12, 3:3, 3:16, 14:6; Acts 4:12; Ephesians 2:8-9; and Hebrews 9:27. As I pray the following prayer out loud before my Creator, this document now becomes my '*New Birth Certificate*'.

I, ______________________, choose to accept you, Jesus, into my life today. I confess that I have broken God's laws, and I deserve his judgment. But today, I repent of my sins and receive your forgiveness, knowing that the precious gift of eternal life is promised to all who believe. I am now just beginning a purpose-filled life of devotion and service to God, and I look forward to becoming a true disciple as I gradually grow in the knowledge and fear of God's Word and His ways. Today, ___________________, is my spiritual birthday!

If you have prayed this prayer for the first time, consider the words of 2 Corinthians 5:17: "If anyone is in Christ, he is a new creation; old things have passed away; behold, all things have become new." I hope you have made the best decision of your life— tell someone else about it!

Creation is the foundation doctrine for all biblical truth. On the other hand, evolutionary theory, and the humanist worldview has laid a sure foundation for all kinds of evil, sin, and corruption. Consider these final words from Dr. Henry M. Morris, PhD: "Evolutionism has been the chief opponent of the saving gospel of Christ, undermining the faith of multitudes in the Bible and its promises. It is not too much to say that evolutionary theory, in one form or another, has provided the pseudo-rationale for all that is false and harmful in the world (the real cause, of course, is the innate sinfulness of the human heart, with its rebellion against the Word of God) ... all who opposed the true God must always resort to some kind of evolution, for this is the only possible alternative to 'special creation' by a transcendent God."[6]

"Lift your eyes and look to the heavens: Who created all these things? He who brings out the starry host one by one, and calls them each by name. Because of his great power and mighty strength, not one of them is missing ... Do you not know? Have you not heard? The Lord is the everlasting God, the Creator of the ends of the earth. He will not grow tired or weary, and his understanding no one can fathom. He gives strength to the weary and increases the power of the weak. Even youths grow tired and weary, and young men stumble and fall; but those who hope in the Lord will renew their strength. They will soar on wings like eagles; they will run and not grow weary, they will walk and not faint" (Isaiah 40:26, 28-31 NIV).

Can we really use evolution theory or any other human

generated ideology to hide from God? Not so, according to Hebrews 4:13, where we read: "And there is no creature hidden from His sight, but all things are naked and open to the eyes of Him to whom we must give account."

Yes, of course, we are accountable before a transcendent God, but our Creator offers amazing grace, unconditional love, and unimaginable mercy to those who seek him. Everyone wants to be loved and accepted. The problem is that a Holy God cannot accept our sin with impunity. God's justice requires that the debt of sin must be paid. That is why he sent his Son to the cross to pay the debt for us— so that we can be delivered from the wrath of God to be poured out on the unbelieving world. That is how much he loves us! He was willing to pay the sin debt on our behalf— taking our punishment upon himself and allowing us to be set free from the penalty of sin and to become members his eternal family. Our acceptance of this unspeakable gift of unconditional love is certainly not a difficult decision to make. Do not let pride and unbelief stand in the way of God's plan to save you and to give you the certain hope of spending an eternity of pleasure, joy, and fulfillment in His presence. Yes, you and I were created for the purpose of relationship. We were designed to have relationship with God and each other. If in our effort to find meaning in life, we only discover personal relationship with one another and do not seek to know the loving God who created us, then we are sure to miss out on the very best that life has to offer, and the hope of an endless eternity in the presence of our Maker. We must acknowledge him, seek him, and come to know him!

This book is a contribution to the assault that is being waged against the dangerous deception of secular materialistic atheism— a movement that was born out of the false and misleading ideology of evolutionary humanism.

Please remember that God loves you more than you can possibly comprehend and He has done everything that was necessary to rescue us from the terrible curse of death and decay that remains upon this world.

With this said, I am reminded once again that God's image-bearing sons and daughters all carry free will, and can certainly choose a worldview that denies the powerful truths I have presented in these many pages. But one thing is certain if they do— they can no longer depend on science, history, and logic to justify their position.

THE HUMAN RACE

I looked into the world I see and
searched for answers desperately.
I found the truth and set my course
to serve the God who is my source.
And though the days grow weary
as I face a world of strife,
I have my Lord's assurance
that I possess eternal life.
Temptation came to knock
Upon the door-gate of my heart.
I had the choice to live for self
or as God's servant do my part.
So by his grace I stood my ground
refusing compromise,
to live for self would rob me
of a great eternal prize.

So when I stand before my God
and look upon his face,
I will have to give account
of how I ran *the human race*.
If I live my life unto myself
a failure I would be.
But if I can touch a dying world,
perhaps I'll help to set men free.
Few people understand this life
and that seems very odd.
My purpose is to glorify
the true and living God.

DAVID E. PEEPLES
(Hebrews 12:1)

EPILOGUE

In researching and writing *The Mystery of Life* I have been compelled to make a strong case for the undeniable connection between science and faith. The narrative addresses the greatest culture war of our time and proposes a worldview that places the sovereign, eternal God of the Bible at the center of the discussion. This benevolent Creator God of heaven is the designing agent and sustaining power behind a material world in which human life is at center stage. Any knowledge or understanding we have gained in our struggle for life is because of our complete and absolute dependence on Him and all that He provides for us.

Evolutionary thought has led mankind directly toward humanism, a man-centered religious ideology that became the breeding ground for all sorts of evil, racism, corruption, immorality, hatred, and violence— not to mention mass genocide. After all, if there is no God, as humanistic atheism proclaims, that means there is no sin, no heaven or hell, and no human accountability to a powerful authority greater than ourselves. If this is the case, then what possible difference does it make if I take whatever I want in life at the expense of those around me? Who restrains me if I decide to covet my neighbor's wife or embezzle money in a shady business deal? After all, it's all about "survival of the fittest," right? Years ago, I can recall briefly observing a display of popular publications at a local book store. On this occasion I was surprised to observe that one of the best-sellers that year was a book entitled *Taking Care of Number One.* This is what humanism does to people: it gives them license to act upon selfish impulses without restraint— to consistently place one's own needs, wants, and desires before those around us leading to self-centered lives. This has led fallen humanity into deeper corruption, as morally impoverished mankind makes his own rules; and these rules are typically designed to satisfy the desires of the flesh— desires and impulses that are at war with God.

The greatest danger for humanity imposed by the false and misleading doctrine of evolutionary humanism has been the widespread generation of doubt and skepticism toward biblical truth. Spiritual skepticism resulting from evolutionary thought has generated much rebellion in the heart of man toward the authority of God. This false religion has become the greatest obstacle to the advance of the gospel of Jesus Christ over the last century.

The message of the gospel seems utterly fantastic and unbelievable in the minds of many people in this generation.

What we experience in the physical world through our five basic senses is not total reality. Everything we think, experience, and feel in the material realm is influenced by the curse upon a fallen, three-dimensional world. The attributes of this present age do not even come close to reflecting the quality, beauty and perfection that the earth once had when God declared that his creative handiwork as "very good"! We are born, we live out our days, but in the end, we will all face disease, pain, and loneliness followed by certain death. This is the nature of life in the present world. Evolution has failed to deliver on its grandiose prediction of a more advanced, stronger, and more disease-resistant humanity. Many people have been duped into believing that this pain-ridden world is all there is. So, the best they can hope for, in the materialistic view, is to convince themselves that there is no greater power to which we must give account when their mortal lives have ended. Many of the elite within modern culture have denied divine authority. By denying any possibility of the supernatural, people struggle to make sense out of lives without purpose and meaning. The gospel of Jesus Christ is commonly rejected when people embrace evolutionary thought. What we know and experience today is only an intermission in the reality that began with a perfect creation and will continue when the same benevolent creator completes his plan and purpose for the ages.

By believing in the supernatural and submitting our lives to the Word of God, we are able to comprehend the deep mysteries of life that have baffled the world's greatest analytical minds for thousands of years. As we examine the scriptures, we begin to understand why there is life and death, love and hate, war and peace, marriage and divorce, truth and falsehood, heaven and hell. God intended for his Word to be an instruction manual to guide us through life. With this said, the vast majority of humanity either knows or suspects that the Holy Bible is a true and reliable source of divinely inspired revelation. Even though many people seldom open their Bible, most suspect that it contains wisdom and provides understanding for the most difficult challenges we face. The grueling problems mankind has created for himself have resulted from an unwillingness to submit our lives to God's truth, which represents the authority of the creator over his creation. When we reject his Word, we find ourselves in rebellion against that authority.

The trustworthiness and authenticity of the biblical text is not the central theme of *The Mystery of Life* project, and this topic would certainly require many bound volumes of its own in order to adequately address this crucial subject. However, because the biblical narrative is paramount to the primary theme of this publication, I feel obligated to address this issue at least to some small degree in order to give the reader confidence that the Bible can be trusted. Therefore, in this epilogue section I will provide many reasons why we can trust the Bible and confidently believe it is a reliable revelation of God's wisdom and will for his image-bearing sons and daughters.

In nature, we see the brilliant handiwork of a master design engineer. Our instinct and personal conscience reveal that we were created as moral beings that are accountable to an authority much greater than ourselves. However, these avenues of understanding can only give us a general revelation of who God is and what He is like. Even though the primary evidence of a benevolent creator is very impressive, it does not provide us with detailed knowledge about the personal nature of God and why He has placed such high value on human life. It tells us nothing about his love, grace, mercy, and forgiveness, for example. We need a more specific revelation from above. If God is real, and desires a relationship with man, then it is entirely up to him to communicate his word and his will to us. Therefore, the question is, has God taken the initiative to communicate divine truth to mankind? Is the Bible a revelation of divine origin? Can we trust the scriptures as the complete, authoritative Word of God? What evidence supports the bible as authentic and reliable, as a communication from the sovereign creator of heaven and earth?

God is a personal being who desires a relationship with all those whom he lovingly created in his own image. We need greater knowledge of the sovereign Lord in order to relate to him on a personal level. Man was designed for a relationship with God and his fellow man. God gave us the bible to understand who we are, and how to live the right way. We need to understand the character of the one who created us, and how to relate to Him. Without the Bible, we would never understand what it means to *know God.* Because you and I were created to experience personal relationship, it has always been God's plan to have lasting friendship with man.

The manufacturer of any complicated devise will provide an instruction manual to explain how their product works and how to troubleshoot if problems arise. The Holy Bible is our instruction manual for life. It reveals in detail who God is, why we are here, how we can find purpose for living, and what will happen when we die. The most challenging and difficult questions man has struggled with are addressed in God's revelation. Yet people often ignore its truth and go on living troubled lives without real meaning and without purpose. Why not just read the instruction manual so that we can live out our lives the way they were meant to be lived? In considering *the mystery of life*, we must start where the Bible starts, with creation. The value, dignity, and great worth God places on human life is deeply rooted in the fact that we were created in his image. God has chosen humans to represent his character on the earth. This high-calling and privilege makes sin all the more tragic as separation from God, through disobedience prevents man from fulfilling his intended creative purpose. Earlier in the book, I mentioned that evolutionary thought undermines the message of the cross. Let me now explain more clearly what I meant by that. Satan has attempted to convince humankind that random chance events are the only realistic way to explain why we exist. This is an all-out assault on the fundamental Christian doctrine of creation. Satan has targeted Genesis because it is the foundational truth upon which all other Christian theology rests. The rational goes something like this: If there is no creation by intelligent design, then there is no creator. If there is no creator, then there is no sin-induced separation from the creator by mankind. If there is no separation from a supreme transcendent being greater than ourselves, then there is no need for a savior to rescue mankind from his fallen condition. And if there is no human need for a redeeming savior, then the message of the cross of Jesus Christ amounts to pure nonsense in the minds of those who have embraced a humanistic worldview.

In her popular book *Total Truth*, Nancy Pearcey explains it like this: "It is because humans are the masterpiece of God's creation that the destructiveness of sin produces such horror and sorrow."[1] Only when man comprehends that sin has separated him from his Creator, will he begin to recognize the need for a savior. Because we have become the sons and daughters of God by creation initially, you and I were designed to reflect the character of God in the physical dimension, and to give back to Him the glory that He alone deserves. To so many Americans living in a secularized, hedonistic culture, the concept of sin makes no sense whatsoever. This is why the deceitful notion of evolution theory must be thoroughly exposed for the satanic lie that it is. The pathway to redemption, opened wide by the death and resurrection of Jesus Christ, restores men and women to their original high-calling as sons and daughters of God. Only then will we discover our true identity and fully appreciate the immense value that God places on our lives. As the image of God is fully renewed in us, we can once again reflect God's character on the earth as his crowning creation. Because we have all sinned against a Holy God, we need instruction about what God has done for us and how man is to be reconciled to Him. This is why we have the Bible. It is all the instruction we need to relate to God across the unsurmountable divide between the physical and spiritual worlds.

Because evolutionary thought is so widespread in American culture and has infiltrated every institution of modern society, many people see no need for the Bible. They simply do not believe in a personal God that every person will answer to. The devil typically does his conniving work through pure deception. By twisting and distorting the truth, he controls the masses in an all-out effort to thwart the plan of God to save mankind. I encourage you to *examine the evidence* and decide for yourself if the Bible we hold in our hands can be trusted. What evidence was left for us to solve this great mystery? If God is all-powerful and all-knowing then, perhaps, he would leave his divine signature upon his written revelation to humanity. With eternal lives at stake, a personal loving God would surely leave little room for doubt that divine intelligence was at work in recording the sacred Scriptures. The truth about biblical inspiration is the final battleground of engagement between the church and the world. The Christian faith will stand or fall upon the question of the divine inspiration and authority of the Bible.

How did the Bible get from the mind of God to the heart of man?

How did we get the bible? We must determine for ourselves if the Bible is the authentic Word of God. The veracity of scripture is paramount in our search for meaning, truth, and understanding. If the Bible is the inerrant Word of God, its message is surely of eternal value.

First of all, God chose to use man to reveal his thoughts, will, and intentions to those with a willingness to hear them, and a heart that was open to receive them. The truth of scripture was delivered from the mind of God, by way of the Holy Spirit, to the heart of certain men whom God had chosen. This is Revelation! In Ezra 1:1, we read how God directed man: “In the first year of Cyrus king of Persia, in order to fulfill the word of the Lord spoken by Jeremiah, *the Lord moved the heart* of Cyrus king of Persia to make a proclamation throughout his realm and put it in writing." And in Jeremiah 1:9 we read: "Then the Lord put forth His hand and touched my mouth, and the Lord said to me: behold I have put my words in your mouth." Another example is found in 2 Peter 1:20,21 which explains that "above all, you must understand that no prophecy of Scripture came about by the prophet's own interpretation. For prophecy never had its origin in the will of man, but men spoke from God as they were *carried along by the Holy Spirit*."

Secondly, God called these specific individuals by divine appointment to document in writing the revelation he had given them for the benefit of all humankind. We call this Inspiration!

And, finally, the record of God's revelation in written form was preserved and safe-guarded by a chosen nation of people, while the divine message was made available to the multitudes. As mankind would hear, learn, and understand the heart of God through his written revelation, the Holy Spirit allows the message to penetrate the heart of man and transform him from within. We call this Illumination!

There are a number of supernatural characteristics that have been recognized in the scriptures that provide all the proof we need to verify the authenticity of the bible as God's eternal truth. We will take a look at these supernatural elements that sets the Bible apart from all other books.

Supernatural Element 1: God's instrument of revelation

God chose a unique nation of people through whom he would directly reveal his thoughts, words, and actions. This imperfect nation was entrusted with the responsibility of recording God's Word, and were commissioned with the awesome task of protecting, preserving, and accurately passing on this sacred narrative to each generation until the whole world could hear the word and will of God. In chapter 8, we considered the miraculous beginning and survival of the Jewish people whom God set apart to be an instrument of divine revelation to the whole earth. Through one miracle after another God established this unique nation, commissioning them to receive his divine oracles and, ultimately, they would introduce the promised Messiah into the world. Without God's miracles, there would be no nation of Israel, and without Israel we would have no Bible and no Messiah.

Supernatural Element 2: Unique origin-amazing continuity

The Bible is a unified message system with '*redemption*' as its central theme. As you will see, God left no possibility for collusion between the writers of Scripture. The Bible was written in three languages— Hebrew, Aramaic, and Greek. It was recorded on whatever writing material was available at the time, such as clay tablets, stone, papyrus, parchment, leather, and even certain metals. Its forty-four separate writers came from very diverse backgrounds—they were kings, fisherman, shepherds, a statesman, a cupbearer, a tax collector, a physician, and many prophets of Israel, for example. They wrote in different centuries, at diverse locations, on three continents with little or no knowledge of one another, spanning 1,600 years and some forty generations of human history. How is it possible that forty-four isolated writers could maintain a unifying theme over a period of 1,600 years? This could only be done by divine supervision with inspired writers revealing God's truth as it was illuminated to their hearts and minds. With divine oversight, Providence insured that this combined volume of sixty-six books would form a unified message system—a fully interwoven theme that completes one continuous revelation of God's eternal truth. No other religious writings come remotely close to the unique circumstances God used to record his revelation to man. The Koran was inspired by one person— Mohamed. The religious works of Buddha were fashioned by one man. The Book of Mormon was compiled and published by one man— Joseph Smith. We would certainly expect religious books that come from solitary authorship to be consistent in the theme they present, but God chose many men under extremely diverse circumstances to deliver a unified message to mankind in a manner that defies collusion and clearly requires an explanation far beyond human intellect.

With unique characteristics, God himself insured that the continuity and thematic consistency found from Genesis to Revelation could only point to one thing: divine origin. The Bible speaks with amazing unity, conformity, and exceptional harmony on the most controversial topics from cover to cover. Moses recorded his contribution around 1500 BC, and the last writer of the Old Testament, Malachi, recorded his revelation about 400 BC. This left a wide gap between the last writer of the Old Testament and the first writer of the New Testament of somewhere between four and five hundred years, leaving a revelation time lapse between the Old Testament and the appearance of Messiah, who was the primary subject of their prophecies. This was also part of God's plan. This gap in time insured there could be no collusion between the prophets of Old Testament revelation and the appearance of Messiah. The revelation of God's eternal Word was recorded beginning around 1492 BC and was finally completed with the last book of New Testament around AD 80 or slightly before. These eternal words have revealed to us the conflict between God and Satan over the allegiance of mankind. Perhaps the most amazing thing about the Bible is that the contribution of each writer dove-tails cohesively with all others, forming one continuous revelation. It is not a heterogeneous jumble of myths, legends, and superstitions as many atheists would have us believe, but a harmonious collection of knowledge and details that obviously originated from a single divine source. There is no other way to explain its unique origin than to acknowledge that the Holy Spirit has orchestrated all of it.

Considering the Bible addresses the most sensitive and controversial topics from Genesis to Revelation with harmony and continuity throughout, gives us our initial evidence that the text was the work of God and not that of man. Imagine, if you will, the jumbled mess of conflicting thoughts and opinions that would appear in a blended book narrative written under similar circumstances today by a select group of widely diverse writers working independently of one another. Let us say, for example, individual writers were assigned to create a collection of family resource material addressing the most challenging moral and social issues of our time. Does that objective not seem like an overwhelming task? And yet our unique biblical collection of sixty-six books compiled by forty-four writers over 1,600 years focuses on virtually every moral issue of our time with astonishing harmony from cover to cover as the work of each individual writer reinforces the message of the others and the theme of the Old Testament is dramatically fulfilled in the New. It becomes very apparent then that man could never have compiled such a large volume of related books with the unifying qualities of the Bible by human effort alone.

Supernatural Element 3: Divine perspective of the Bible

The Bible is a progressive revelation that begins with the creation of heaven and earth, and ends with the creation of a new heaven and new earth (Revelation 21:1). The Bible gives us God's description of the origin of the universe and reveals to us how life began. It provides a detailed account of the nature of God, the fall of man, the way of redemption and God's plan and purpose for the ages. The New Testament was never meant to replace the Old. It was meant to offer a fulfillment of the underlying theme of the ancient Hebrew text.

In many cases, the New Testament is needed to fully understand the Old. For example, we cannot grasp Leviticus without Hebrews, or Daniel without Revelation. Also, Isaiah 53 would be difficult to understand without the gospels of Matthew, Mark, Luke, and John. But if the Bible is really from the mind of God, then we could expect the scriptures to reveal God's point of view in the narrative. Is the Bible written from the standpoint of God speaking to man? If the scriptures were merely the result of human achievement, as the skeptics claim, it is quite unlikely that the Bible would reflect God's perspective. Phrases like "Thus saith the Lord" and "The Word of the Lord came unto me saying" have appeared over two thousand times in the Old Testament (see Jeremiah 1:9). Also, keep in mind that the Old Testament refers to itself as being the Word of God some 3,800 times, and the New Testament writers, including Jesus, refer to Old Testament events at least 160 times by quoting directly from more than 240 Old Testament passages. This is the type of narrative we would expect to find in sacred writings that originated from the mind of God. 2 Timothy 3:16 tells us that "all scripture is God-breathed and is useful for teaching, for rebuking, for correcting and for training in righteousness." God chose and divinely directed certain individuals to receive and record divine laws, historical facts, poetry, and the prophetic warnings that he intended for human knowledge. The divine providence of God insured that His message to man, although written by human hands, would reflect divine truth, and could be trusted as reliable. The writers of scripture used their personal faculties, and wrote in their own style; yet God allowed them to express his words without distorting the divine message.

We must keep in mind that words recorded by human intellect are always limited to the author's personal knowledge, understanding, and unique experience. However, from what we read in the Bible, it is apparent that the writers of Scripture did not always understand the significance of what they were writing. This is strong evidence that the Bible is not the result of human thought and planning. "This salvation was something the prophets did not fully understand. Though they wrote about it, they had many questions as to what it could all mean. They wondered what the spirit of Christ within them was talking about, for he told them to write down the events which, since then, have happened to Christ: His suffering and his great glory afterwards. And they wondered when and to whom all this would happen" (1 Peter 1:10,11 TLB). The fact that God revealed his word to men and instructed them to record it is clearly seen in the words of the prophet Jeremiah: "In the fourth year of Jehoiakim, son of Josiah, king of Judah, this word came to Jeremiah from the Lord: "Take a scroll and write on it all the words I have spoken to you concerning Israel, Judah and all the other nations from the time I began speaking to you in the reign of Josiah till now" (Jeremiah 36:1, 2).

God was so serious about maintaining the accuracy of his revelation that he left a stern warning to anyone who would add or take away from his divine message. Scripture records this divine warning at least three separate times. "You shall not add to the word which I command you, or take from it, that you may keep the commandments of the Lord your God which I command you" (Deuteronomy 4:2). "Every word of God is pure; He is a shield to those who put their trust in Him. Do not add to His words, lest He rebuke you, and you be found a liar" (Proverbs 30:5, 6). "For I testify to everyone who hears the words of the prophecy of this book; If anyone adds to these things,

God will add to him the plagues that are written in this book ..." (Revelation 22:18).

Man is incomplete and unfulfilled without a personal relationship with the God who created him. If the Bible is the authoritative Word of God, it is essential that we feed upon its divine truth if we hope to understand what life is all about. "But He answered and said, it is written, 'Man shall not live by bread alone, but by every word the proceeds from the mouth of God." (Matthew 4:4)

Supernatural Element 4: Impartial content of the Bible

People often wonder why God favors Israel over other nations. Is this true? Definitely not! God chose Israel and he used them in a remarkable way to accomplish his divine purpose. But he loves all people the same— all were created in his image. A closer look at the Bible text reveals that God was completely impartial when he instructed the recording of details concerning the history of both nations and individuals. For example, the written history of the Jewish nation recorded in scripture is one of stubborn rebellion, idolatry, and deceitfulness. Now, since all the writers of the Old Testament were of Hebrew ancestry, it does not make sense that they would record their national history as a despicable record of ingratitude, idolatry, and unbelief. This naked honesty could only have been inspired by the sovereign mind of an impartial God who favors no living soul over another. The Bible speaks with impartiality and a total absence of favoritism when it comes to the heart of God for humanity. The biblical narrative provides a detailed and unmerciful account of the personal sins of those who wrote it. For example, the sins and personal failures of some of the greatest men in world history are clearly exposed for the entire world to know of. The worst sins of Adam, Abraham, Isaac, Jacob, Moses, David, Solomon, Hezekiah, Elijah, Peter, and Paul, for example, are exposed; yet through repentance and the merciful forgiveness of Providence, these men accomplished more for God in their lifetimes than perhaps anyone in world history. It does not make sense that the writers of the Scriptures would emphasize their own personal sins and failures, as well as their national sin, if they were writing from their own individual thoughts and perspective.

Supernatural Element 5: All-inclusive nature of the Bible

The Bible starts at the beginning with the creation of heaven and earth, and continues through nine human dispensations in which God is dealing with humanity. This includes the final act when God recreates the heavens and the earth in preparation for the perfect age in which the redeemed of God will enjoy the presence of their Maker uninterrupted forever.

Human dispensations include the following from the perspective of God's dealing with man.

1. "The Edenic Dispensation" the *Age of Innocence* that ends with the fall of man.

2. "The Antediluvian Dispensation" the *Age of Conscience* prior to the great flood.

3. "The Postdiluvian Dispensation" the *Age of Human Government* after the flood.

4. "The Patriarchal Dispensation" *the Age of God's covenant promises* to the patriarchs.

5. "The Legal Dispensation" *the Age of Divine Law* given through God's servant Moses.

6. "The Ecclesiastical Dispensation" the *Age of God's Grace* also called the 'Church Age'.

7. "The Tribulation Dispensation" *the Age of God's Wrath* poured out on the earth.

8. "The Messianic Dispensation" *the Kingdom Age*, known as the 'Millennial Reign of Christ'.

9. "The Dispensation of the Fullness of Times" *the Perfect Age* that will never end.

Supernatural Element 6: The remarkable transmission

Let us consider for a moment whether the Bible, as we know it today, reflects the same thought and intent of the original autographs.

Considering the turbulent history of the Hebrew nation, to whom the sacred oracles of God were entrusted, it seems as though the Bible would have had little chance to survive with its original language intact. Most original manuscripts of the ancient Hebrew text have not survived to the present day for a couple of reasons that we will discuss in a moment. However, this is not the case with the New Testament documents. We still retain within church archives 24,633 original handwritten copies of either all or part of the New Testament, which survived from antiquity. By comparison, the next closest surviving literature from ancient times, *Homer's Iliad*, can claim only 643 handwritten copies. With a surplus of ancient NT copies on hand, scholars can easily compare and contrast the textual purity of contemporary translations, of the same narrative, that are currently in wide circulation. The sacred documents forming the New Testament had to be copied and recopied for centuries until the advent of the printing press. After centuries of labor-intense copying, one might expect the Bible to contain many errors and omissions, resulting in a serious loss of its original meaning and vitality. However, with so many surviving manuscripts with which to compare current translations, the textual purity of the New Testament today has been firmly established and this recognized inerrancy of the sacred Scriptures from ancient time until the present day cannot be denied. Thus, the abundance of manuscript evidence attesting to the accuracy of current New Testament documents makes our modern translations very reliable renderings of the original language.

The case of the Old Testament transmission of the Holy Scriptures presents a quite different story altogether. The reason we do not have many copies of the ancient Hebrew text today is primarily two-fold. First of all, these documents were copied onto perishable material such as parchments, papyri and animal skins, which were terribly inferior to the reliable writing material we possess today. This would certainly account, in part, for a lack of preservation through the centuries. However, a more resounding reason for the significant lack of original OT copies is found in the way these documents were handed down from ancient time. In fact, in the case of the Old Testament, the lack of original copies is actually a testament to the accuracy with which these manuscripts were transmitted to future generations. The explanation for this is found in the extreme diligence and meticulous care with which the Old Testament documents were copied and passed on. Since ancient documents could only be preserved by handwritten copy, the accuracy of textual transmission was solely dependent upon the special care with which these ancient manuscripts were handled. Because early writing materials were constantly subject to decomposition, a special class of workers was employed to constantly replace the worn and dilapidated copies with newer ones. Within Hebrew culture the extreme care used to preserve the sacred writings of the Old Testament was such an important part of their national pride and cultural identity that an entire class of men were trained, whose sole duty was to accurately preserve and to transmit the Holy Scriptures with the most precise exactitude humanly possible. This class of professionals are identified as the Talmudic and Masoretic scribes. In the copying process, scribes were pledged to follow twenty separate grammatical requirements in completing their work to ensure the precise accuracy of textual transmission. For example, the number of lines, words per line and letters per line were counted to match the original. This is only a fraction of the many detailed tests that each newly copied manuscript was subjected to. At the completion of each copy, if more than three misplaced strokes of the pen were found, the copy was immediately destroyed. The primary reason that we have so few handwritten copies of the ancient Hebrew Scriptures is because these scribes did not keep the worn-out copies from which the new documents were produced. The reason older documents were destroyed or burned after copying was completed is because of the scribe's commitment to textual purity. They could not allow any worn-out copies to remain in circulation that might be misread at some point resulting from blurred and indistinct lettering. This could possibly lead to miss-copied documents over time. These dedicated scribes were determined that not the least stroke of the pen would be lost from the original narrative.

God chose the Hebrew nation, not only to record his sacred Word, but he also entrusted and divinely charged them with the responsibility of protecting, safe-guarding, and preserving the accuracy of his divine message. The question is, how well did these scribes do at fulfilling their great commission to pass along divinely accurate manuscripts to future generations without compromising the original language? The accuracy of hand-copied Hebrew documents was in serious doubt by many scholars until the accidental discovery of a treasure trove of ancient scrolls in caves near the Dead Sea by a Bedouin shepherd boy in 1947. While searching for a lost goat, a young shepherd boy tossed a rock into a cave on the west side of the Dead Sea and heard a sound of breaking pottery. Upon closer inspection, he discovered some large ancient-looking jars containing leather scrolls carefully wrapped in linen cloth. The jars were sealed in a way that would protect the scrolls from decay for a very long time. As these ancient manuscripts were diligently studied and carefully photographed it became apparent to Hebrew scholars that this find was the most significant archeological discovery in the Holy Land in modern times.

The Dead Sea scrolls have provided scholars with an extremely rare opportunity of comparing ancient hand-written documents with the oldest Hebrew manuscripts available to them at that time. This fascinating discovery would surely help scholars determine the accuracy of textual transmission from ancient times to the present. These scrolls contained fragments from every book of the Old Testament, except the book of Esther. Also included were some extra-biblical works of great worth and a few valuable biblical commentaries as well. Through the providence of God, these rare documents had survived totally undisturbed for nearly 2,000 years. Up until this amazing discovery the oldest Old Testament manuscript in existence was dated at around AD 900. The Dead Sea scrolls have afforded textual scholars the unique opportunity to compare these AD 900 documents with their ancient counterparts that have been dated back as far as 125 BC. With this momentous opportunity scholars were able to test the accuracy of textual transmission of the Hebrew Old Testament since before the time of Christ. The transmission accuracy of the Old Testament could now be examined as never before in human history. One of the ancient leather scrolls, dated at 125 BC, included a complete manuscript of the Hebrew book of Isaiah, one of the most beloved books of the Old Testament cannon. When this book was compared to the Isaiah manuscript dated at AD 916, scholars were amazed at the results. Even though over 1,000 years had separated the two documents, they were virtually identical when compared word for word. For example, of the 166 words comprising the fifty-third chapter, only one single word was in question— a word that did not affect the meaning of the original language. The Dead Sea scrolls have erased any lingering doubt as to whether or not the Old Testament of today can be considered accurate when compared to documents dating before the time of Christ. Outside of some insignificant variation in spelling and grammar, we can be certain that today's modern copies of the biblical text do not differ substantially from the content of the original narrative. By divine providence, a sovereign eternal God has clearly protected and preserved his revelation to humanity for the beneficence of all generations. Of those who deny the trustworthiness of the Bible and are skeptical about its reliability today, perhaps they simply do not want it to be a truthful revelation of God's heart for mankind.

Supernatural Element 7: Miraculous survival of the Bible

Even though the works of many non-biblical writers lend full support to the trustworthiness of the Scriptures, many bold attempts have been made to discredit the authority of the Bible. Through the modern view of textual criticism, many liberal-minded theologians and biblical critics have attempted to undermine the veracity of the Scriptures. They question the authorship and the dates of the biblical writings, as well as many historical details– wanting to discredit its written content in an effort to cast doubt on the reality of the people, places, and events it describes. Historically, the New Testament gospels are primarily a collection of first-hand accounts written by men who witnessed the events they wrote about and received first-hand the powerful words of Jesus through intimate personal contact. Needless to say, the Bible has survived the attacks of the higher critics, becoming more thoroughly vindicated today than ever before in history. The Bible has faced more criticism than any other known work of literature. It has been defamed, burned, banned, and outlawed from the early days of the Roman Empire up until the recent attacks of secular liberalism, communism, and a radical element within modern-day Islam. For example, in AD 303 the wicked Roman emperor, Diocletian, had proclaimed in an edict that all Christians and their sacred book were to be destroyed. He sent Roman soldiers throughout the empire burning churches to the ground and destroyed all the sacred Scriptures they could find. However, this tyrannical edict certainly did not last long, for it was only twenty-five years later that Constantine succeeded this evil despot and commissioned copies of the Holy Scriptures be printed at government expense, while declaring Christianity the official religion of the Roman Empire. During these dark ages of religious persecution Christian monks were forced to sew copies of the Scriptures into the lining of their clothing to avoid prosecution as they moved from place to place sharing the hope of God's love. In the 1700s, an infidel Frenchman by the name of Voltaire became a bitter enemy of Christianity and the Bible. As he began to persecute and rail against the church and the sacred Scriptures, Voltaire vowed that Christianity and the Bible would be eradicated from the earth by the end of the nineteenth century. However, as divine providence would have it, some fifty years after he died in 1778, the house of Voltaire became occupied by the Geneva Bible Society and was transformed into a Bible printing and distribution center. Ironically, they used the infidel's own printing press to produce copy after copy of the sacred Scriptures for wide distribution. Men like Adolf Hitler, Joseph Stalin and many others have tried to rid the world of the church and the Bible, but this persecution only made the Scriptures more valued and treasured. Evil men have sought to destroy God's Word, and to persecute those who profess Christianity; but every attempt has failed while the Bible has endured to become influential worldwide despite the efforts of a few infidels. The enemies of truth pound away until their hammers break, but God's anvil still stands firm. Satan would not go to such great lengths to destroy the Holy Bible unless it posed a serious threat to his control over the hearts of men. If the Bible is not God's Word, it would have been forgotten long ago.

The Bible is the most despised book ever written, as hatred for the Word of God runs deep in modern cultures where many people wish to rid themselves of burdensome moral constraints. It is illegal, in some countries, to own, sell, or distribute the Bible. Should we be surprised that the devil does not want people to understand and obey the inspired Scriptures. The Holy Bible has inspired great culture through music, art, literature, and architecture. Without biblical influence Western culture would not know the greatest advancement, and achievement in the history of the world. The Bible in our lap is the only physical thing on earth that is eternal in nature. It has survived persecution and despotism. And it will remain after the world we know passes into history. "The grass withers, the flower fades, but the word of our God stands forever" (Isaiah 40:8). "Heaven and earth shall pass away, but my words will by no means pass away" (Jesus– Matthew 24:35). The Bible is the most intimate possible connection between heaven and earth.

Supernatural Element 8: Universal circulation & influence

No other book in the history of mankind can compare with the Bible in terms of its universal influence and circulation throughout the world. The Bible is by far the most dearly beloved book of all time. As the world's all-time best seller, it is the most widely read and the most extensively translated literature in all human history, having a worldwide circulation that is now approaching one copy for every person on the planet. The Bible has received more public attention and has been quoted more often than any other work of literature, by far. The Bible was the first book to come off a printing press, and the first book to be translated from one language into another. By 1979, the Bible had been translated into more than 1,500 known languages and dialects. At that time, the translation rate included 95 percent of all human languages. Today there are very few languages, dialects, and remote tribes that do not have at least some of the Bible available to them in a form they can understand. It is important to note that the Bible can be translated into any language without losing its vitality and spiritual power. The message of the Bible has so inspired the thinking of man that the number of books and commentaries written in response has been equally impressive. And the profound influence the Bible has had upon the civil affairs of mankind has been nothing short of astonishing. This is, of course, what we would expect if it was a divinely inspired message of hope and promise that originated from the mind of God.

Supernatural Element 9: Textual harmony with science

The Bible has proven to be scientifically accurate in its entire content. This supernatural quality of the Scriptures is powerful evidence that God's hand has directed the writers he chose. They recorded great knowledge and information that was beyond the understanding of mankind at the time— details only an omniscient God would have known. For many years, claims were made that the text of scripture was in direct conflict with what we understood about science. As it turns out, the problem was not with the Bible but man's lack of knowledge about the physical world. Mankind's curious discoveries had not yet caught up to the Bible. For example, for a very long time, men believed the earth was flat, when all along, the Bible said it was circular (see Isaiah 40:22). Under the influence of Aristotle, it was believed that the sun rotated around the earth. This was the prevailing view of the day until about the sixteenth century, when Kepler and Galileo put an end to that outdated nonsense. Johannes Kepler was a strong believer in God and the Bible. He once declared that "science is simply thinking God's thoughts after him."[2] Mankind did not even know that blood circulates throughout the body until around 1650 when William Harvey discovered the circulatory system. During the Middle Ages, the Bubonic plague ravaged Europe killing one-fourth of the entire population. This death plague occurred because mankind did not adequately understand germs, contagious disease, and quarantine. People had failed to appreciate the importance of cleanliness and the appropriate washing in running water. Victims of the plague were not even isolated from healthy individuals, so others quickly became infected. But the book of Leviticus chapter 13, written several thousand years before these most tragic circumstances unfolded was packed with knowledge that could only have come from a personal loving God, and would have saved millions of lives then, as it does today. The wise-men of ancient Egypt believed that the earth was held up by five pillars. This was the prevailing view in that culture, during the Egyptian dynasty. However, they did not read Job 26:7. For centuries it was believed that there were only a few more than one thousand stars in the universe. Now we know there are an infinite number of stars that cannot even be counted. They did not read Jeremiah 33:22. For hundreds of years, it was thought that bleeding a sick person would improve health when people became gravely ill. It was called bloodletting. This procedure is what led to the death of George Washington. Leviticus 17:11, however, tells us that the life of every creature is in its blood. How could Moses possibly have known that if it had not been revealed to him by the Creator? Now, we give people blood transfusions to save lives.

The Bible records only the truth. It is pure nonsense to say the Bible is scientifically inaccurate. People who say this know little about science, and even less about the Bible. Today, we know that every known fact of science mankind has discovered to date is in complete harmony with Scripture. God established the laws of nature to rule over his creation (Psalm 148). What can we take away from this? The harmony between science and Scripture reveals that only God would have known these facts before mankind confirmed the truthfulness of scientific revelation.

Supernatural Element 10:
Historical accuracy of the Bible

Among other things, the Bible is a work of historical literature. Because of this, we can now compare the biblical record with many known facts of history obtained through scientific and historical research and investigation. If the Bible is truly the Word of God, we would expect it to be very accurate in every recorded detail of history. The following material provides a small sampling of the many names, places, and historical details mentioned in the biblical narrative that have been clearly identified through historical research. This includes non-biblical sources like ancient inscriptions, historical manuscripts, and engraved tablets, pots, and monuments. For more than a century and a half, scholars from public and private universities the world over have led research teams on archeological expeditions throughout the biblical Middle East. The fascinating discoveries these scholars have uncovered proves far beyond doubt that the Bible is reliable and accurate in its historical content whenever it has been put to the test.

Many attempts have been made to discredit the veracity of the Bible's historical record. One by one, these erroneous claims have been debunked in the light of biblical archeology. Are the historical figures, events, and places recorded in Scripture to be accepted as real history in a literal sense? When Jesus was entering the city of Jerusalem on the back of a donkey, as recorded in Luke 19: 36-40, the people began to shout their praise. The pious Pharisees were indignant, and told him to silence the crowd by rebuking his disciples. Jesus gave this response: "If these people should hold their tongues, the stones will immediately cry out." Well, the stones have cried out, but not immediately, yet their voice has declared a resounding cry throughout the world for over two centuries. The facts uncovered by the scientific discipline of archeology in the last couple of centuries are nothing short of astonishing. Many prominent authorities agree that historical research has added convincing support to the biblical account of history. Let us consider here just a few biblical references receiving archeological and historical confirmation.

Some Bible critics have in fact attempted to discount the life and journeys of the Old Testament patriarch Abraham. However, extensive recent excavation at Ur of the Chaldees has uncovered positive proof of his life and activity. Apparently, Abraham was well-known throughout ancient Mesopotamia as a record of his activity was also discovered in the excavation of Babylonia as well. Other discoveries have helped man to confirm the migration of Abraham from Ur to Canaan as recorded in the book of Genesis. Other excavation projects have confirmed the historical record of many kings and rulers associated with the biblical narrative. Particularly impressive is the confirmation of the reign of *King Solomon*. The Old Testament record of his extensive building projects has been validated by an abundance archeological evidence. For example, according to 1 Kings 10:26, we are told that King Solomon commanded a staggering number of horses and chariots. The modern excavation of the ancient fortified city of Megiddo has resulted in the positive identification of Solomon's stables that were extensive enough to accommodate an armory of horses and chariots. The excavation of another location under King Solomon's reign revealed an ancient blast furnace, which undoubtedly has a positive connection with the record in 2 Kings that mentions Solomon's casting of metals. Also, regarding Solomon, Middle-eastern archeology expert Dr. Nelson Glueck has explained that "the whereabouts of Solomon's lost city of Ezion-Geber was for centuries an unfathomable mystery, because no one paid attention to the biblical statement that it was located 'beside Eloth, on the shore of the Red Sea, in the land of Edom'"[3] (1 Kings 9:26). The city was found right where the Bible said it was.

On December 18, 1995, a featured article of *Time* magazine was a narrative entitled "Is the Bible Fact or Fiction?" This article, from a popular publisher that typically leans toward the liberal view of humanistic evolution, presents the Bible as containing "some truth." Although the writer acknowledged some archeological discoveries that validate the historical accuracy of Scripture, it left the reader with a feeling of skepticism by suggesting that unless a biblical event can be proven through historical research, it cannot be considered trusted and reliable information.

Hebrews 11:6 tells us that without faith, it is impossible to please God. We must reject the idea that we should only believe what we can conclusively prove. None of us live that way. We all exercise a degree of faith in what we do and why we do it. You and I will only get on a plane if we have good reason to believe that the pilots will keep us safe; and we would certainly never jump out of an airplane unless we had the faith to believe that whoever packed our parachute knew exactly what they were doing. Therefore, when we look at a topic like the veracity of Scripture, we are looking for evidence pointing us toward belief that is "beyond reasonable doubt." If the Creator is the sovereign Lord of the Bible, then we cannot discount his ability to preserve the integrity of Scripture. With this said, let us take a look at some more historical facts, and determine if the mounting evidence favors the divine origin of the Scriptures.

The archeological evidence supporting the historicity of the Bible is so vast in terms of volume of documentation that we can go all the way back to the Garden of Eden to begin our overview of material that has been uncovered by historical research. The following information can be found in Haley's Bible handbook between pages 68 and 90.

The Temptation Seal-Found in Babylonia: this engraved relief depicts the temptation of Eve by the serpent in the Garden of Eden.

The Adam and Eve Seal– From Assyrian archives: this ancient artifact depicts the sorrow and dejection of Adam and Eve (walking with heads down) after their devastating fall to temptation and subsequently being forced out of the Garden of Eden.

The Weld Prism: As the oldest known outline of world history from around 2170 BC it lists eight rulers before the great flood and reveals knowledge of long-life spans during that time.

The Gilgamesh Seal: Archeological evidence abounds for a cataclysmic worldwide flood and can be found in the historical traditions of every major culture in early human history. This is especially true upon examination of the ancient Babylonian civilization as seen in the Epic of Gilgamesh revealed by the inscription text found on the Gilgamesh Seal.

Pre-flood cities uncovered: The excavation of pre-flood cities in Mesopotamia (Ur, Fara, Susa, and Kish among other sites) has revealed something very interesting. At many ancient sites in the Middle East, archeologists commonly find that one city was built on top of the ruins of a previous one. In 1929, a joint expedition commenced between the University of Pennsylvania and the British Museum under the direction of Dr. C. L. Woolley. This research team set out to complete a thorough search through the ruins of Ur, the original home of the biblical patriarch Abraham. They uncovered numerous layers of human occupation beginning near ground level, followed by an eight-foot-thick layer of water laid sediment containing no human artifacts at all. As they continued to dig down through the sediment layer, they eventually uncovered another distinct layer of human occupation. What they found below the sediment layer was a pre-flood civilization that had been buried by a great cataclysmic event. According to Dr. Woolley, "that eight feet of sediment implied a very great depth and long period of water; it could not have been left by any ordinary overflow of rivers, but only by some vast inundation of water such as the Biblical Flood."[4]

Further excavation of the general area also revealed that the eight feet of sediment was universal throughout the entire region, as confirmed in the uncovering of other pre-flood locations. Halley's Bible handbook reports the following information based upon the excavation of multiple pre-flood sites in the region of ancient Mesopotamia: "The fact of a vast flood covering the whole area of early civilization is established by the eight foot layer of silt which cuts through the 'culture levels of all the Euphrates Valley sites."[5] This fascinating revelation provides some very impressive evidence favoring the authenticity of the ancient biblical flood of Noah as recorded in Genesis 6-8. At the original pre-flood site of Ur, Woolley reported that some magnificent discoveries were made among the human artifacts uncovered there, indicating that early man possessed a very high level of intelligence, and artistic skill.

Tower of Babel Ruins: More than two dozen very early Ziggurat-type tower structures from ancient Mesopotamia have been excavated, including that of the Genesis "Tower of Babel", where, according to the Bible, God confused the speech of man, forcing him to scatter abroad across the face of the earth.

Ur of the Chaldees: As stated above, world-renowned archeologist C. L. Woolley excavated the Mesopotamian birthplace of the Bible patriarch Abraham beginning in 1922. He uncovered one of the best-known ancient sites of all time— a city that was at the height of its splendor when Abraham left to follow God's command leading him to Canaan and a new land of promise.

The Code of Hammurabi: Discovered at ancient Susa by a Frenchman in 1902, this very early code of inscribed laws identifies Hammurabi, the most famous of the early Babylonian kings, and it dates all the way back to the time of Abraham. This discovery is the oldest legal system in the world and demonstrates that early culture was already far advanced as a civilization.

Other archeological research has identified evidence confirming the following biblical events:

Abraham's sojourn into Egypt (Gen. 12), Abraham's battle with the Kings (Gen. 14), the divine destruction of Sodom and Gomorrah (Gen. 18-19), The events involving Joseph and Potiphar's wife in Egypt (Gen. 39), The seven years of famine in Egypt while Joseph ruled over the house of Pharaoh (Gen. 41), Israel's sojourn into Egypt to escape famine in Canaan (Exodus 1), the birth of Moses (Exodus 2), the death of Pharaoh's first-born (Exodus 12), Israel's Exodus from Egypt under Moses (Exodus 12), the defeat of Jericho (Joshua 6), the reign of King Saul (1Sam. 1), the prosperous reign of King David and his many conquests (2 Sam. 1), and, as mentioned, the reign of King Solomon, including his gold, stables, furnace, and naval fleet (1 Kings 7-14). Other sites include the building of Samaria, the rebuilding of Jericho, Shishak's invasion, Ahab's House of Ivory, Jezebel's cosmetic box, Assyrian captivity of the Northern ten tribes of Israel, Hezekiah's tunnel, Manasseh's reign, Queen Ester's palace, Babylonian captivity of Judah, the reign of Belshazzar, the fall of Babylon, the Edict of king Cyrus, and the repentance of Nineveh.

Again, and again independent sources have corroborated historical facts of the Old Testament text. According to Dr. Nelson

Gluck, the most eminent Jewish archeologist of the last century, over twenty-five thousand names and locations documented in the Old Testament have been identified. He made this statement in 1955. This number has certainly grown substantially over the years as new Bible facts are being confirmed throughout the Middle-east at an accelerated pace. Research scholars have never identified a single ancient artifact that contradicts or that discredits any statement recorded in the Scriptures. Dr. Nelson Gluck explains it this way: "It may be stated categorically that no archeological discovery has ever controverted a Biblical reference. Scores of archeological findings have been made which confirm in clear outline or in exact detail historical statements in the Bible. And by the same token, proper evaluation of Biblical descriptions has often led to amazing discoveries."[6] Glueck remained in the forefront of biblical research his entire lifetime. After an outstanding career of accumulating and categorizing first-hand knowledge from archeological research throughout the region of the Middle East, the world-famous archeologist documented his work in a book entitled *Rivers in the Desert*.

Another respected scholar, Dr. J. O. Kinnaman, said this about Middle Eastern archeology: "Of the hundreds of thousands of artifacts found by the archeologists, not one has ever been discovered that contradicts or denies one word, phrase, clause, or sentence of the Bible, but always confirms and verifies the facts of the biblical record."[7] As a brilliant language scholar and former professor of linguistics at Princeton Theological Seminary, Dr. Robert Dick Wilson had mastered forty-five different languages and dialects during his long career. Dr. Wilson was, by far, the most qualified Old Testament linguist of the twentieth century.

He made the following statement after a full career of studying the biblical text from a linguistic point of view: "After forty-five years of scholarly research in biblical textual studies and in language study, I have come now to the conviction that no man knows enough to assail the truthfulness of the Old Testament. When there is sufficient documentary evidence to make an investigation, the statement of the Bible, in the original text, has stood the test."[8]

As already mentioned, the Dead Sea caves have yielded a treasure trove of well-preserved ancient Bible manuscripts. And of the many caves discovered, cave #4 alone housed fifteen thousand fragments from 574 manuscripts, predating any previous copies of Old Testament documents by as much as one thousand years. These valuable manuscripts were believed to be hidden by a community of Jewish scribes called the Essenes, an isolated religious sect living and working at Qumran near the shores of the Dead Sea from 200 BC until well after the life of Jesus Christ played out in dramatic fashion on the ancient Jewish landscape. The Hebrew Old Testament preservation tradition of the Essene community is described in the writings of the ancient historians Josephus and Philo.

The March 5, 1990, issue of *Time* magazine published an article entitled "Score One for the Bible" in which documentation was reported on the research excavation of ancient Jericho in 1930 and 1936. Joshua 6:20 records that at the direction of God, the people gave a loud shout and the walls of the city fell flat— meaning they fell outward and down the hillside instead of collapsing inward upon themselves as was typically the case whenever fortified city walls were breached by battering rams in ancient times. By falling flat, this allowed the Jewish invasion force to proceed rapidly up over the rubble and into the city with minimal resistance to secure an easy victory. These two excavations proved historically that the walls of ancient Jericho fell flat, outward, and away from the city, exactly how the Bible so accurately described.

In 1846, historical explorer Austen Henry Layard discovered a black obelisk among the ruins of ancient Nimrud, located in what is now present-day Iraq. This exciting discovery recorded the military exploits of the biblical King Shalmaneser I, who historically chronicled his army's defeat of King Jehu of Israel. In another incident Hezekiah, King of Israel, had ordered emergency fortification of the walls of Jerusalem in order to repel an invasion from Sennacherib the King of Assyria in 701 BC. This event is recorded in 2 Chronicles 32:1, 2, and 5. After considerable damage and destruction from the famous six-day war of 1967, a team of researchers from Israel's Hebrew University were successful in locating the ancient ruins of this wall construction that King Hezekiah had ordered in urgent preparation for the expected Assyrian invasion.

Another nineteenth century archeological discovery found in Iraq yielded a clay cylinder bearing the decree of King Cyrus allowing the Jews to return to their homeland, ending the seventy-year captivity imposed by God for Jewish rebellion. The event was recorded in Ezra 1:1-3. This clay cylinder provided a clear confirmation of one of the most astonishing events recorded in the Scriptures. It is fascinating as well, concerning

Cyrus, how God used a pagan ruler to fulfill his divine purpose. I think we can all agree that only God knows the future before it happens. In the case of King Cyrus, God predicted, through the prophet Isaiah, the birth of this man, including an accurate foretelling of his very name and what was predetermined for him to accomplish nearly 150 years before he was born. King Cyrus was a very important person in Scripture, whom God used in a powerful way. His name is recorded some twenty-two times in the Bible. This is certainly fascinating proof that God spoke very clearly to the men he chose to record his divine revelation. King Cyrus is the man who founded the Persian Empire. In Isaiah 44: 24, 26-28, God declares himself the Creator and Redeemer and predicts very clearly that he will use a (yet to be born) man named Cyrus to restore the ruined city of Jerusalem. However, when this prophecy was spoken, Jerusalem was still a prospering and thriving city that would not even be defeated until more than one-hundred years later, by the Babylonian king Nebuchadnezzar. This historically confirmed fulfillment of Bible prophecy is proof that the God of the bible reigns.

The Roman historian Tacitus, known to be hostile toward Christianity, recorded in the Annals of Tacitus xv. 44. a clear record of the life of Jesus, Pontius Pilate, and Tiberius Ceasar that dates between AD 115-117. The Jewish Rabbinical writings of the Sanhedrin 43a, written between AD 70-200, records the event of Jesus' death on Passover eve as described in the New Testament.

King David's rule has been confirmed by the discovery of some remarkable engravings on a basalt monument unearthed at

Tel Dan in 1993. The fact that ancient Israelites were in bondage to Egypt, as the Bible records, was confirmed by a discovery in Thebes in 1895. The Merneptah Stele that was uncovered in Pharaoh Merneptah's tomb dates to the Jewish slavery in Egypt, and records the presence of the Israelites during his reign. Pharaoh Merneptah is believed by some to be the Pharaoh of the Exodus. In fact, an autopsy of his remains confirmed the cause of death was drowning. Could it also be that the multiple broken bones identified along with the drowning death conformation resulted from a deadly pursuit of the Israelites into the Red Sea, where Pharaoh's entire army perished according to Scripture?

Dr. Robert Dick Wilson's book *A Scientific Investigation of the Old Testament* documents the historical identification of twenty-nine kings. Included among them were some rulers over Egypt, Assyria, and Babylon. In the discovery of ancient monuments honoring their reign, every name transliterates as it appears in the Old Testament.

The *Moabite Stone*, also called the *Mesha Inscription*, which was discovered in 1868, dates to 830 BC. In 2 Kings 3, we read of ruler Mesha, a vassal king of the Moabites who paid tribute to Israel. This monument records some of his extensive military campaigns. Still other inscriptions uncovered in the Middle East confirm the reign of various biblical kings, along with other nations and places as recorded in Scripture. The following is only a short list of some of those biblical confirmations (just to name a few).

Sargon II—Sennacherib—Nebuchadnezzar—Belshazzar—

Ahab—Jehu—Ahaz—Omri—

Hezekiah—Hoshea—Jeroboam—Manasseh—Pekah—Joash— Menehem—Ahaziah—

Uzziah—Jehoiakim—Shalmaneser III—Nahor—Haran—Nineveh—Elam

The science of archeology has gradually confirmed the ancient historical record set forth in the Bible. The accuracy of the Old Testament has now become well-established by the continuous excavation of ancient ruins. The basis for the faith of millions of Christians worldwide depends upon the reliability of modern New Testament documents. For this reason, the monumental effort of Bible scholars, researchers, and even scientists to validate the accuracy of the New Testament is unparalleled in the history of ancient literature. Keep in mind that thousands of people, who were eye witnesses of the events surrounding the life, death, and resurrection of Jesus Christ, were still living when the gospels were written. If the truth had not been recorded accurately, they would have refuted those documents as fraudulent. Thousands of early saints were tortured and martyred for their faith, yet no record exists that any of them had ever denied the accuracy of the Scriptures; and once the Bible was translated and copied for widespread distribution and usage, it would have been impossible to corrupt the text or to introduce any fraudulent misinformation.

Thanks to modern archeology, nearly all the ancient cities recorded in the book of Acts have now been located and positively identified. Other New Testament sites that have been identified include the court where Jesus was tried by Pilate, the Pool of Bethesda, the ancient temple area, and the Pool of Siloam. These are some of the exact locations where many profound New Testament events transpired nearly two thousand ago. Even though modern archeology has greatly helped to validate the New Testament record, its most impressive support comes from non-biblical written documents that have survived to this day. Researchers have identified numerous extra-biblical historical texts from ancient times referring to the person of Jesus Christ, including the many events surrounding his life and ministry as an itinerate preacher, his trial and crucifixion, his resurrection from the dead, and the early Christian movement that ensued. A number of these nonbiblical ancient scholars and early historians are listed here for your consideration: Ignatius, Polycarp, Irenaeus, Clement of Alexandria, Tertullian, Justin Martyr, Eusebias, Tacitus, Lucian, Josephus, Seutonius, Origen, Aristides, Philo, Thallus, Phlegon, and Plinius. According to Dr. Gary Habermas, author of *The Historical Jesus*, when the writings of these and a few other ancient texts are combined, we have a very accurate accounting of the life of Jesus without even referencing the New Testament record at all. The historical existence of Jesus Christ, and the movement he began with his sacrificial death and resurrection, is not in question. There has been no other person in history whose existence has been so thoroughly documented. Only infidels bent on refuting Christian faith have had the nerve to question the historical evidence verifying the life and times of Jesus of Nazareth.

According to the world-famous biblical scholar and

researcher Sir Frederick Kenyon, "The interval then between the dates of original composition and the earliest extant evidence becomes so small as to be in fact negligible, and the last foundation for any doubt that the Scriptures have come down to us substantially as they were written has now been removed. Both the authenticity and the general integrity of the books of the New Testament may be regarded as finally established."[9] Because of historical and archeological research, we are rapidly reaching the point where the query is no longer: what Bible facts have been historically identified? The query is now rapidly becoming: what biblical names, events, and places have not been identified? The convincing evidence of modern archeology certainly does not prove conclusively the divine origin of the Bible. However, what it does prove is that our Bibles are trustworthy and historically reliable. If we can trust the biblical record regarding the historical figures, events, and places detailed, this would certainly be a good indication that it should be equally reliable and exceptionally trustworthy in other areas as well. Considering the amazing harmony between the Bible and the vast treasure trove of uncovered historical evidence from the past, we must conclude that the Bible easily passes the test of historical accuracy.

Supernatural Element 11: Prophetic purity of the Bible

If God wanted to communicate a solemn message to mankind, how would he do that in such a way that man would know for certain the information had a divine origin? Well, for one thing, he could record history in advance, hundreds, or perhaps a thousand years or more before the events recorded came to pass. It is certain that only an omniscient God knows the future before it happens. This is exactly what God did with the Bible. His prophetic fingerprints are all over its pages, from Genesis to Revelation. This very impressive supernatural quality of Scripture is perhaps the most obvious reason why we know that the Bible came from the all-knowing mind of God. It would be risky to be a prophet in ancient Israel. According to Hebrew law you would have to be accurate 100% of the time or you were branded a false prophet and put to death.

We can trust the Bible because it is prophetically accurate. Of the many predictions recorded in the Old Testament, over three hundred refer directly or indirectly to the promise of a coming Messiah dating as far back as 1,500 years before his incarnation. We will examine some of the more familiar Messianic passages of Scripture later in this section. The Bible not only presents to us a revelation of the past, but is also the only book ever written that goes well beyond our current world and predicts major events that will come to pass in the future. Perhaps then the strongest evidence ever presented to verify the supernatural origin of the Bible comes from the study of Bible prophecy. If God is the source of all biblical revelation, and God is 'all-knowing', then we could expect the prophetic passages of Scripture to receive a very precise and literal fulfillment. According to Dictionary.com, the word *prophecy* means "a foretelling or prediction of what is to come— a divinely inspired utterance or revelation." As we briefly consider this topic, we should keep in mind that the God of the Bible, who declares himself the creator of heaven and earth, is the only one with all knowledge. This omniscience means God has total unlimited knowledge of the past, present, and future. When his revealed word to man makes so many predictions of what is to come, and those predictions receive a precise fulfillment in the course of time, there is no doubt that God has spoken, and his signature of prophetic revelation is all over the Bible. "As for the prophet who prophesies of peace, when the word of the prophet comes to pass the prophet will be known as one whom the Lord has truly sent" (Jeremiah 28:9). The fascination of our world with the practices of the occult stems from the age-old desire of man to contact the spiritual world and to know the future before it happens. The best way to relate to the spiritual world is to read the Bible. Few people realize that the content of the Bible, according to scholars, is upward of 30 percent prophetic in nature. With nearly one-third of the biblical text consisting of prophetic revelation, most of which having already been fulfilled, we can examine the outcome of many of these ancient predictions and see how they compare with historical records. The Bible text contains approximately 2,700 detailed predictions throughout its pages. Of these vast predictions, over 2,200 have already received complete fulfillment. We can be confident that the remaining 500 or so prophecies, concerning the end times and the second coming of Christ, will also be fulfilled, and, perhaps in our lifetime. "For I am the Lord. I speak, and the word which I speak will come to pass...I will say the word and perform it, says the Lord God" (Ezekiel 12:25). "For assuredly, I say to you, till heaven and earth pass away, one jot or one tittle will by no means pass from the law till all is fulfilled" (Jesus— Matthew 5:18).

Many of the predictions that were revealed by God through the ancient Hebrew prophets were directed toward ancient biblical cities of the Middle East. Here we shall examine one of these ancient passages of scripture in detail to see how well the Bible corresponds with recorded history. Approximately six centuries before Christ, God revealed through the prophet Ezekiel, a judgment declared upon the ancient city of Tyre— a city that was located in what we know as modern-day Lebanon. In the ancient world, Tyre was one of the great trade and commercial centers. The citizens of Tyre excelled as merchants, mariners, and explorers. As a strongly fortified city along the western shore of the Mediterranean Sea, this ancient metropolis was thriving during the time that Joshua led the children of Israel into the conquest of Canaan. Centuries later, during the time of the prophet Ezekiel, God declared divine judgment on Tyre because her people had rejoiced at the fall of Jerusalem. This ancient prophecy by Ezekiel is one of the most detailed predictions recorded in the Old Testament. For centuries, it seemed as though it would never see a clear fulfillment. In the twenty-sixth chapter of Ezekiel's writing, a prophetic pronouncement against Tyre is recorded. The prophet revealed that several nations would rise against Tyre's fortifications, and they would ultimately destroy her massive walls and break down her gates and towers. God's word, penned by the prophet, predicted that a foreign adversary would remove her timbers and foundation stones and actually scrape away the debris and rubble to make the city like a bare rock. It predicts that the rubble and debris from the fallen city, and even the dirt, would be cast into the sea. This unique prophecy specifically stated that Nebuchadnezzar, king of Babylon, would lay siege to Tyre from the north, and would eventually conquer the city. The prophet declared that other nations would participate in the destruction of Tyre, and that a great spoil would be taken from the wealth of this city, causing many other nations to fear greatly at her downfall. The prophecy also recorded that ancient Tyre would become a place for the spreading of nets in the midst of the sea, and foretold that the original site of the city would never be rebuilt. This prophetic passage included some very strange details that seemed rather unlikely to ever be accurately fulfilled.

This wealthy city along the Mediterranean coast was a prominent commercial trading center inhabited by the sea-faring Phoenicians. In the year 586 BC, Nebuchadnezzar, King of Babylon, laid siege to the fortified Phoenician capitol. His arduous campaign lasted for thirteen years until finally, in 573 BC, the city yielded to his control. However, no treasure of consequence fell into enemy hands because the citizens of Tyre had secretly transported their merchandise and wealth onto a small island fortress about a half mile out into the Mediterranean Sea. Although the city was destroyed and left in ruins, Nebuchadnezzar profited little and was unable to pursue the Phoenician treasure to its new strategic location. Nevertheless, a portion of the ancient prophecy had been fulfilled. With the wealth of the mainland city now deposited safely on the nearby island, the Phoenician citizens went about the task of building their island citadel into one of the great trade centers of the ancient world. For some 241 years, the mainland city of Tyre remained in a heap of rubble just as the Babylonian army had left it. Later, however, when Alexander the Great began his extensive military campaign, Greek legions swept across the land conquering and subduing everyone they faced.

Although fearing the strong Phoenician naval fleet, Alexander moved south to conquer the wealthy island fortress of Tyre. In 322 BC, his Grecian forces approached the fortified island citadel. Alexander called upon the merchant rulers of Tyre to open their gates to him at once. Upon their refusal, Alexander became enraged and determined to overtake the city by any means possible. Not having a naval fleet, however, left him at a disadvantage. Alexander struggled to devise a plan that would allow him conquest of the Phoenician treasures. Finally, he formulated a plan by which he believed the conquest of the wealthy island fortress would be within his grasp. Alexander decided that he would build a land bridge from the mainland shore to the island fortress, thereby allowing his troops to invade the wealthy Phoenician capital. Even though the island inhabitants boasted a superior naval fleet with which to harass and keep the enemy at bay, Alexander commanded his legions to remove all of the rubble and debris from the overthrown mainland city to be used for constructing a causeway that would connect the mainland to the island fortress. The debris of stones, timbers, pillars, and even the dust of the original mainland city was cast into the sea leaving the once-thriving metropolis as bare earth.

After months of difficult labor, marred by costly setbacks imposed by the Phoenician naval forces, the causeway was finally completed, and Alexander marched his army across dry ground to decisively conquer the wealthy metropolis. The island citadel of Tyre was reduced to rubble, and the financial wealth of the once great city fell into the hands of the Greeks. The fall of Tyre caused such fear in neighboring cities that they opened their gates to Alexander, and, without opposition, submitted to his conquest. The island city had eventually recovered from the Greek invasion, but years later it was once again destroyed by another foreign invader by the name of Antigonus. In time, a new commercial harbor was opened along the Red Sea and the trade routes shifted so that the great wealth that had formerly flowed into the port of Tyre now found its way to Alexandria instead. Thus, Tyre had received a final blow. The once flourishing ancient metropolis never again regained the prominence of her past glory.

Today the ancient trade center has become a popular haven for fishing boats, and, upon the very ground where the Phoenician merchants once traded their fine goods, fishermen are now spreading their nets to dry in the afternoon sun. Tyre, the great trade and commercial center of the ancient world for centuries, was subdued by the hand of Almighty God, and the freshwater springs that once supplied the needs of the original mainland city still flow today. Although it remains an excellent location with enough fresh water to supply a large modern city, the site has never been rebuilt in more than 2,600 years as Ezekiel accurately predicted. History confirms this prophecy was accurately fulfilled down to the minutest detail. Thus, we have convincing proof of the accuracy of God's Word spoken through His ancient prophet Ezekiel.

If the prophet had recorded this series of predictions using only human foresight, without divine impartation of future events, what is the remote possibility that every detail of this prophecy would have randomly, by chance, come to pass in the course of time? Based on probability estimates, it has been calculated that the chance fulfillment of these unusual prophetic details is fixed at one chance in seventy-five million.[10] It is, therefore, certain that only an omniscient God could foretell, years in advance, events such as these with such amazing precision.

Many more detailed Bible prophecies were directed at other cities of the ancient world, such as Babylon, Edom, Sidon, Ammon, Samaria, Nineveh, Jericho, as well as Jerusalem for example. These other prophecies have also been fulfilled with amazing accuracy. The absolute fulfillment of these unusual prophetic details concerning Tyre, along with many other ancient cities of the Middle-east stands as indisputable evidence of the divine origin of the Old Testament record. Since we know for certain that the prophetic accuracy of the Bible has been well established, we can only conclude that the Bible is truly the result of God's own planning and design.

An Address in History

If you were trying to identify me, for example, how would you do it? Well, you could start with my name, check out my birth certificate, and come upon my address, narrowing it down to street number, street name, city, county, state, and country. You could search for my social security number, a photo perhaps, employer, phone number and email address. This would pretty much complete the search. The Bible is about the Messiah, who would come to bridge the wide gulf between a Holy God and fallen humanity. Therefore, it is rather important that God identifies for us who the Messiah will be. What is his address in history? To start with, the Bible says that he will be the offspring of a woman (an allusion to the virgin birth). We know this detail because, throughout the Bible, genealogies always use the father to document lineage. This statement in Genesis 3:15 is a very unusual exception to that standard practice. He would be of the lineage of Shem, of the seed of Abraham, the son of Isaac, of the line of Jacob, and the tribe of Judah, from the family of Jesse, and of the house of David. The gospel of St. Luke chapter 3 traces the lineage of Jesus all the way back to Adam, the first man. This gives us a pretty good start in identifying the Messiah, sent by God, who was to bare the sins of the world. But this was not enough identification for God. He chose to pinpoint the Messiah for us so precisely that there could be no doubt whatsoever who he was. That is why the God of Heaven revealed over three hundred prophecies in the Old Testament that would unmistakably identify the true Messiah when he came into the world. The following is only a short list of these many prophecies that God used to identify the Messiah. They were all precisely fulfilled in the life of one man. Each prediction lists the Old Testament prophecy followed by its fulfillment in the New Testament.

Messianic Prophecy	**Old testament Reference**	**New Testament Fulfillment**
Born in Bethlehem of Judea	Micah 5:2	Matthew 2:1
To be born of a virgin	Isaiah 7:14	Luke 1:26-31
He would be called Immanuel	Isaiah 7:14	Matthew 1:23
Slaughter of the innocence	Jeremiah 31:15	Matthew 2:16-18
He would come out of Egypt	Hosea 1:1	Matthew 2:14, 15
To be called a Nazarene	Isaiah 11:1	Matthew 2:23

Be a light to the gentiles	Isaiah 9:1, 2	Matthew 4:13-16
Receive praise of children	Psalm 8:2	Matthew 21:15, 16
Preceded by a forerunner	Isaiah 40:3, Malachi 3:1	Matthew 3:1, 2
He would be a prophet	Deuteronomy 18:15	Acts 3:20-22
He would speak in parables	Psalm 78:2	Matthew 13:34, 35
Enter Jerusalem on a colt	Zechariah 9:9	Luke 19:35-37
Be betrayed by a friend	Psalm 41:9	Matthew 10:4
Sold for thirty pieces of silver	Zechariah 11:12	Matthew 26:14, 15
Blood money thrown - temple	Zechariah 11:13	Matthew 27:5-7
Used to buy potter's field	Zechariah 11:13	Matthew 27:5-7
His betrayer to be replaced	Psalm 109:7,8	Acts 1:16-20
Forsaken by his disciples	Zechariah 13:7	Mark 14:50
Accused by false witnesses	Psalm 35:11	Matthew 26:59, 60
Silent before his accusers	Isaiah 53:7	Matthew 27:26
Would be beaten and bruised	Isaiah 53:5	Matthew 27:26
Would be smitten and spat on	Isaiah 50:6, Micah 5:1	Matthew 26:67, 27:30
Mocked by the crowd	Psalm 22:7, 8	Matthew 27:29
His hands and feet nailed	Psalm 22:16	Luke 23:33
His side would be pierced	Zechariah 12:10	John 19:34
Would die by crucifixion	Psalm 22:14, 15	Luke 23:33
Be crucified among thieves	Isaiah 53:12	Matthew 27:38
Intercedes for his persecutors	Isaiah 53:12	Luke 23:34
Rejected by His own people	Isaiah 53:3	John 7:5, John 1:11

Hated without a cause	Psalm 69:4	John 15:24,25
Garments parted – lots cast	Psalm 22:18	John 19:23,24
Gall and vinegar to drink	Psalm 69:21	Matthew 17:34
Would be forsaken by God	Psalm 22:1	Matthew 27:26
Would commit himself to God	Psalm 31:5	Luke 23:46
No bones would be broken	Psalm 34:20	John 19:32,33
His heart would be ruptured	Psalm 22:14	John 19:34
His death would be voluntary	Isaiah 53:12	John 10:11
He would die for others	Daniel 9:26	Matthew 20:28
His death, a sacrifice for sin	Isaiah 53:3-12	Romans 5:6-8

According to Peter Stoner, a mathematical professor, and the author of *Science Speaks*,[10] the chance fulfillment of just eight of these major Messianic prophecies (if written by the wisdom of man without divine input) is estimated to have a combined chance occurrence of 1 x 10^{28}, and the chance that all eight prophecies would be fulfilled in only one man he places at 1 x 10^{17}. He goes on to estimate the chance fulfillment of sixteen of these prophecies in the life of one man at 1 x 10^{45} where the chance occurrence of 1 x 10^{50} is considered to be scientifically impossible. He doesn't stop there, however, but also gives us the estimated chance fulfillment of forty eight major prophecies coming to pass in the life of one man at 1 x 10^{157}. To give some perspective to these probability figures, Professor Stoner relates that if you covered the entire state of Texas two feet deep in silver dollars that includes only one checked coin in the entire collection, the chance that a blindfolded man could randomly select the single checked coin on the first try is estimated at 1 chance in 10^{17}. This is the same chance that only eight of these prophecies would be fulfilled randomly in the life of one man. If we then consider the probability of forty-eight major prophecies being randomly fulfilled in one man with a chance of 1 x 10^{157}, we are talking about mathematical nonsense here with a chance fulfillment far exceeding the probability that a blindfolded man could find a single checked electron in an entire universe completely full of the same. And yet God has revealed over three hundred prophetic predictions to identify exactly who the Messiah would be. God has left absolutely no room for doubt in this matter.

The outline above includes about forty of the several hundred prophecies written to identify a coming Messiah by the prophets of the Old Testament– prophecies written five hundred to a thousand or more years before the actual birth of Christ, and yet every one of these detailed predictions was amazingly fulfilled in the life of only one man– Jesus. Could anyone possibly believe that this was coincidental? This is why God used prophecy throughout the Bible to foretell future events. We could also list hundreds of more prophecies that were directed at various people, places, cities, nations, and rulers that were also fulfilled with precise accuracy. The only logical way to explain history written five hundred to fifteen hundred years in advance of the events they describe is to admit to ourselves that this unique information required a divine impartation of knowledge from the Holy Spirit of God to the faithful writers of the OT Scriptures.

It is certainly very interesting to note here that King David, who wrote the 22nd chapter of Psalms more than a thousand years before Christ, gives us an excellent description of the agony and suffering of one who is put to death by hanging on a cross. There is certainly no way that David could have known of this event in advance, because at the time he recorded these details, a sentence of death by crucifixion as a means of capital punishment was unknown to the Jewish nation. During David's time, death was routinely carried out by stoning. Crucifixion, as a form of lawful execution for a capital crime, was introduced by the Romans who gained dominance over the Jewish nation nearly a thousand years later, well before the time of Jesus. These prophetic details required a divine impartation of knowledge unavailable to the writer by any other means. Christian faith was founded on a vast body of evidence, explainable only by a divine revelation. The study of Bible prophecy demonstrates that there is abundant support for the authenticity of God's Word. This evidence is available to all who are willing to consider it. However, no amount of evidence will ever be sufficient to persuade those who are unwilling to believe.

As a young man, William Ramsay, an English research scholar, traveled to the Middle East to disprove, based on historical research, the accuracy of Luke's gospel and the book of Acts as recorded in the New Testament. As a young skeptic, who set out to discredit the Bible, he spent many years digging up artifacts and recording research data throughout the vast region of Asia Minor. After many decades of exploration, however, Dr. Ramsay was finally forced to conclude that the detailed accounts and statements documented by Saint Luke in the New Testament were discovered to be astonishingly accurate down to the smallest detail. Based upon his own discoveries and the research of others confirming the recorded facts from the New Testament, Dr. William Ramsay became a Christian believer and ultimately a great Bible scholar. His scholarly books have become classics on the historicity of the New Testament and he was eventually knighted in his home country of Great Britain, becoming Sir William Ramsay.

Dr. William F. Albright, whose expertise ranks him as one of the world's most renowned biblical archeologists, has claimed that there is no basis at all for dating any book of the New Testament after AD 80. Further evidence suggests that they were most likely all completed by AD 75. This means that the New Testament documents were recorded by the pen of those who had first-hand knowledge of the events they were describing.

Confirmation of the life of Joseph in ancient Egypt, as recorded in the historical account of Genesis, has come down to us in a most interesting fashion. A previously overlooked cache of ancient Egyptian coins, stashed in the vaults of a museum, have now confirmed the biblical record of Joseph, and the role he played in government service to the pharaoh of Egypt long ago. Although originally discovered a century earlier, in 2009 this cache of ancient coins was rediscovered by authorities at the Egyptian National Museum. Shortly afterward, a local Cairo newspaper disclosed that the coins in question reportedly bear the name and image of the biblical Joseph. The conclusion of the team of antiquities researchers was that the coins had originated at the time when Joseph served as Pharaoh's treasurer during the seven years of plenty, followed by the seven years of famine, as predicted by Joseph when he was called upon to interpret Pharaoh's dream recorded in Genesis chapter 41. One of the coins discovered bore Joseph's effigy as the minister of the treasury in Pharaoh's court. Inscribed on another old coin, bearing Joseph's image, his name appears twice; written in hieroglyphs, his Hebrew name is inscribed and also the Egyptian name that was given to him by Pharaoh when he was appointed to be the treasurer of Egypt. Still another coin displayed an inscription on it and bore the image of a cow, which would have, most likely, symbolized Pharaoh's famous dream of the seven fat and seven lean cows. What an amazing historical confirmation this has provided of the biblical events documenting the life of Joseph, dating back to around 2000 BC. Further confirmation of the life of Joseph, and the seven years of plenty followed by the seven years of famine in Egypt was discovered on a marble tablet inscription. This fascinating marble tablet was uncovered in the nineteenth century within the ruins of an ancient fortress in what is now present-day Yemen. It was dated around eighteen centuries before the birth of Christ. Abundant valid evidence also confirms the life story of Moses as a leader of the Hebrew throng in ancient Egypt. And Flavius Josephus records that two Egyptian priest-scholars, named Manetho and Cheremon, left a record identifying the Hebrew presence in Egypt and their sudden exodus under Moses.

There is clearly no shortage of astonishing biblical confirmation available today. Everyone who is willing should consider the vast evidence suggesting that the Holy Scriptures are and were divinely inspired, and that the real architect of the Bible was none other than the Holy Spirit. He had selectively illuminated the willing hearts and minds of specifically chosen men in ancient times, who faithfully recorded God's divine revelation. Archeologists have discovered that ancient governing authorizes, historians, kings, and rulers will often leave behind accurate cultural records in the form of ancient seals and stone inscriptions. These and other written documents often record the military exploits, decrees, proclamations, and additional historical narratives that bear witness to human exploits throughout the Middle East region. According to the biblical scholar and historical expert Mr. Grant Jeffrey, "Anyone who carefully examines the ancient, extra-biblical historical records with an open mind cannot honestly hold on to the belief that the Bible is nothing more than a collection of legends, fables, and myths."[11]

Witness of the Disciples and Apostles

The resurrection of Jesus was the pivotal event in the formation of the Christian church that began in the first century and continues to expand rapidly the world over. The documents and evidence surrounding the resurrection of Jesus involve facts that have never been refuted. We will not spend additional time on this topic here as this could certainly fill an entire book. What I would like to call to your attention concerns the lives of the disciples of Jesus and their dramatic transformation after his resurrection. The disciples of Christ knew him better than anyone else who ever lived. After the crucifixion, the Scriptures paint a bleak picture of the Lord's disciples cowering behind locked doors, in fear for their lives. However, after the resurrection we observe a remarkable transformation of these men into bold ambassadors for Christ, whose witness could not be silenced by any law, power, or authority. As they went forth with supernatural courage to share the truth about what they saw and heard, no force on earth could deter their bold proclamation. The result was the beginning of a rapidly spreading fire of passionate faith that ignited throughout the ancient world and would never be quenched by the forces of evil on this earth. According to the writer Eusebius, an early church historian of the fourth century, the apostles and disciples of Jesus became scattered across the ancient world, spreading the gospel of Jesus Christ everywhere they went. Except for the apostle John, these brave souls would all face a terrible martyr's death for the righteous cause of the resurrected Christ. The only way to explain the passionate zeal that led to their premature deaths is to understand that their motivation for missionary service was based on the facts they witnessed concerning the resurrection of Jesus. The historical records confirm that these men all faced their fate without ever denying faith in the risen Lord and what he accomplished on behalf of fallen humanity.

Only a fool would die for a lie if they knew it was a lie, but a good man might dare to die for the truth, if they knew it was the truth. James (the son of Zebedee) was beheaded at Jerusalem for his faith. Andrew was crucified in Greece after being beaten. Bartholomew called Nathanael preached the good news throughout Asia Minor and was flayed to death with a whip in Armenia. Thomas was speared to death in India after a very successful missionary journey. Peter was crucified upside down, believing he was unworthy to die in the same manner as his Lord. Phillip was hung to death. Thaddeus was killed by arrows in Armenia. Jude was put to death by arrows in Persia. James (the brother of Jesus) was thrown off the pinnacle of the temple in Jerusalem. After discovering that he had survived the awful fall, his enemies beat him to death with a club. Matthew was killed with a sword in Ethiopia. Matthias, the disciple chosen to replace Judas Iscariot, was stoned, and then beheaded. Mark was dragged through the streets of Alexandria with a rope until he died. Luke was hanged to death in Greece. Barnabas was stoned by the Jews in Cyprus. The apostle Paul was imprisoned and tortured by the Roman Emperor Nero. After enduring a lengthy imprisonment, during which time he wrote important parts of the New Testament, he was beheaded in AD 67. John, whom Jesus loved, was the only survivor of the severe persecution faced by the early church leaders. He was thrown alive into a pot of boiling oil and expected to die. After surviving the ordeal, which left him severely disfigured for life, he was banished to the isle of Patmos, to live like an animal. It was there that this great apostle wrote the prophetic book of Revelation. He was the only disciple to die of old age. The disciples of Jesus, and the followers to whom they witnessed, would ultimately turn the ancient world upside down by introducing a living faith. Christianity was birthed and began to spread like wild-fire from the cradle of human civilization after the life, death, and resurrection of Jesus Christ. This hope and promise was for all who would dare to believe that the Lord God had provided the redemption for mankind as promised so long before.

Supernatural Element 12:
Miraculous events of the Bible

One cannot believe the Bible is God's divine Word, and in the same breath, deny the miracles and supernatural events it records. To deny miraculous events is simply to deny God. If

God is who he claims to be, then the recording of supernatural events is what we would expect to find. It would be inconceivable to believe in God and, at the same time, to deny any possibility of the supernatural. The God of the Bible is a God of the miraculous. Skeptics will routinely scoff at and ridicule anyone who actually believes in the great miracles of Scripture. Many people claim the Bible is full of myths, legends, and fables. They hide behind pseudoscience and the bold claim of naturalists that anyone who believes in miracles should have their head examined. But I do not believe this debate should start with the question: are miracles possible? When it comes to miraculous events, the only real question we need to ask is: does the evidence indicate that miracles have occurred? If no inexplicable supernatural events have ever been recorded, then the atheist might have a leg to stand on. Unfortunately for humanism, this is not the case. The Bible records many supernatural events in documenting God's interaction with mankind over the course of history. If physical and historical evidence confirms that miraculous events have occurred in history, then the question "Are miracles possible?" becomes frivolous. Throughout this book, the goal has been to present valid evidence worthy of consideration. In our brief deliberation into the critical topic of the veracity of Scripture, again, we must follow the physical and historical evidence and we will, ultimately, be led to the truth. With the powerful evidence we have examined throughout this epilogue section, many would likely contend that the Bible itself is a miracle. A strong consensus of evidence, supported by the 'supernatural' elements I have documented, suggest that the physical world and the Bible have both originated from the mind and will of God. When we seriously consider the topic of miracles, there is no single event in the annals of mankind that has been more thoroughly examined and scrutinized than the 'resurrection' of Christ. There are a number of very helpful books in print that have presented all the arguments concerning the resurrection. The evidence they present strongly confirms that this event is an undisputed fact in history. Everything I have encountered in my research over the last thirty years has confirmed to me that the truthfulness and rationality of the resurrection of Jesus Christ has been thoroughly established. Every attempt presented to deny this event has been met with a thorough defeat because the monumental evidence that was left behind to validate this historical event is indisputable.

Consider this brief overview of the evidence. For example, the death of Jesus was thoroughly confirmed at the cross. He was placed in hewn-out tomb and the stone placed over the entrance was secured with a Roman seal— to break the seal was a capital offence. It has been estimated that moving the massive stone to open the tomb would have required the combined strength of a dozen or more men. A Roman unit of trained soldiers (12-16 men) was posted at the scene to ensure that nobody had access to the grave site and the penalty for sleeping on duty was death. There was an earthquake and a bright flash of light that astonished the soldiers. Afterward, the tomb was found to be empty. Nobody could ever produce the body. The chief priests bribed the soldiers to lie about what they saw. Jesus appeared multiple times to his disciples after he rose from the grave, and on one occasion he was seen by five hundred people at the same time. His followers were so thoroughly transformed by what they saw, they went out with supernatural boldness and changed the world. These men were so sure of the truth about the resurrection that they were willing to face death before denying what they saw. The resurrection is a well-documented historical event, and was ultimately confirmed by non-biblical sources as well.

Another miracle in history that has been scoffed at relentlessly for centuries is the biblical record of a worldwide flood in the days of Noah, as recorded in Genesis chapter 6-8. With perceived evidence for evolution theory rapidly disappearing based on empirical science and documented historical findings, many scientists are beginning to take another look at the biblical claim of a universal flood. They are discovering that the monumental devastation described in the biblical text is perfectly compatible with historical, geological, and anthropological research findings encountered over the last century. Archeological evidence alone reveals more than two hundred accounts of a one-time massive worldwide flood that wiped out nearly all of humanity and the land animals. For a thorough understanding of how God could have preserved animal species aboard Noah's Ark during the universal flood, do a little research of your own and visit the 'Ark Encounter' in Williamstown, Kentucky where you will discover a life-size replica of Noah's ark and learn how it was used to preserve life through a devastating worldwide deluge.

Israel is a miracle nation, as we have already discovered in a previous chapter. Still other supernatural events recorded in the Scriptures have also received confirmation based on the findings of historical and archeological research. These events would include, for example, the diversity of human language that is associated with the Tower of Babel incident. A clay tablet discovered in Babylon references a temple that was once destroyed with the people scattered and their "speech made strange." The famous Oriental language scholar, Max Mueller, claims that all human languages can be traced to an original. This professional testimony suggests that the confusion of language by God at the Tower of Babel is a feasible explanation for why there are so many languages in the world today. Other verified miracles of antiquity would include a three-hour period of darkness at the death of Christ, the unusual manner in which the walls of Jericho fell, and the supernatural destruction of Sodom and Gomorrah. In addition to these events, there is also extra-biblical evidence supporting the biblical account of the "ten plagues of Egypt," as recorded in the book of Exodus. The Ipuwer Papyrus, which is housed in the Dutch National Museum of Antiquities, is something of a poetic lamentation. This ancient document describes great disasters that took place in ancient Egypt. The record is dated at around 1400 BC, placing it close to the time of the Exodus (circa 1446 BC). This ancient papyrus makes a fascinating reference to the plague of blood throughout the land, the death of the livestock, the plague of hail and fire, the extended days of darkness, and the death of the first born of Egypt.

Supernatural Element 13: Its life-changing power

Does the Bible have modern-day relevance to our lives? If it is the Word of God, we can expect it to address the needs of people in every generation with life-changing power. In the midst of the increasing confusion and uncertainty all around us, we can still turn to an open Bible, and find the words of help and encouragement we so desperately need. The Bible clearly addresses normal human needs, actions, and attitudes, while expressing above all God's special love and impartial devotion for every human soul He created. It disturbs the complacent, comforts the sorrowful, rebukes the proud, reforms the sinful, strengthens the weak, and brings hope to broken lives. The Bible is uniquely different than all other books. It is very much alive and fully empowered by the Spirit of God for the life-changing benefit of every man, woman, and child. Its powerful God-breathed narrative seems to have a remarkable way of penetrating the very soul of man to reveal his spiritual need in the sight of a loving personal God. "For the word of God is living and powerful, and sharper than any two-edged sword, piercing even to the division of soul and spirit, and of joints and marrow, and is a discerner of the thoughts and intents of the heart" (Hebrews 4:12). The Bible is somewhat like a mirror into which we look to observe ourselves the way we really are, the way God sees us. Unfortunately though, throughout history men have tried to judge the Bible rather than allowing it to judge their hearts. The Bible contains all the information necessary to fill the deepest longing within every human soul— that longing for a relationship with our eternal Maker that gives lasting peace to the inner man and a certain hope beyond this life. What could be more important than that? As we have previously discussed, the central theme of God's book is the redemption of mankind. It offers a personal petition to the heart of every individual, and requires a response on the part of each one of us. Everyone must choose to accept or reject God's plan of redemption. Nobody can abstain; we either choose life or we choose death depending on how we respond to the free gift of God's grace that is offered to every man, woman, and child through the blood of Jesus Christ, by his personal sacrifice on our behalf. God's eternal hope that is offered for every member of the human family required the voluntary death of his only Son on Calvary. The biblical narrative has never become outdated. Its message is as fresh, meaningful, and life-changing today as it ever has been.

The earth-shaking and world-altering power of the Holy Scriptures is clearly evident as we look at the significant impact it has had on great personalities and world leaders. No evil tyrant or murderous dictator has ever been a friend to or an advocate of the Bible. On the other hand, history reveals that no successful, highly revered world leader has ever been an enemy of the Scriptures. Therefore, to deny the authority and authenticity of the Bible is to set oneself firmly against the values and principles held by many great leaders of Western Civilization. The following is a very brief list of what world leaders have had to say about the Bible's influence.[12]

George Washington: "It is impossible to rightly govern the world without the Bible."

Benjamin Franklin: "Young men, my advice to you is that you cultivate an acquaintance with, and a firm belief in, the Holy Scriptures."

Patrick Henry: "This is a Book worth more than all the others that were ever printed."

Andrew Jackson: "That book, sir, is the rock on which our Republic rests."

Abraham Lincoln: "But for this Book we could not know right from wrong. I believe the Bible is the best gift God has ever given to man."

Theodore Roosevelt: "No educated man can afford to be ignorant of the Bible."

Woodrow Wilson: "The Bible is the one supreme source of revelation of the meaning of Life."

Sir Isaac Newton: "We account the Scriptures of God to be the most sublime philosophy. I find more sure marks of authenticity in the Bible than in any profane history whatsoever."

Michael Faraday: "Why will people go astray when they have this blessed Book to guide them?"

Chiang Kai-Shek: "The Bible is the voice of the Holy Spirit."

Winston Churchill: "We rest with assurance upon the impregnable rock of Holy Scriptures."

H. S. Wells: "The Bible has been the Book that held together the fabric of Western civilization ... The civilization we possess could not have come into existence and could not have been sustained without it."

"The law of the Lord is perfect, converting the soul; the testimony of the Lord is sure, making wise the simple." (Psalm 19:7)

No Lack of Communication

The Bible has revealed to us invaluable information that we would not otherwise know. It gives us knowledge about the origin of life and the physical world; it reveals to us the nature of man whom God created in his own image, and it relates the existence and character qualities of God and the relationship between God and man with competition from the devil's schemes. It also reveals to us God's perfect plan and purpose for the ages, as he interacts with humankind, who is the centerpiece of his vast creative works. Because of sin, mankind became separated from his creator in terms of relationship. This left mankind totally at the mercy of God to bridge the wide gulf that now separated the spiritual and the physical worlds. The purpose, therefore, of the divine communication in the Bible is to reveal to us how well God has accomplished bridging this gap, and what it means to mankind in terms of our response to the redemptive work God has performed on our behalf. Of all God's vast creation, only mankind is equipped to relate to him on a personal level. God has always desired to have a loving, intimate relationship with man without violating the free will with which man was created. We all know for certain that a personal loving relationship between mutual friends cannot exist without freedom of choice.

In communicating his plan and will for mankind, it became necessary for God to authenticate his revealed Word in a way that would distinguish it from all other written works that might lay claim to divine origin. Our Christian faith is built upon the Holy Bible; without it we would be wandering souls without hope and purpose. In examining just how God has thoroughly authenticated his communication with man, we soon discover that the Bible is a fully integrated message system that was carefully designed and engineered by divine instruction. The evidence of design throughout the Bible is a clear demonstration of divine authorship. Not only can we trust the Bible, but because of the immense effort with which God has undertaken in communicating it to us, you and I will also be held accountable for what it has to say. (see John 12:48)

When we think about it from a logical point of view, we should each ask ourselves the following important question: if Jesus was not the Son of God and the promised Messiah of Israel, how could he possibly command such wide influence and powerful magnetism over human lives two thousand years after his time on earth? How could a poor wandering Jew, from an obscure village of the ancient world, in the first century, gain such a profoundly strong influence over the world's population two thousand years removed from his lifetime? The man called Jesus had no political power. He had no formal education. He never wrote a book, led an army, or held public office. And yet after only three short years of public ministry two thousand years ago, he has commanded a dynamic Christian movement that continues to multiply with every generation. Today, he is reported to have several billion followers worldwide. There is absolutely no logical explanation for this fact, unless he is exactly who he claimed to be— the eternal Son of God.

The vast body of evidence that supports the Bible as a trustworthy and authentic message of divine origin is rock solid. There should be no valid reason for anyone to deny that the Bible is an inspired communication of divine impartation. The Bible text embodies all the knowledge and information man needs in order to fill the deep longing of his eternal soul— to know the truth about the Creator and find answers to the most challenging questions he will face in this world. It is the blueprint owner's manual of the Master Architect, and only by following its instruction, can we have the full and satisfying life we were created for. The orthodox view of the Christian faith holds that the biblical text alone is the inspired, illuminated revelation of truth from God to mankind, and is the sole basis of divine authority for all true believers. The Bible introduces us to the Christian faith, which is not just one of many worldview philosophies we may choose from. When people reject the Holy Bible and Christianity, they are not rejecting a philosophy— they are rejecting a person. And this person is God's only provision for the problem of sin that has ravaged our world with death, disease, sorrow, and unimaginable pain and suffering. This is our only way out of the terrible mess we are in! Religion is all about man's effort to gain favor with God. Christianity, on the other hand, is about God's effort to provide fallen humanity with the means by which to be restored to a right-standing relationship with his Maker. All we must do is to accept God's generous offer of redemption. The rest is up to Him! We cannot afford to be ignorant of the Word of God, and we will pay a terrible price for ignoring it.

The Hidden Gospel Message

Did you know that the message of the Christian gospel is hidden within the pages of the first book of the Bible? Once again, we see the providence of God at work as the Holy Spirit has orchestrated another surprise appearance of the fingerprints of God upon his divine Word. The gospel message is found in the genealogy of Noah. This was no accident! It is hard to believe that any Hebrew writer would dare conspire to hide the 'Christian Gospel' within the pages of the venerated Torah! But there it is in Genesis chapter 5. There are ten names listed here that complete the genealogy of Noah, beginning at creation, and ending with preparation for the great flood. Most people skip right over the genealogies of the Bible, but they are important and we need to take a closer look at this one in Genesis 5. In the Bible names are significant and their meaning is often very relevant. Abraham, for example, means "Father of Many." and he became the father of many nations. Jesus means "Savior," Christ means "Anointed one," and He was the Messiah– the anointed savior of the world and God's appointed atonement for sin.

Now, what do the ten names in Genesis chapter 5 actually mean, and what is their significance in the grand scheme of God's design? The very first name we encounter is Adam, which means "man" Adam's son carried the name Seth, which means "appointed." Seth had a son and named him Enosh, which means "mortal." Enosh bore a son whom he named Kenan. The name Kenan denotes "sorrow." Kenan bore a son whom he named Mahalalel which translates "the Blessed God." Mahalalel's son was named Jared, which means "shall come down." Jared bore a son and named him "Enoch." which means "teaching." Enoch's son was named Methuselah. This name translates "His death shall bring." Methuselah begot a son whom he named "Lamech." His name means "the Despairing." Lamech bore a son and called him "Noah." The word translates "comfort" or "rest." Now, when we put the meanings of these ten names altogether, we find the hidden message: Man (is) appointed mortal sorrow: (but) the Blessed God shall come down teaching (that) His death shall bring the despairing rest." That is fascinating!

So, 'man' was appointed 'mortal sorrow' (because sin brought death); but 'the Blessed God' (Jesus) 'came down' (from heaven), teaching us that 'his death would bring those living in despair' the 'rest' they were longing for – giving peace and comfort to their troubled souls. Remember Jesus said: "Come unto me all of you who are weary and heavy laden (weighed down by the burden of sin) and I will give you 'rest' (for your soul).

Testing for Divine Revelation

Suppose that a friend wanted to introduce someone to a book that he believes contains inspired revelation from the mind of God. They might say, okay, I will take a look, but first let me tell you the rigorous tests that book will have to be subjected to in order to qualify as divine revelation.

1. It must be written by multiple individuals over many hundreds of years, with each contributing writer working independently of the others, thus preventing any collusion between them.

2. This revelation must address, with consistent unity throughout the combined narrative, the most challenging and controversial topics that have ever come before the councils of men.

3. The collective works of these many diverse writers must be in complete harmony and agreement with one another on each topic addressed.

4. It must address with impartial honesty the sins and failures of those who have penned the words, including their national sin and personal record of disobedience to God's commands.

5. It must reach out to the intended audience in a style of writing that would be indicative of God speaking personally and directly to his human creation.

6. It must speak forth the truth concerning every issue of consequence to man and be in complete agreement with the Bible on every matter addressed.

7. It must have survived to this day with its divine message intact and basically unaltered as compared with the original written documents. No editing changes shall be found when compared to the original written autographs.

8. It must have suffered an incredible history of hatred and persecution at the hands of the enemies of God without ever succumbing to the ignorance of men. It must, ultimately, rival the Bible as the most prolific, beloved, and popular literature

in the history of mankind.

9. Its message must be so profound that it becomes translated into every known human language for the benefit of all mankind over the course of time.

10. It must become the very foundation upon which societies of free men are established and the affairs of honest men are based. This profound influence must, ultimately, be reflected in government, education, culture, and civil law.

11. It will not contain a single detail of informative revelation that can be proven to be false by the progressive advance of human knowledge through history, science, and education.

12. It must be saturated with prophetic words written hundreds of years or more in advance of the precise fulfillment of each and every prediction. This will make it obvious to all that the message could only have come from an omniscient God.

13. It must record profound supernatural events, which can only be explained by the work of an omnipotent Creator, and a significant number of these profound events must be verifiable by means of eye-witness accounts, physical evidence, and historical documentation.

14. The narrative must clearly state that its origin is of a divine nature and this fact will be demonstrated by textual evidence upon scrutiny and examination.

15. This book must profoundly and forever transform the lives of those who have accepted and embraced it as a message of divine truth.

16. This book must point mankind to the redemptive plan and purpose of a loving, personal God and its theme must be consistent with the theme recorded in the biblical narrative.

These are the criteria God has used to deliver his eternal truth to the human lives that were created in his image and likeness. The Bible, and no other published work, fulfills the detailed criterion outlined here. The profound magnitude of the biblical narrative and the life-changing qualities that have made it the most beloved and controversial document in the history of man have been presented for your careful thought and sincere consideration. Nothing else even comes close to matching the obvious fingerprints and signature of the Holy Spirit that are all over this divinely inspired classic.

A Verdict is Reached

People often disbelieve things that are true because they do not want them to be true. The real question we should be asking ourselves is this: in my current life circumstance, do I believe in something that represents the final authority in my life, and upon what foundation can I stand to give me hope for a promising future when my days on earth are over? If what you believe is in direct conflict with the Bible when it is taken a face value, then do not trust it. Do not trust your feelings and emotions. They can mislead you. The powerful evidence that favors the Bible alone as the inerrant Word of God is what makes it so worth believing. We are always admonished in scriptures to "test all things; holding fast to what is good" (1 Thess. 5:21). Will you trust in the Word or the world? I believe the Bible is the only anchor you can always count on in the storms of life. The self-centered human heart often says, "I want to be the boss. I want to do things my way and not be restricted by any burdensome moral constraints." In our rebellious nature, we want to exclude anyone else from being the lord over our lives. We want to be in control and do things our own way. This is why God has given us the Bible. He knew we would need a good instruction manual, or we would struggle to figure out how to live the way we were intended to— allowing us to experience the fulfillment we all desire. Do not let your free will get in the way of knowing the God of the universe and living out your intended creative purpose on this earth. Who is the final authority in your life? Is it God or self? If you are not currently experiencing the fulfillment you desire, then take a stand on the promises of God. The Bible cannot help you if you refuse to believe it and remain skeptical, and the death of Christ cannot save you if you refuse to accept that it was God's only means of atonement for your sin. If the Bible is not the truth, then you and I have no hope in a fallen world that is full of death, disease, pain, and suffering, with no way out from under the curse upon this planet. The Bible gives us very clear instruction on the sure way out of the mess we humans are in. Accept it as the inspired Word of a loving God, and make it the final authority in your life. As you love it, learn it, and live it, the Bible will radically transform who you are from the inside out (see Romans 12:2). We could pass a law that says you must love your neighbor, but as creatures exercising free will, we will only love our neighbor if it is in our hearts to do so. Only God can give us unconditional love for others, and only the Bible can tell us how get it! (John 8: 31,32).

In this brief summary, we have barely touched on the broad topic of biblical veracity. However, the evidence we have examined here overwhelmingly supports the firm belief of many that the Bible alone is God's divine revelation to mankind. It is a widely used resource that, when taken seriously, will transform human lives. Perhaps the most profound evidence that the Bible is God's Word is that when we read it with an open mind and open heart, these powerful words somehow jump off the page to penetrate our very souls, transforming our thinking, conduct, and worldview. Ultimately, it will dictate the lifestyle we live as the Holy Spirit illuminates its message to our hearts. God took the initiative to communicate with mankind (his crowning creation), and in doing so, he provided us with a divine standard to live by — an instruction manual that will lead us into a fulfilling relationship with himself. Even though it has always been treasured as a great work of literature, the true value of the Bible text is found in its message, not its style. Those of us who have accepted the authority of biblical truth have found it to satisfy our deep desire to answer the most troubling questions about life; and more importantly we have discovered the fulfillment of a personal relationship with our Creator. Yes, the Bible is truly the authentic, inspired Word of God. It has provided us with a progressive, harmonious collection of divine thought and knowledge, which constitutes a complete, unified message system. The Bible is reliable, and we can stake our lives on it. We should embrace it with great appreciation and reverently fear the God who gave it, because this will be the standard by which our lives will be judged when each of us faces our Creator one day. If we believe and hold fast to the promises of God, we will never have to struggle with the thought of whether we are good enough to make it. Our salvation is not and never will be based on what we have done, but our hope is based solely on what He has done for us.

The extensive documentation of truly reliable evidence that we have examined in this epilogue section overwhelmingly supports the unwavering conviction of many that the Holy Bible is God's divine revelation to mankind.

The Bible says that God cannot lie, and the word of the Lord is true (Hebrews 6:18, Psalm 33:4, Proverbs 30:5, and, Psalm 12:6). I think we would be wise to trust it!

"Beware lest anyone cheat you through philosophy and empty deceit, according to the traditions of men, according to the basic principles of the world, and not according to Christ. For in Him dwells all the fullness of the Godhead bodily; and you are complete in Him, who is the head of all principality and power" (Colossians 2:8-10).

"You are worthy, O Lord, to receive glory and honor and power. For you created all things, and by your will they exist and were created" (Revelation 4:11)

Endnotes

Chapter One

1 Dr. Lee M. Spetner, (American Physicist- MIT graduate) "Not By Chance" (Judaica Press, Inc: Brooklyn, New York 1997) preface viii.

2 Charles Darwin, (1809-1882) (British Student of Theology who left Cambridge to become a self-taught naturalist and became instrumental in popularizing the theory of evolution through reliance on natural selection) "The Origin of Species" (John Murray Publishing: London, England 1859) p. 84. see "The Battle for Truth" David A. Noebel (Harvest House Publishers: Eugene, Oregon 2001) p.120,121.

3 Charles Darwin, "The Origin of Species" (P.F. Collier and Son Corp: New York, NY 1909) p.20.

4 Ibid. p.169. also see "Examine the Evidence" Ralph Muncaster (Harvest House Publishers: Eugene, Oregon 2004) p. 48.

5 Ibid. p.170. also see "Examine the Evidence" (op cit) p.48.

6 Ibid. p.169.

7 Charles Darwin, "The Origin of Species" (John Murray Publishing: London, England 1859) p.292. "The Origin of Species" (London, England 6th edition, 1902) p.341-342. see "The Battle for Truth" (op cit) p.342. and

"Dawkins God" (Blackwell: Oxford, England 2005) p.81.

8 Desmond Morris (British Zoologist) "The Naked Ape: A Zoologist's Study of the Human Animal" (1st American ed. McGraw-Hill: New York 1967) see the book review condensed in Reader's Digest magazine, July 1968.

9 Time Magazine: Cover story "How Man Became Man" (November 7, 1977).

10 Time Magazine: "Puzzling Out Man's Ascent" (November 7, 1977) p.77.

11 Julian Huxley (1887-1975) (British evolutionary biologist and a strong proponent of the natural selection hypothesis) "At Random" a television preview, which aired on November 21, 1959. see Sol Tax, "Evolution of Life" (University of Chicago Press: Chicago, Illinois 1960) p.1. see "The Battle for Truth" David Noebel, (op cit) p.118.

12 Julian Huxley, "Evolution: The Modern Synthesis" (Harper and Brothers Publishers: New York 1942) p.457. "The Long War Against God" Henry Morris, (Master Books: Green Forrest, AR. 2008) p.322.

13 Carl Sagan (1934-1996) (American evolutionary astronomer and cosmologist), "Cosmos" (Random House Publishers: New York 1980) p. 27. see "The Battle for Truth" (op cit) p.118.

14 Carl Sagan, "The Dragons of Eden: Speculations on the Evolution of Human Intelligence" (Random House Publishers: New York 1977) p.6. see "The Battle for Truth" (op cit) p.120.

15 Isaac Asimov (1920-1992) (a former professor of biochemistry at the Boston University school of medicine) "Science and Creationism" ed. Ashley Montague, (Oxford University Press: Oxford, England 1984) p.183. see "The Battle for Truth" (op cit) p.118.

16 Corliss Lamont (1902-1995) (a left-wing socialist philosopher from Columbia University), "The Philosophy of Humanism", rev. ed. (Frederick Ungar: New York 1949, 1982) p.120. see "The Battle for Truth" (op cit) p.121.

17 George Gaylord Simpson (1902-1984) (American born paleontologist from Columbia University and a major participant in the modern evolutionary synthesis) "The Meaning of Evolution" original 1949. (Yale

University Press: New Haven, Ct. 1971) p.345. see "The Battle for Truth" (op cit) p.117.

18 Rene Dubos (1901-1982) (French born American microbiologist and a strong advocate for humanist evolutionary philosophy) "Humanistic Biology" (the American Scientist, vol. 53, March, 1965) p.6. see "Scientific Creationism" Dr. Henry Morris, (Creation-Life Publishers: San Diego, California 1978) p.10.

19 Julian Huxley, Education and Humanism, "Essays of a Humanist" 1964 p.125.

20 Humanist Manifesto I (Prometheus Books: Buffalo, New York. 1933, 1980) p.8. see "The Battle for Truth" (op cit) p.117.

21 Humanist Manifesto II (Prometheus Books: Buffalo, New York 1973, 1980) p.17. see "The Battle for Truth" (op cit) p.117.

22 Carl Sagan, "Cosmos" (Random House: New York 1980) p.4. see "God's Undertaker" John Lennox, (Lion Hudson Pub: Oxford, England 2009) p.30.

23 Michael Denton, (British-Australian biochemist & author) "Evolution: A Theory in Crisis" (Adler & Adler Pub: Bethesda, Maryland 1985,1986) p.342. (British printing, Burnett Books Ltd,: London, England 1985) see "The Long War Against God" Henry Morris, (Master Books: Green Forest 2008) p.277.

24 Richard Dawkins, (British biologist and prominent evolutionary philosopher. Oxford, England) "The Blind Watchmaker" (W. W. Norton & Co: New York 1986) p.43. see "Is There Really a God"? Ken Ham booklet (Answers in Genesis: Florence, Ky. 1998) p.7.

25 Michael Behe, (American biochemist and biomolecular researcher from Lehigh University, known for introducing the design principle of 'Irriducible Complexity': a single system which is composed of several to many well-matched interacting parts that contribute to the basic function, where upon the removal of any one of the parts causes

the system to effectively cease functioning. “Darwin’s Black Box: The Biochemical Challenge to Evolution” (The Free Press: New York 1996) p.252-253. “Darwin’s Black Box” 2006 edition, p.39. J. Buell & G. Hearn, eds. “Darwinism: Science or Philosophy”? (Dallas, Texas: Foundation for Thought and Ethics, 1994) p.68-69. see “The Case for Faith” Lee Strobel (Zondervan Publishing: Grand Rapids, Michigan 2000) p.101.

26 Charles Darwin, “On The Origin of Species” (John Murray Pub: London, England 1859) p.219. facsimile of the 1st ed. (Harvard University Press: Cambridge, Mass. and London, England 1964) p.189. also see “The Origin of Species” 6th ed. (New York University Press: New York 1988) p.154. cited in “Total Truth” Nancy Pearcey, (Crossway Books: Wheaton, IL. 2005) p.187.

27 Charles Darwin, “On the Origin of Species” (J.M. Dent & Sons Ltd: London, England 1971) p.167. see the discussion on adaptations in “Creation Facts of Life” Gary Parker, (Master Books: Green Forrest, AR. 1994) p.80-81. this material from “The Origin of Species” is found in Darwin’s chapter entitled ‘Difficulties With the Theory’.

28 Michael Behe, “The Edge of Evolution: the search for the limits of Darwinism” (Free Press: New York 2007) p.166. cited in “God’s Undertaker” John Lennox, (Lion Hudson Publishing: Oxford, England 2009) p.120.

29 Bertrand Russell, (1872-1970) (British writer & philosopher - Nobel prize for Literature in 1950) see the “History of Western Philosophy” (Routledge: London, England 2000) p.570. see “God’s Undertaker” (op cit) p.84

30 Colin Patterson, (1933-1998) (a former paleontologist at the British Museum of Natural History-London) spoken in an address at the American Museum of Natural history, (New York, Nov 5, 1981) see “Creation Facts of Life” Gary Parker, (Master books, Green Forrest, Ar. 1994) p.107.

31 Duane T. Gish, (1921-2013) (former biochemist - UC Berkley, an

officer of the Institute for Creation Research and a strong prominent of creationism during the 20th century. Known for confronting many evolutionary biologists in formal debate) "Evolution: The Fossils Say No"! (ICR Publishing Co: San Diego, California 1973) p.3.

32 Henry Morris, (1918-2006) (a highly regarded earth scientist, and a former college professor with multiple PhDs who became a young earth creationist and directed the Institute for Creation Research for many years while authoring more than twenty books as a highly regarded scientific Christian apologist) "Many Infallible Proofs" (Master Books: El Cajon, California 1974) p.249

33 Ibid. p. 249.

34 Galileo Galilei, (1564-1642) (early scientific pioneer - physicist & astronomer) In a letter to Grand Duchess Christina, 1615. cited in "The Language of God" Francis Collins (Free Press: New York 2006) p.158.

35 James Tour, (American scientist highly regarded as a synthetic organic chemist, specializing in nanotechnology-Rice University) see Candace Adams article, "Leading Nanoscientist Builds Big Faith" (The Baptist Standard, March 15, 2000) cited in "The Case for Faith", Lee Strobel, (Zondervan Publishing: Grand Rapids, Michigan 2000) p.111.

Chapter Two

1 Ernst Mayr, (1904-2005)(a leading evolutionary biologist of the twentieth century) "Populations, Species and Evolution" (Harvard U. Press: Cambridge, Mass. 1970) p.102. cited in "Scientific Creationism" Henry Morris, (Creation-Life Publishers: San Diego, Ca. 1978) p.54.

2 Corliss Lamont, see chapter one, reference 16, (2nd citation) for emphasis.

3 Werner Gitt (German engineering professor and a young earth creationist) "In the Beginning was Information" (Master Books: Green Forrest, AR. 2005) p.93.

4 George Gaylord Simpson, "The Non-prevalence of Humanoids" (Science vol. 143, 1964) p.772. (pathlights.com/creation - evolution encyclopedia) 2014.

5 Lee Spetner, "Not By Chance: Shattering the Modern Theory of Evolution" (The Judaica Press: Brooklyn, New York 1997) p.131.

6 Ibid. p.138.

7 Pierre-Paul Grasse, (1895-1985) (French zoologist of the 20th century and author of over three hundred publications) "Evolution of Living Organisms: Evidence for a New Theory of Transformation" (Academic Press: New York 1977) p.87. see "Creation Facts of Life" Gary Parker, (Master Books, Green Forrest, AR. 1994) p.109.

8 Ibid, p.87. (op cit) Gary Parker, p.109

9 Pierre-Paul Grasse, "Evolution of Living Organisms" (Academic Press: New York 1977) p.103. (soulwinners.com.au/8.html) 2014

10 Pierre-Paul Grasse, "Evolution of Living Organisms" (Academic Press, New York 1977) p.38. (pathlights.com /creation - cvolution cncyclopedia) also see (2christ.org/quotes) 2014

11 Jonathan Sarfati (PhD in chemistry and a young earth creationist/author working with Creation Ministries International (CMI) a non-profit Christian apologetics ministry) " Refuting Evolution", (Master Books, Green Forrest, AR. 1999) p.40,41. (antibiotic resistance) see also C. Wieland on, 'Superbugs: Not Super After All', Creation Ex Nihilo, 20(1): p.10-13 (June-August 1992).

12 Albert Szent-Gyorgyi, (1893-1986) (the Cambridge educated Nobel Prize winner –biochemist/physiologist who discovered vitamin C) see "Creation Facts of Life" Gary Parker, (Master Books: Green Forrest, AR. 1994) p.109.

13 Gary Parker, (PhD in biology with a specialty in geology and paleontology. biology professor, author and former atheist) "Creation Facts of Life", (Master books: Green Forrest, AR. 1994) p.122.

14 Thomas Nagel, (American Philosopher) 'Philosophy and Public Affairs'

Wiley Inter Science, vol. 36, (2), 20, 2008 p.199. see "God's Undertaker" John Lennox (op cit) p.120.

15 Notice the genetics discussion – Richard Goldschmidt/Norman Macbeth's "Darwin Retired" (Dell publishers: New York, 1971) p.33,154. see "Total Truth" Nancy Pearcey, (Crossway Books: Wheaton, Illinois 2005) p.160.

16 Richard Goldschmidt, (Zoologist/Geneticist-UC Berkley) in Norman Macbeth's "Darwin Retired" (Dell publishers: New York, 1971) p.33,154. also see ('Material Basis of Evolution' Yale University Press) (2christ.org/quotes) 2014.

17 Jeremy Rifkin, (social theorist- specializing in technological and environmental change) 'Algeny' (Viking Press: New York 1983) p. 134. see "The Long War Against God" Henry M. Morris, (Master Books: Green Forrest, Ar. 2000) p.160.

18 Gordon Taylor, (American microbiologist) "The Great Evolution Mystery" (Harper & Row Publishers: New York 1983) p.34,38. (2christ.org/quotes) 2014.

19 Michael Behe, (professional observations cited by John Lennox "God's Undertaker" 2007) p.110.

20 Stephen J. Gould, (1941-2002) (former professor of Evolutionary Biology & Paleontology, Harvard University – a leading evolutionary spokesman of the 20th century) "The Return of Hopeful Monsters" Natural History, vol. LXXXVI(6), June-July 1977, p.28. (creation.org/articles/quotes) 2014.

21 Nils Heribert-Nilsson, (1883-1955) (Swedish Botanist/Geneticist) 'Synthetische Artbildung' (Verlag CWK Gleerup Press: Lund, Sweden 1953) p.174. (2christ.org/quotes) 2014.

22 Stephen J. Gould, 'The Return of Hopeful Monsters' article published in: 'Natural History' June/July, 1977 also see "The Panda's Thumb: more reflections in natural history", (Penguin Press: London, England

1980) (with reprint 1990) p.158. (pathlights.com/creation-evolution encyclopedia) 2014.

23 F.H.T. Rhodes, H.S. Zim & P.R. Shaffer, "Fossils" (Golden Press: New York 1962) p.10. see "Scientific Creationism" Henry Morris, (op cit) p.97

24 John Woodmorappe, (American geologist/biologist and young earth creationist) see discussion- CRS Quarterly vol. 18, No. 1, p.46-71. (2christ.org/quotes) 2014.

25 David Raup, (1933-2015) (Former University of Chicago paleontologist and curator at the Field Museum of Natural History, Chicago) quoted from 'Conflicts Between Darwin and Paleontology' (Field Museum of Natural History Bulletin, vol. 50 (1), 1979) see "Creation Facts of Life", G. Parker, (op cit) p.143.

26 Duane T. Gish, "Evolution—The Fossils Say No"! (ICR Publishing Co: San Diego, Ca. 1973) p.17.

27 Ibid. p.41.

28 Henry M. Morris, "The Remarkable Birth of Planet Earth" (ICR Publishing Co: San Diego, Ca. 1972) p.22.

29 David B. Kitts, (1923-2010) (professor of geology-University of Oklahoma, Head Curator with the dept. of geology, Stoval Museum and a well known evolutionary paleontologist) "Paleontology and Evolutionary Theory" 'Evolution', vol. 28, Sept. 1974. p. 467. see Genesis Park used by permission ("Examine the Evidence" Ralph O. Muncaster, (Harvest House: Eugene, Oregon, 2004) p. 53. cited at (2christ.org/quotes) 2014.

30 Stephen J. Gould, "The Panda's Thumb" (W.W. Norton: New York 1980) p.238-239.

31 Stephen J. Gould, The Episodic Nature of Evolutionary Change "The Panda's Thumb" (W.W. Norton: New York 1985) see "God's Undertaker" John Lennox, (op cit) p.114.

32 Gary Parker, "Creation Facts of Life" (Master Books: Green Forrest, AR. 1994) p.208.

33 Niles Eldredge, (biologist/paleontologist - a curator at the American Museum of Natural History - dept. of invertebrates) 'Progress in Evolution'? New Scientist, vol. 110, 1986, p.55. (quote mine project/eldredge) 2014.

34 Peter L. Forey, (1945-2016) (paleontologist with the British Museum of Natural History) 'Neontological Analysis versus Paleontological Stories' (Academic Press: London, England 1982) p.120-121.

35 Brian Sykes, (professor emeritus of Human Genetics, Oxford) 'Nature', vol. 352, August 1, 1991, p.381.

36 Carl Sagan, "The Dragons of Eden" (Random House Publishers: New York 1977) p.6. see "The Battle for Truth" D. Noebel, (op cit) p.122.

37 Steven M. Stanley, (American paleontologist and evolutionary biologist – U. of Hawaii) "Macroevolution: Pattern and Process" (W. M. Freeman pub: San Francisco, California 1979) p.39. see "The Long War Against God" Henry Morris, (Master Books: Green Forrest, AR. 2008) p.27.

38 David Woodruff, (1943-2015) (biology professor - UC San Diego) 'Science' vol. 208, 1980, p.716.

39 Mark Ridley, (British zoologist, Oxford University /doctoral advisor: Richard Dawkins) 'Who Doubts Evolution'? New Scientist, vol. 90. June 25, 1981) p.831.

40 Gary Parker, see (ideacenter.org), (2christ.org/quotes) and (goodreads.com/author/quotes/garyparker) 2014.

Chapter Three

1 John Lennox, (Oxford Professor of mathematics, philosopher of Science - a prominent Christian apologist who has lectured and debated throughout the world on 'the origin of the physical dimension' and 'the existence of God') see his book "God's Undertaker, has Science Buried God"? Lion Hudson Pub: Oxford, England 2009.

2 Jerry Adler & John Carey: 'Is Man a Subtle Accident'? Newsweek, vol. 96,

No. 18 Nov. 3, 1980 p.95. (2christ.org/quotes) 2014

3 Newsweek, Jerry Adler & John Carey, 'Is Man a Subtle Accident'? November 3, 1980 p.95-96. Reporting on the: 'Macroevolution' conference -- Chicago Field Museum of Natural History

4 Chester A. Arnold, (1901-1977) (professor of botany and curator of fossil plants- University of Michigan) in "An Introduction to Paleobotany" (McGraw-Hill Publishing Co: New York, 1947) p.7. (2christ.org/quotes) see at "Scientific Creationism" Henry Morris, (Creation-life Pub: San Diego, Ca. 1974) p.86, 87.

5 Eldred Corner, (1906-1996) (professor of botany- Cambridge) A.M. MacLeod & L.S. Cobley, eds. "Evolution in Contemporary Botanical Thought" (Quadrangle Books: Chicago, IL. 1961) p. 95, 97. from Bird, L. p. 234. see "Creation Facts of Life" Gary Parker, (op cit) p.148.

6 Steven M. Stanley, "The New Evolutionary Timetable: Fossils, Genes, and the Origin of Species" (Basic Books: New York 1981) p.95.

7 Douglas Futuyma, (American evolutionary biologist) "Science on Trial: The Case for Evolution" (Pantheon Books: New York 1983) p.82.

8 Colin Patterson, (in a letter to Luther D. Sunderland, April 10, 1979) as appearing in "Darwin's Enigma: Fossils and Other Problems" by Luther Sunderland (Master Book Publishers: Green Forrest, AR) (4th edition printing El Cajon, Ca. 1988) p.89. see "Refuting Evolution" Jonathan Sarfati, (Master Books: Green Forest, AR. 1999) p.48.

9 Albert Fleischmann, (1862-1942) (former German professor of zoologist and comparative anatomy) 'The Doctrine of Organic Evolution in the Light of Modern Research' Journal of the Transactions of the Victoria Institute 65, 1933. p.194-95, 205-6, 208-9. (nwcreation.net/evolutionquotes) 2014.

10 Stephen J. Gould, 'Is a New and General Theory of Evolution Emerging'? an article appearing in Paleobiology, vol. 6(1) Jan. 1980, p.127; p. 40 as cited in Genesis Park. see "Examine the Evidence" Ralph O. Muncaster, (Harvest House Publishers; Eugene, Oregon 2004) p.54.

11 Java-man fraud/Eugene Dubois retraction, "Evolution- The Fossils Say No!" Duane T. Gish (ICR Publishing: San Diego, California 1973) p.85-88.
12 Duane T. Gish, (expert on the fossil record exposes the fraud of Piltdown Man) "Evolution-The Fossils Say No!" (ICR Publishing: San Diego, California 1973) p.92
13 Solly Zuckerman, (1904-1993) "The Amazing Story of Creation" Gish (ICR Pub: El Cajon, Ca.1990) p.84.
14 Solly Zuckerman, "Beyond the Ivory Tower" (Taplinger Publishing Co: New York 1971) see "The Amazing Story of Creation" Duane T. Gish (op cit), p.84.
15 Richard Leakey, (paleoanthropologist son of Louis Leakey and director of the National Museums of Kenya) The Weekend Australian magazine May 7&8 1983, p.3. (creation.org/articles/quotes) 2014
16 Charles Oxnard, Reference 13 above, p.84. also see quotation in "Creation Facts of Life" Gary Parker, (Master Books: Green Forest, AR. 1994) p.162-3.
17 Duane T. Gish, Reference 12 above, p.103.
18 Timothy White, (paleoanthropologist- University of California, Berkeley) see Ian Anderson article 'Hominid Collarbone Exposed as Dolphin's Rib' 'New Scientist' April 28, 1983. p.199. (creation.org/articles/quotes) 2015.
19 Greg Kirby, (Senior lecturer in population biology at Flinders University) quoted from an address to the biology teachers Association, South Australia. 1976. see (chick.com/evolution) by Chick Publications, Jan/Feb, 1987.
20 Stephen J. Gould, "The Panda's Thumb" (W.W. Norton publishers: New York 1980) p.126.
21 Henry M. Morris, "Science and the Bible" (Moody Press: Chicago, Illinois 1986) p.61
22 Robert Wiedersheim, (1848-1923) (German Anatomist) writing in 'The Structure of Man: An Index to His Past History' 1893. 2nd edition. (translated by H. and M. Bernard) (Macmillan and Co: London, England 1895) see (wikipedia.org/wiki/robert_wiedersheim) 2014.

23 Horatio Newman, (1875-1957) (a zoologist from the University of Chicago who offered a written statement read into evidence at the famous Scopes Trial of 1925) Darrow, Clarence & William J. Bryan (1997) 'The World's most Famous Court Trial: The Tennessee Evolution Case Published' The Lawbook Exchange, Ltd. p.268. found at (wikipedia.org/wiki/robert_wiedersheim) 2014.

24 Henry Morris, "The Long War Against God" (Master Books: Green Forrest, AR. 2000) (4th printing) 2008 p.113

25 Wolfgang Smith, (mathematician, physicist, philosopher- MIT, UCLA, OSU) 'Teilhardism and the New Religion: A Thorough Analysis of the Teachings of Pierre Teilhard de Chardin' (Tan Books, Rockford, IL. 1988) p.5-6. see "The Long War Against God" H. Morris (op cit) p.161.

26 David A. Noebel, (American religious leader, Christian philosopher and avid writer) "The Battle for Truth" (Harvest House Publishers: Eugene, Oregon 2001) p.146.

27 Lee Strobel, (American journalist, Christian author and apologist) "The Case for Faith" (Zondervan Publishing House: Grand Rapids, Michigan, 2000) p.91

Chapter Four

1 John Lennox, (Oxford mathematician and philosopher of science) "God's Undertaker: Has Science Buried God" (Lion Hudson Pub: Oxford, England 2nd edition 2009) p.127.

2 Ibid. p.129

3 Michael Behe, J. Buell & G Hearn, eds. Darwinism: Science or Philosophy ? Dallas, Texas: Foundation for Thought and Ethics, 1994. quoted in: Stephen C. Meyer, 'The Explanatory Power of Design', p.126. see "The Case for Faith" Lee Strobel, (Zondervan Publishing: Grand Rapids, Michigan 2000) p.101.

4 I.L. Cohen, (engineer & mathematician/ New York Academy of Science) "Darwin Was Wrong: A Study in Probabilities" (New Research publications: 1984) p.205. (Greenvale, NY. ed. 1985) p.6-8, 209-210, 214-215. see the creation - evolution encyclopedia at (pathlights.com) 2014

5 Fred Hoyle, (1915-2001) (British astronomer and professor, Cambridge University) 'Hoyle on Evolution'. Nature, vol. 294, Nov 12, 1981, p.105. see "The Case for Faith" Lee Strobel (op cit) p.101.

6 Fred Hoyle, ' The Universe, Past and Present Reflections' Annual Review of Astronomy and Astrophysics: vol 20, 1982. p.16. Engineering and Science 1981 p.12, quoted in Holder, 'Is the Universe Designed'? 'Nature' Nov 12, 1981 vol 294: p.105. partial quote, see (godandscience.org/apologetics/quotes) 2014.

7 Christian de Duve, (1917-2013) (Nobel Laureate, Belgian Cytologist/Biochemist) 'A Guided Tour of the Living Cell' Scientific American Library Series Nov. 1984. see science quotes at (simpletoremember.com) 2014

8 Kenneth Ham, (conference speaker, educator, author and Christian Apologist) 'Is There Really a God?' booklet (Answers in Genesis, Florence, KY. 1998) p.23.

9 Edgar Mitchell, (1930-2016) (American Pilot- NASA: decorated Apollo 14 Astronaut and the sixth person to walk on the moon) see (simpletoremember.com/science-quotes) 2014.

10 Francis Collins, "The Language of God" (Free Press: New York 2006) p.1-2.

11 Richard Dawkins, "Genetics: Why Prince Charles Is So Wrong" (checkbiotech.org Jan. 28 2003) see: "Total Truth" Nancy Pearcey (Crossway Books: Wheaton, IL. 2005) p.191.

12 Nancy Pearcey, (evangelical author, Frances Schaeffer scholar, Christian philosopher and apologist serving at the Discovery Institute Center for Science and Culture) "Total Truth" (Crossway Books Wheaton, IL. 2005) p.197.

13 Ibid. p.198.

14 Francis Collins, (director of the Human Genome Project) "The Language of God", (Free Press: New York 2006) p.1. (paraphrase) cited by Marilyn Adamson "Does God Exist- Six Reasons to Believe" p.5. (everystudent.com) 2013.

15 Francis Collins, "The Language of God" (Free Press: New York 2006) p.3. Nancy Pearcey, 'Copying the Human Script: Genome Project Raises Hopes, Fears' World July 8, 2000. see "God's Undertaker" J. Lennox (op cit) p.176..

16 Michael Denton, (Btitish/Australian microbiologist and author) "Evolution: A Theory in Crisis" (Burnett Books Ltd: London, England, 1985) p.334. see quotes by men of science (simpletoremember.com/science-quotes) 2013.

17 Werner Gitt, (director: German Institute of Physics and Technology) "In the Beginning was Information" CLV, (Bielefeld, Germany 1997) p.79. see "Is There Really a God"? Ken Ham booklet (Answers in Genesis 1998) p.26.

18 Ibid. p.107. see "God's Undertaker" John Lennox (op cit) p.145.

19 Bill Gates, (Chairman- Microsoft Corporation) "The Road Ahead" 1995, (Penguin: London, revised) (Blue Penguin: Boulder 1996) p.228. see "God's Undertaker" J. Lennox (op cit) p.145.

20 Walter T. Brown, (American engineer and young earth creationist-MIT) "In the Beginning: Compelling Evidence for Creation" 7th edition (Phoenix: Center for Scientific Creation 1986) p.6. see "The Battle for Truth" David Noebel (Harvest House 2001) p.141.

21 Molecular Configuration: 'Chirality' requirements in organic physiology. see "Examine the Evidence" Ralph O. Muncaster (Harvest House: Eugene, Oregon 2004) p.110-115.

22 Francis Collins, "The Language of God" (Free Press: New York 2006) p.90-91.

23 Michael Denton, "Evolution: A Theory in Crisis" (Adler & Adler, Bethesda, MD. 1986) p.250. see "The Battle for Truth" D. Noebel (op cit) p.140. and "God's Undertaker" J. Lennox (op cit) p.122.

24 Walter L. Bradley, (advocate of Intelligent Design- former professor of engineering: Baylor University) interview with Mr. Lee Strobel "The Case for Faith" (Zondervan Publishing: Grand Rapids, Michigan 2000) p.109.

25 Richard Dawkins, 'The Necessity of Darwinism' New Scientist, 1982 vol. 94 (April 15) p.130. see Ken Ham "Is There Really a God"? booklet by (Answers in Genesis 1998) p.9-10.

26 Stephen C. Meyer, (history/philosophy of science graduate at Cambridge, director of the Center for Science and Culture: The Discovery Institute "The Message of the Microcosm: DNA and the Death of Materialism" Cosmic Pursuit Fall 1997 p.41-42. see Ralph O. Muncaster "Examine the Evidence" (op cit) p.117,123.

27 Michael Behe, "Darwin's Black Box" (The Free Press: New York 1996) p.193,243,251. see "The Case For Faith" Lee Strobel (op cit) p.90-91.

28 Michael Denton, "Evolution: A Theory in Crisis" (Adler & Adler: Chevy Chase, Maryland 1985, 86) p.77. see "The Case For Faith" Lee Strobel (op cit) p. 87.

29 Michael Denton, "Evolution: A Theory in Crisis" (op cit) p.62,358. (nwcreation.net/evolutionquotes) 2014.

30 Ibid. p.249-50. also see "God's Undertaker" J. Lennox (op cit) p.122.

31 Ibid. p.250. and "God's Undertaker" J. Lennox (op cit) p.122,123.

32 Jacques Monod, (1910-1976) (French Biologist: Nobel Prize winner 1965) 'Chance and Necessity: An Essay on the Natural Philosophy of Modern Biology' (Collins Press: London 1972) p.134 see "God's Undertaker" J .Lennox (op cit) p.123.

33 Michael Behe, "Darwin's Black Box" (Simon & Schuster: New York 1996) p.39. see "God's Undertaker" J. Lennox (op cit) p.123,124.

34 Ibid. p.193. see "God's Undertaker" J. Lennox (op cit) p.125.

35 Frederick Hoyle, "The Intelligent Universe" (Michael Joseph pub: London, England) (Holt, Rinehart, Winston: reprint New York 1983) p.20,21,23.

36 Wolfgang Smith, "Teilhardism and the New Religion: A Thorough Analysis of the Teachings of de Chardin" (Tan publishing: Rockford, IL. 1988) p.1-2. "The Long War Against God" Henry Morris (partial quote op cit) p.279.

37 Charles Darwin, 'The Life and Letters of Charles Darwin' 1887, vol. 2. p.229. see famous quotes from famous evolutionists at (anounted-one.net/quotes) 2014.

38 Ilya Prigogine, (1917-2003) (Russian born, Belgian physical chemist and decorated Nobel recipient- 1977) 'Physics Today 25' by Prigogine, Gregair, & Babbyabtz, discussion p.23-28. (simpletoremember.com) 2014.

39 Paul Davies, (award winning British Physicist- professor at Arizona State University) "Superforce: The Search for a Grand Unified Theory of Nature" (Simon & Schuster: New York 1984) p.223. see "The Battle For Truth" David Noebel (op cit) p.140.

40 Ibid. p.243 (godandscience.org/apologetics/quotes) 2014.

41 Richard Dawkins, "The Blind Watchmaker" (W. W. Norton & Co: New York 1986) p.287. see "Total Truth" Nancy Pearcey (op cit) p.168.

42 S C Todd, 'A View from Kansas on the Evolution Debate', Nature 401 Sept. 30, 1999 p.423. see "Total Truth" Nancy Pearcey (Crossway Books: Wheaton, Ill. 2005) p.168.

43 Richard Dawkins, "The Blind Watchmaker" (W.W. Norton & Co: New York 1986) p.1. (emphasis added) see "Total Truth" N. Pearcey (op cit) p.183.

44 George Wald, (1906-1997) (American evolutionary scientist, professor emeritus of Biology/Harvard- Nobel recipient specializing in Neurobiology) 'Innovation and Biology' Scientific American, vol. 199,

Sept. 1958, p.100. (famous quotes from famous evolutionists, groups. google.com/forum) (the quote mine project) 2018.

45 Theodosius Dobyhansky, (1900-1975) (Ukrainian-American geneticist and evolutionary biologist) Review of Evolution, 'Creation and Science' by F.L. Marsh 1945. The American Naturalist p.75,79.

46 Nancy Pearcey, "Total Truth: Liberating Christianity from its Cultural Captivity" (Crossway Books: Wheaton, IL. 2005) (study guide edition) p.168-169.

Chapter Five

1 For detailed information on the use of radiometric dating methodology refer to: "Scientific Creationism" by Dr. Henry Morris (Creation Life Publishers: San Diego, Ca. 1974) beginning with p.137.

2 Robert H. Rastall, (economic Geology- Cambridge) Encyclopedia Britannica, vol. 10 (William Benton Publishing: Chicago, IL. 1956) p.168. (2christ.org/quotes) 2014.

3 Ronald R. West, (assistant professor of paleobiology, Kansas St. University) 'Paleoecology and Uniformitarianism' Compass, vol. 45, May 1968, p.216. see (creationism.org/articles/quotes) 2014.

4 Niles Eldredge, "Time Frames: The Rethinking of Darwinian Evolution and the Theory of Punctuated Equilibia" (Simon & Schuster: New York 1985) p.51,52 see the Quote Mine Project: Gould, Eldredge, etc. quotes 2014.

5 A. E. Wilder-Smith, (British organic chemist, and former atheist) "Man's Origin, Man's Destiny" (Harold Shaw Publishers: Jan. 1968) (9) p.127-128. (paperback by Bethany House publishing, 1975) (2christ.org/quotes) 2014.

6 Henry M. Morris, "Scientific Creationism" (Creation Life Publishers: San Diego, Ca. 1974) p.155. see also: Stuart E. Nevins article: 'Evolution: The

Oceans Say No'! ICR Acts and Facts, Impact Series #8, October, 1973.

7 Ibid, p.158.

8 Ibid, p.157-158. see also: Dr. Thomas G. Barnes article: 'Origin and Destiny of the Earth's Magnetic Field' (ICR publishing: San Diego, Ca. 1973) p.64.

9 Ibid, p.151-153. see also: Hans Pettersson, (Swedish oceanographer) 'Cosmic Spherules and Meteoric Dust', Scientific American, vol. 202 February 1960) p.132.

10 Ibid, p.153. (ocean salt content)

11 Ibid, p.167-169. (see the mathematical calculation of population dynamics based on the old earth scenario)

12 Ibid, p.169 for a further study of evidence supporting a young earth as compared to the billions of years demanded by evolution theory, I refer the reader to: Dr. Henry M. Morris, "The Biblical Basis for Modern Science" (Baker Publishing, 1985) p.477-480

13 Colin Patterson, (1933-1998) "Evolution" (Cornell University Press: Ithaca, New York 1978) p.150. (pathlights.com creation/evolution encyclopedia) 2014

14 L. H. Mathews, (evolutionary biologist) 'forward' to Charles Darwin, 'Origin of Species" 1971 edition, (J. M. Dent & Sons, Ltd: London, England) p. x,xi. see "Scientific Creationism" Henry M. Morris (Creation Life Publishers: San Diego, Ca. 1974) p.6.

15 Pierre-Paul Grasse, "Evolution of Living Organisms" (Academic Press: Nerw York 1977) p.107. he adds: (Natural selection is a conservation mechanism rather that a creative process. It serves to hinder the spread of harmful mutations through a population, keeping the status quo) as cited in "The Long War Against God" Dr. Henry Morris (Master Books: Green Forest, Ar. 2000) p.161.

16 Theodore Roszak, (1933-2011) (professor emeritus of History) California St. University 'Unfinished Animal: the Aquarian Frontier

and the Evolution of Consciousness' 1975, p.101-102. (nwcreation.net/evolutionquotes) 2014.

17 Dean Kenyon, (Biochemist: San Francisco St. University) see Walter T. Brown, "In the Beginning" (Phoenix: Center for Scientific Creation, 1986) p.6. cited in "The Battle For Truth" D. Noebel (op cit) p.142.

18 Wolfgang Smith, 'The Universe is Ultimately to be explained in Terms of a Metacosmic Reality' Margenau and Varghese (eds): Cosmos, Bios, & Theos p.113 quotes from famous evolutionists (anointed-one.net/quotes) 2014

19 Louis T. More, (Professor of Paleontology- Princeton) "The Dogma of Evolution" (Princeton University Press: Princeton, New Jersey 1925) 2nd printing p.160. also see "Darwin's Enigma: Fossils and Other Problems" Luther D. Sunderland (Master Books: Santee, Ca. 1988) p.27. and "Mind Siege" Tim Lahaye (Word Publishing, 2000) p.140.

20 Julian Huxley, Humanist Assoc. membership brochure (San Jose, California) 'What is Humanism'? Humanist Community of San Jose. features Julian Huxley. see also "Scientific Creationism" (op cit) p.197.

21 Henry Morris "Scientific Creationism" (op cit) p.197. (see Justice Abe Fortas comment in connection with US Supreme Court ruling striking down Arkansas anti-evolution law).

22 David A. Noebel, "The Battle for Truth" (Harvest House Publishers: Eugene, Oregon. 2001) p.147.

23 Michael Ruse, (science philosopher- Florida State University and a leading anti-creationist advocate) (1) 'Saving Darwinism from the Darwinians - How evolution became a religion: creationists correct'? The National Post May 13, 2000, p.B1,B3,B7. (2) 'Mystery of Mysteries: Is Evolution a Social Construction'? (Harvard University Press, Cambridge, Mass. 1999) cited by Nancy Pearcey "Total Truth" (Crossway Books, Wheaton, Illinois. 2005) p.172. These statements by Dr. Ruse are in stark contrast to his testimony at the Arkansas 'Balanced Treatment Bill Trial

of 1981/82' where he and other anti-creationists denied the claim that evolution should be accurately viewed as an anti-God religious dogma.

24 Ibid. (1) (op. cit) (2) p.172.

25 Dr. R. Kirk, as cited in 'The Rediscovery of Creation' (National Review, May 27, 1983) p.641. (google: quotes reveal the credulity of evolutionists) also see S. Jaki, 'Cosmos and Creator' 1982. (nwcreation.net/evolution) 2015.

26 Marjorie Grene, (American philosopher of science with a particular emphasis on biology) 'The Faith of Darwinism' Encounter, November 1959, p.49. reprinted in Grene, Marjorie "The knower and the known" (University of California Press: Berkley, Ca. 1974) p.187. retrieved – google books Sept. 17, 2014.

27 David Coppedge, 'Speaking of Science, Creation Matters', May/June 2003. (debunking evolution) (newgeology.us/presentation32.html) 2015.

28 Colin Patterson, (quoted from a letter to Luther D. Sunderland, April 10, 1979) (talkorigins.org) 2014. see "Darwin's Enigma: Fossils and other Problems" Luther D. Sunderland, (Master Books: San Diego, Ca. 1984) p.89.

29 Richard Bliss, (Professor of Biology-Christian Heritage College) 'It Takes a Miracle for Evolution' Christian Heritage Courier 1 (March 1979) p.4. cited by Tim Lahaye in "Mind Siege" Word Publishing (op cit) p.142.

30 Klaus Dose, (director of the Institute for Biochemistry, Gutenberg University, West Germany) 'The Origin of Life: More Questions than Answers', Inter-disciplinary Science Review, 13, 1988. p.348. see "God's Undertaker" J. Lennox (op cit) p. 133 & "The Case For Faith" Lee Strobel (op cit) p.107

31 Walter L. Bradley & Charles B. Thaxton, "Information and The Origin of Life" The Creation Hypothesis (Intervarsity Press: Douners Grove, Illinois 1994) p.194. see also 'Life from the Heavens? Not This Way' Fazale Rana & Hugh Ross, Facts and Faith, Quarter 1, 2000. cited by Lee

Strobel in "The Case For Faith" (op cit) p.107.

32 Julian Huxley, (evolutionist and former president of UNESCO) (statement taken from a television interview questioning J. Huxley about the rapid rise in Darwinian thought in the 20th century. (late 1950's) cited by Dr. Henry M. Morris in "The Troubled Water of Evolution" 1974, p.58.

33 Nancy Pearcey, "Total Truth" (Crossway Books, Wheaton, IL. 2005) p.176.

34 Richard Lewontin, (research professor – Museum of Comparative Zoology, Harvard University) "Billions and Billions of Demons" (New York Review of Books, Jan. 9, 1997) p.28,31. cited at (ucsb.edu/fscf/FAQ/evolution) review of Carl Sagan's book "The Demon Haunted World: Science as a Candle in the Dark", (New York Review of books, Jan 9, 1997) also see "God's Undertaker" J. Lennox, (op cit) p.35,36.

35 Julian Huxley, "Evolution: The Modern Synthesis", (Harper & Bros. Publishers: New York 1942) p. 457. see "The Battle for Truth" D. Noebel (op cit) p.118.

36 Gary Parker, "Creation Facts of Life", (Master Books: Green Forrest, AR. 1994) p.122.

37 Norman MacBeth, (Harvard trained attorney and out-spoken critic of the neo-Darwinism) "Darwin Retired", (Gambit Books: Boston, Mass. 1971) (forward) also see "Creation Facts of Life" Gary Parker (op cit) p.122.

38 Phillip Johnson, (former law professor: University of California-Berkley) "Darwin on Trial", (Regnery Gateway: Washington DC 1991)

39 Norman MacBeth, "Darwin Retired: An Appeal to Reason" (Gambit Books: Boston, Mass. 1971) (forward)

40 Johannes Kepler, (1571-1630) cited in Morris Kline "Mathematics: The Loss of Certainty" (Oxford University Press: New York 1980) p.31, as cited by J. Lennox , "God's Undertaker: Has Science Buried God"? (Lion Hudson Publishing: Oxford, England 2009) p.21.

41 Galileo, (1564-1642) 'GG ll Saggiatore', (1623) as translated in 'The Philosophy of the 16th & 17th Centuries' 1966 by Richard Henry Popkin, p.65. see Morris Kline's book "Mathematics: The Loss of Certainty" (1980) cited by John Lennox, "God's Undertaker" 2009 p.24.

42 Keith Ward, (British philosopher, theologian and scholar) "God, Chance, and Necessity" (One World Publications: Oxford, England 1996) see "God's Undertaker" J. Lennox (op cit) p.63.

43 Henry F. Schaefer, (highly decorated professor-University of Georgia - computational and theoretical chemist, a five time nominee for the Nobel Prize) J.L. Sheler & J.M. Schrof, 'The Creation', U.S. News & World Report (December 23, 1991) p.56-64. (godandscience.org/apologetics/quotes) 2014.

44 Kurt Godel, (world renown Austrian mathematician, logician, & philosopher) Palle Yourgrau, 'A World Without Time: The Forgotten Legacy of Godel and Einstein' (Basic Books: New York 2005) p.104-105. as cited by Antony Flew in "There is a God" (Harper Collins Pub: New York 2007) p.167.

45 John Lennox, "God's Undertaker" (Lion Books: Oxford, England 2009) p.63.

46 John Lennox, Ibid. p.65.

47 John Lennox, 'U-tube debates' an illustration often used by J. Lennox in many of his debates concerning the question: 'has science eliminated the need for God'?

48 Stephen Hawking, 'ABC Television's 20/20' 1989 as cited by J. Lennox "God's Undertaker" (op cit) p.63.

49 Arno Penzias, (America physicist and radio astronomer who won a Nobel Prize in 1978) In Denis Brian, Genius Talk, (Plenum: New York, 1995) see "God's Undertaker" J. Lennox, (op cit) p.75.

50 Charles Townes, (highly decorated American Nobel Prize winning physicist 1964) 'Making Waves' American Physical Society, 1995. see

"God's Undertaker" J. Lennox (op cit) p.69.

51 Julian Huxley, 'At Random' a television preview aired on Nov. 21, 1959. cited in "The Battle For Truth" David A. Noebel, (op cit) p.124.

52 Junior High biology textbook, "Matter- Its forms and changes" 2nd edition (Brandwein, Stollberg, Greenstone, Yasso, Brovey) Junior High School textbook, Los Angeles City Schools (Harcourt Brace Jovanovich, Inc: (New York, Chicago, San Francisco, Atlanta, & Dallas) 1972.

53 High school textbook, "BSCS Biology: A Molecular Approach" 8th edition, Jon Greenberg (revision editor) (Everyday Learning Corp: 2001) p.446. as cited in "Total Truth" N. Pearcey, (op cit) p.169.

54 High school textbook, (Neil A. Campbell, Jane B. Reece, & Lawrence G. Mitchell) "Biology" 5th edition. (Addison Wesley: Reading, Mass. 1999) p.426. see "Total Truth" N. Pearcey, (op cit) p.169.

55 College level textbook, Douglas Futuyma, "Evolutionary Biology" 3rd edition textbook, (Sinauer Co: Sunderland, Mass. 1998) p.15. see "Total Truth" N. Pearcey, (op cit) p.169.

56 University textbook, "Evolution" 2nd edition (Sudbury, Jones & Bartlett, 1996) p.62. see "God's Undertaker" J. Lennox, (op cit) p. 87.

57 Douglas Futuyma, (Evolutionary Biologist) "Science on Trial: The Case for Evolution" 1983, p.82. also see: "Evolutionary Biology" 2nd edition (Sinauer Co: Sunderland, Mass. 1986) p.3. as cited in 'God's Undertaker' John Lennox, (op cit) p.87, & "Total Truth" Nancy Pearcey, (op cit) p.200.

58 Richard Lewontin, cited by D. Futuyma "Science On Trial" (Sinauer Co: Sunderland, Mass. 1995) p. 161. see "God's Undertaker" J. Lennox, (op cit) p.99.

59 The Humanist magazine, 'A Religion for a New Age' J. Dunphy The Humanist Jan-Feb 1983, p.23,26. (emphasis added) cited by Wendell R. Bird, Origin of the Species – Revisited, vol. 2 p. 257. see "Refuting Evolution" J. Sarfati, p.20-21, and "Mind Seige" Lahaye p.170.

Chapter Six

1 The documentary film "Expelled: No Intelligence Allowed" was directed by Nathan Frankowski and hosted by Ben Stein. 2008.

2 A Professional 'textbook critique' by Mills, Lancaster & Bradley: 'Origin of Life- Evolution in Biology Textbooks- A Critique' The American Biology Teacher, vol 55, No. 2, February, 1993, p.83.

3 Tim Lahaye, (Christian minister and apologist, distinguished author, expert educator in Biblical escatology) "Mind Siege-The Battle for Truth in the New Millennium" (Word Publishing: Nashville, TN. 2000) p.153,154.

4 M.H. Armitage and K.L. Anderson, Acta Histochemica 115, 603-608: (intellihub.com/Scientist-terminated) 2013.

5 Nancy Pearcey, 'Scopes in Reverse' Washington Times, July 24, 2000. cited in "Total Truth" Nancy Pearcey (Crossway Books: Wheaton, IL. 2005) p.173.

6 Reader's Digest magazine- a response from the editors to an ICR article publication request. (Institute for Creation Research News Letter Jan 1982. El Cajon, Ca.)

7 Luther D. Sunderland, "Darwin's Enigma: Fossils and Other Problems", 4th edition. (Master Books: Santee, Ca. 1988) p.7,8. (2christ.org/quotes) 2014.

8 Henry Morris, (author of more than twenty books on science and faith) "Many Infallible Proofs" (Master Books: El Cajon, California 1974) p.246.

9 Albert Einstein, Max Jammer, "Einstein and Religion" (Princeton University Press: Princeton, NJ, 1999) p.48. see "There Is a God" Antony Flew, (Harper & Collins Publications: New York 2007) p.99.

10 Antony Flew, (Professor of Philosophy/Oxford) 'Atheist Becomes Theist: Exclusive Interview with Former Atheist Antony Flew' at Biola University. see interview at (godandscience.org/apologetics/quotes) 2015.

11 Wernher Von Braun, (rocket scientist and space pioneer) as quoted by James Perloff, "Tornado in a Junkyard" (Refuge Books: Arlington, Mass. 1999) p.253. see (nwcreation.net/evolution quotes) 2015. also see T. McIver, 'Ancient Tales and Space-Age Myths of Creationist Evangelism' 1986. The Skeptical Inquirer vol. 10: p.258-276. 'Gone Bananas' World Sept. 7, 2002. (godandscience.org/apologetics/quotes) 2015.

12 Wernher Von Braun, This Week Magazine Jan 24, 1960. p.2.

13 Kenneth Ham, 'Is There Really a God'? 1998. booklet by Answers in Genesis, Florence, KY. p.33.

14 Albert Fleischmann, "Witnesses Against Evolution" by Mr. John F. Meldau (Christian Victory Publishing: Denver, Co. 1968) p.13. see citation at (nwcreation.net/evolutionquotes) 2014.

15 Henry M. Morris, "The Long War Against God", (Master Books: Green Forrest, AR. 2008) p.256 (4th printing)

16 Ibid, p.32.

17 Frank Tipler, (mathematical physicist and cosmologist - Tulane University) "The Physics of Christianity" (Doubleday publishing: New York 2007)

18 William Paley (1743-1805) (Christian apologist and philosopher) This is a variation of an original quote from his book "A View of the Evidences of Christianity" 1794. (this quotation is often confused and has sometimes been mistakenly credited to a Mr. Herbert Spencer.)

19 Niles Eldredge, "Time Frames: The Rethinking of Darwinian Evolution and the Theory of Punctuated Equilibria" (Simon & Schuster: New York 1985) p.44. "Time Frames: The Evolution of Punctuated Equilibria" (Princeton University Press: Princeton, New Jersey 1985) p.144-45. cited in "God's Undertaker" J. Lennox (op cit) p.114. (famous quotes from famous evolutionists) 2014.

20 Arthur Keith, (anti-Christian physical anthropologist) writing in the 'forward' to the 100th anniversary edition of Charles Darwin's 'Origin

of Species, 1959. see Google Groups- famous quotes from famous evolutionists. cited at (soulwinners.com.au/8.html) 2014.

21 Frances Crick, (British molecular biologist & Nobel Laureate for his epic DNA discoveries/ research) "Life Itself, Its Origin and Nature" (Futura pub: London, England 1981,82) p.88 (Simon & Schuster: New York 1981) p.88. cited in "God's Undertaker" J. Lennox, (op cit) p.133.

22 Roger Haines Jr. (research attorney- Ca. Third District Court of Appeals) 'Macroevolution Questioned' Creation Research Society Quarterly, Dec 1976, p.169.

23 Edmund J. Ambrose, (professor of cellular biology, University of London) "The Nature and Origin of the Biological World", (John Wiley & Sons, 1982) p. 164. (anointed-one.net/quotes) 2014.

24 Mark Czarnecki, (evolutionary paleontologist) 'The Revival of the Creationist Crusade' MacLeans's Jan. 19, 1981, p.56. (The TalkOrigins Archive- Quote Mine Project) also see (anointed-one.net/quotes) 2014

25 Robert Jastrow, (American astronomer, physicist, cosmologist and a leading NASA scientist) "The Enchanted Loom: Mind In the Universe" (Simon & Schuster: New York, 1981) p.19. (anointed-one.net/quotes) 2013.

26 Stephen J. Gould, 'Evolution's Erratic Pace' Natural History vol. 86. (May 1977) p.14. (The Talk Origins Archive -Quote Mine Project) (talkorigins.org) 2014.

27 Allan Sandage, (widely regarded as the world's greatest living cosmologist: winner of the Crawford prize in astronomy) J.N. Willford: March 12, 1991. 'Sizing up the Cosmos: An Astronomers Quest' New York Times, p.B9. also see "God's Undertaker" J. Lennox (op cit) p.66. (godandscience.org/apologetics/quotes) 2014.

28 H. J. Lipson, (Professor of Physics, University of Manchester, UK) 'A Physicist Looks at Evolution' Physics Bulletin, vol. 31. 1980 p.138. (emphasis his) Is evolution religion? (nwcreation.net/evolution) 2014.

29 Wolfgang Smith, "Teilhardism and the New Religion: A Thorough Analysis of the Teachings of de Chardin" (Tan Books & Publishers: Rockford, IL. 1988) p.1-2. (2christ.org/quotes)(wordpress.com) 2014.

30 Frederick B. Burnham, (theologian and professor-history of science) quoted from an appearance on ABC's Nightline with Ted Koppel, April 24, 1992. quotes for creation (english.sdaglobal.org/research/qotcratn.htm)2014.

31 Sir William Dawson, "The Story of Earth and Man" (Harper & Brothers: New York, 1887) p.317,322,330,339. (nwcreation.net/evolutionquotes) 2014.

32 Henry M. Morris, "The Long War Against God" (Master Books: Green Forest, AR. 2008) p.199.

33 Ernst Chain, (1945 Nobel Prize winner- Isolated and Purified Penicillin) (Jerry Bergman, Ph.D. April 2008. 'Ernst Chain: Antibiotic Pioneer' Acts&Facts, vol. 37, No. 4, p.10-12.) quotes 1 & 4 (A) Clark, R.W. 1985. 'The Life of Ernst Chain: Penicillin and Beyond' New York: St. Martin's Press, p.147. quotes 2&3 (B) Chain, E. 1970. 'Social Responsibility and the Scientist in Modern Western Society' (Robert Waley Cohen memorial lecture) London: 'The Council of Christians and Jews' p.25. Dr. Ernst Chain, along with Alexander Fleming and Howard Florey, all received the Nobel Prize for their discovery and development of penicillin—a medical breakthrough that would prove to be an invaluable asset to the world community at large.

34 Robert Jastrow, "God and the Astronomers" (W.W. Norton & Co: New York, 1978) p.115-116. in Moreland J.P. ed., 'The Creation Hypothesis', 1994, p.292-293 (godandscience.org/apologetics/quotes) 2014.

35 Arno Penzias,"Science and Christianity" Henry F. Schaefer p.49. also see A. Penzias quoted by Malcolm Browne, 'Clues to the Universe's Origin Expected' New York Times, March 12, 1978, p. 1. see "The Language of God" Francis Collins (Free Press: New York 2006) paperback edition p.76.

36 Soren Lovtrup, (Swedish Biologist/Embryologist) "Darwinism: The Refutation of a Myth" (Croom Helm pub: London/New York 1987) p.422. (2christ.org/quotes) and (nwcreation.net/evolutionquotes) 2014.

37 Louis Bounoure, (Professor and Director of Research at the French National Center of Scientific Research) 'The Advocate' March 8, 1984. p.17. (2christ.org/quotes) and (nwcreation.net/evolutionquotes) 2014.

38 T. N. Tahmisian, (former Atomic Energy Commission physiologist) 'The Fresno Bee', August 20, 1959. as quoted by N. J. Mitchell 'Evolution and the Emperor's New Clothes' (Roydon Publications: United Kingdom 1983) title page. (2christ.org/quotes) and (nwcreation.net/evolutionquotes) 2014.

39 John Lennox, "God's Undertaker" (Lion Books: Oxford, England 2009) p.210.

Chapter Seven

1 Julian Huxley, "Evolution after Darwin' Sol Tax. edition, (University of Chicago Press: Chicago, Illinois 1960) cited in "God's Undertaker" J. Lennox, paperback edition (op cit) p.87.

2 Nancy Pearcey, (fellow - Discovery Institute) "Total Truth" (Crossway Books: Wheaton, Illinois. 2005) p.189.

3 Stephen Hawking, (award-winning British astrophysicist) 'A Brief History of Time' (Bantam Press: New York 1998) p.210. (1988 ed. p.175) cited by Francis Collins, "The Language of God" (Free Press- paperback 2007) p.62.

4 Paul Davies, 'A Brief History of the Multiverse' The New York Times, published on April 12, 2003. cited in "Total Truth" Nancy Pearcey, (op cit) p.189.

5 Paul Davies, "God and the New Physics" (Simon & Schuster: New York 1983) p.145. cited in "Mind Siege" Tim Lahaye, (Word publishing:

Nashville,TN 2000) p.143

6 Paul Davies, 'The Just So Universe' Walter L. Bradley, Touchstone, July/August 1999, p.72.

7 John O'Keefe, (planetary scientist and NASA Astronomer) F. Heeren, "Show Me God" (Searchlight Publications: Wheeling, IL. 1995) p.200. see (godandscience.org/apologetics) 2014.

8 John Polkinghorne, (theologian & professor of mathematical physics-Cambridge, a leading voice explaining the relationship between science and religion) 'Science Finds God' Sharon Begley, Newsweek Magazine July 20, 1998 p.48. see "Mind Siege" (op cit) p.138.

9 Sir Frederick Hoyle, "The Intelligent Universe" (Michael Joseph pub: London, England 1983) p.220. see (english.sdglobal.org/research/qotcratn) 2014.

10 Steven Weinberg, (American physicist and Nobel Laureate in High Energy Physics 1979) writing in the journal 'Scientific American' see quotation at (simpletoremember.com/science-quotes) 2015

11 Edward Harrison, (British astronomer and cosmologist) "Masks of the Universe" (Collier Books: Macmillan, 1985) p. 252,263 science quotes (godandscience.org/apologetics/quotes) 2014.

12 George Greenstein, (American Astronomer) "The Symbiotic, Universe: Life and Mind in the Cosmos" (Williams & Morrow: New York, 1988) p.27. see "Total Truth" (op cit) p.189,190.

13 Alexander Polyakov, (Soviet mathematician) S. Gannes, Fortune Magazine October 13, 1986. p. 57. see (godandscience.org/apologetics/quotes) science quotes 2014.

14 John Lennox, U-tube video doc. 'Design of The Universe?' 9/28/2007 John Lennox 'the birth of science' lecture, see also "God's Undertaker" (op cit) p.70. (accuracy comparison by Dr. Paul Davies)

15 John Lennox, "God's Undertaker" (Lion Books: Oxford, England 2nd edition (paperback) 2009) p.70.

16 Arno Penzias, "Cosmos, Bios, and Theos" H. Margenau, R. Varghese, ed. (Open Court pub: La Salle, IL. 1992) p.83. also see 'Our Universe: Accident or Design' (Wits 2050, S. Africa: Starwatch, 1992) p.42. quoted by Walter Bradley in the 'Just So' Universe- the fine-tuning constants and conditions in the cosmos. "God's Undertaker" John Lennox, (op cit) p.58,73.

17 Henry F. Schaefer, (five time Nobel nominee) quoted in his essay: 'Stephen Hawking, The Big Bang, and God' (godevidence.com/2010/08/quotes) 2014.

18 Alan Sandage, (highly decorated American astronomer) "Sizing up the Cosmos: An Astronomers Quest" J.N. Willford, New York Times, March 12, 1991, p.B9.

19 Charles Darwin, (the most predominant founding father of evolution theory, British naturalist and author) "The Autobiography of Charles Darwin" Nora Barlow edition (Collins Co: London 1958) p.92,93. as quoted in Antony Flew's book "There is a God" 2007 p.106. also see Kenneth R. Miller, "Finding Darwin's God" (HarperCollins: New York 1999) p.287. and "The Language of God" Francis Collins, 2007, p.99. (emphasis added)

20 William Paley, (18th century clergyman, naturalist and Christian apologist who first championed the teleological argument for the existence of God long before the recent advent of evolution theory) "Natural Theology: or Evidences of the Existence and Attributes of Deity, Collected from the Appearances of Nature" publish date 1802. (material reprinted in 1972 by (St. Thomas Press: Houston, TX) p.3.

21 Isaac Asimov, 'In the Game of Energy and Thermodynamics You Can't Even Break Even' Smithsonian Institute Journal, June 1970. p.10. (creation.org/articles/quotes) 2014.

22 Antony Flew, (former long-time atheist and advocate of materialism) "There is a God, How the world's most notorious atheist changed his

mind" (Harper Collins pub: New York 2007) p.126.

23 Richard Feynman, (American theoretical physicist- Nobel Prize recipient 1965 for his work in quantum electrodynamics) "The Meaning of it All: Thoughts of a Citizen-Scientist" (Basic Books: New York. 1998) p.43.

24 Julian Huxley, see (gospeloutreach.net) 2014. "Does God Exist" Rolaant L. McKenzie August 2007, p.3.

25 Time Magazine, April 1980. 'Modernizing the Case for God' April 7, 1980. as quoted in "There is a God" Antony Flew, 2007, preface p. xv.

26 Jim Holt 'Science Resurrects God' fascinating article from The Wall Street Journal, (December 24, 1997) Dow Jones & Co., Inc. (quotes from scientists - design of the universe) (godandscience.org/apologetics/quotes) 2014.

27 Sir Isaac Newton, (1643-1727) (widely recognized as one of the most influential scientists of all time and a key figure in the scientific revolution) 'Symmetry in Plants' 1998 Roger Jean & Denis Barabe (part 1) p. xxxvii (a translation of a Latin phrase Newton wrote in his student's notebook) see (wikiquote.org/isaacnewton) 2014. Newton (part 2) ..."Human Nature: An Interdisciplinary Biosocial Perspective' vol 1, Issues 7-12, 1978, p. 47.

28 Paul Davies, (a highly regarded British astrophysicist, professor and author, winner of the Templeton Prize 1995, Faraday Prize 2002) "The Cosmic Blueprint: New Discoveries in Nature's Creative Ability to Order the Universe". (Simon & Schuster: New York, 1988) p.203. (godandscience .org/apologetics/quotes) 2014. see partial quote in: "God's Undertaker" (op cit) p.71.

29 Max Planck, (Nobel Prize winning Physicist and founder of Quantum Theory) 'Religion and Natural Science' (lecture - 1937) Scientific Autobiography and Other Papers. translated by F. Gaynor. New York, 1949. p.184.

30 Antony Hewish, (British physicist and radio astronomer- Nobel Prize recipient-1974, for his work on pulsar research and high energy

physics) Dr. Hewish has consistently argued that religion and science are complementary as noted in the forword to John Polkinghorne's book "Questions of Truth" (Nicholas Beale, Jan 16, 2009) (John Knox Press: Westminster) p.12. (retrieved 7/27/12) (see fifty one responses to questions about God, science, and belief).

31 Tony Rothman, (American physicist, cosmologist and scientific writer) J.L. Casti, "Paradigms Lost" (Avon Books: New York, 1989) p.482-483. (godandscience.org/apologetics/quotes) 2014.

32 Paul Davies, "The Mind of God" (Simon & Schuster: London, England 1992) p.232. see "God's Undertaker" J. Lennox , 2009, p.59.

33 Bertrand Russell, "Religion and Science" (Oxford University Press: Oxford, England 1970) p.243. cited in "God's Undertaker" (op cit) see introduction to chap 2. p.31. & p.40,41.

34 Fyodor Dostoyevsky (Russian writer and intellectual) quoted by John Lennox (Veritas Forum – Duke University) March 1, 2012. Addressing the subject: 'Is God Rendered Irrelevant'? (www.veritas.org) 2012.

35 John Lennox, U-tube lecture 'Common Objections Atheists Make' June 2011.

36 Richard Dawkins, as quoted by John Lennox in a U-tube debate with Dawkins on the topic 'Is There a God'?

37 Alexander Solzhenitsyn, (world famous Russian dissident) as quoted by John Lennox: U-tube debate 'Is There a God'? 2008 apologetics.com.

38 Jurgen Habermas, as quoted in a U-tube lecture by John Lennox 2/15/2011 on the topic 'Is Faith Delusional'?

39 Rolaant L. McKenzie, (business and learning specialist) 'Does God Exist'? August 2007 as cited at (gospeloutreach.net) p.4. 2013.

40 C. S. Lewis, (British novelist, literary critic, and Christian apologist who wrote the highly acclaimed fictional novel series: The Chronicles of Narnia) quotation from (thoughtfulchristianity.net) 2014.

41 William Lane Craig, (Analytical Philosopher and Christian Apologist -

never outmatched in over 100 debates addressing the C/E cultural war in America) as stated in a 1998 debate vs Antony Flew: Madison, Wisconsin.

Chapter Eight

1 Grahame Clark, (British Archaeologist and Ancient Literature specialist) as quoted by John Lennox in a U-tube debate with Victor Singer. 'Is there a God'?

2 Rev. Billy Graham, (highly esteemed as the world's greatest preaching evangelist of the 20th century) "Salvation: The Reason for My Hope" (Thomas Nelson Publishing: Nashville, TN. 2013) p.77. (W Publishing Group)

3 Henry Morris, "The Long War Against God" (Master Books: Green Forrest, AR. 2000) p.277.

4 Peter Hitchens, (British Journalist and influential writer) 'Darwin's bizarre cult' The International Express (UK) Jan 5th 2000. (creation.com/darwin-bizarre-cult) 2015.

5 Rick Warren, 'Meditations on the Purpose-Driven Life' Inspirio, a division of Zondervan publishing. 2002.

6 Frank Tipler, (esteemed professor of mathematical Physics) F.J. Tipler "The Physics Of Immortality" (Doubleday Books: New York, 1994) preface p. ix cited at: (godandscience.org) & (allaboutscience.org) 2014.

7 Frank Tipler, (former atheist) "The Physics Of Christianity" (Doubleday Books, New York 2007) (Introduction) (godandscience.org/apologetics/quotes) 2014.

8 Arthur L. Schawlow, (Nobel Prize winner in 1981 for his work on laser spectroscopy/Stanford University) H. Margenau & R. Varghese, eds. "Cosmos, Bios, and Theos: Scientists Reflect on Science, God, and the Origins of the Universe, Life, and Homo Sapiens" (Open Court Pub: La Salle, IL.1992) p.107 (godandscience.org/apologetics/quotes)

9 Ibid. p.107. also see "God's Undertaker" John Lennox, (Lion-Hudson pub: Oxford, England 2009) p.210.

10 Ravi Zacharias, (Indian-born Christian apologist and an international champion for evangelical truth) as quoted in Servant magazine, spring issue, 1999, p.8. "The Case For Faith" Lee Strobel (Zondervan: Grand Rapids, Mi. 2000) p. 247. (thepoachedegg.net/ravi-zacharias) 2014.

11 Ravi Zacharias, "A Shattered Visage: The Real Face of Atheism" (1994, 2004) as quoted in 'Scorned for God-Part 1' (Let My People Think Podcast).

12 Albert Einstein, Max Jammer, 'Einstein and Religion' (Princeton University Press, Princeton, NJ 1999) p.93. see 'There is a God' Antony Flew, 2007, p.102.

13 Albert Einstein, "The Quotable Einstein", ed. Alice Calaprice, (University Press: Princeton, NJ. 2005) p.195-6. "There is a God" Antony Flew, (HarperCollins pub: New York 2007) p.102,103.

14 Albert Einstein, as quoted in 'The Saturday Evening Post' Oct. 26, 1929.

15 I. L. Cohen, (American engineer and mathematician) "Darwin Was Wrong: A Study in Probabilities" 1984/1985 (Greenvale, New York: New Research Publications, Inc.) p.6-8, 209,10. 214,15.

16 Pierre Paul Grasse, "Evolution of Living Organisms" (Academic Press: New York 1977) p.8. see "The Long War Against God" (op cit) p.278.

17 William Jennings Bryan, (1860-1925) (American Politician and Attorney) 'Quotations on Free Thought and Religion' (about.com) 2015.

18 Martin Luther King Jr., (1929-1968) (famous American Baptist Minister & Civil-Rights Leader/ Nobel Peace Prize Winner 1964) as stated in his Nobel Prize acceptance speech, Olso, Norway 1964. at age 35.

19 John Houghton, "The Search for God – Can Science Help"? (Lion Publishing: Oxford, England 1995) p.54. see "God's Undertaker" (op cit) p.91.

20 Omar Bradley, (famous American military General) quoted from his

1948 Armistice Day Address at the end of World War ll. (christianity.co.nz/science) 2015.

21 William Jennings Bryan, (statement offered into recorded testimony in the famous Scopes monkey trial at Dayton, Tennessee) July 16, 1925.

22 Dr. Wernher von Braun, (former director, Marshall Space Flight Center-NASA) comments included in 'My Faith' American Weekly Feb. 10, 1963, 'a letter to the California St. Bd. of Education' September 14, 1972. and a personal untitled essay on Science and Christian Faith. 1976. (Creationsafaris.com/wgcs_4vonbraun) 2014.

23 Plato, (well-known classical Greek philosopher and mathematician, student of Socrates and teacher of Aristotle) 427 BC - 347 BC see quotation citing at (thoughtfulchristianity.net) 2013.

24 George Wald, partial quotation repeated for emphasis from chapter four, reference 44.

Chapter Nine

1 Albert Einstein, see famous quotation at (goodreads.com/author/quotes/einstein) 2014.

2 James Gorman, (science reporter - the New York Times) 'The Tortoise or the Hare?' Discover, October 1980, p.88 (evolutionfacts.com/evolution-handbook) 2014.

3 Wernher von Braun, (1912-1977) from an untitled essay written as a forward to a paperback book by Harold Hill, "From the Goo to You by Way of the Zoo" (Logos International: Plainfield, New Jersey 1976) (creationsafaris.com/wgcs_4vonbraun)(emphasis added) 2014.

4 Roy Varghese, (in reflecting on the 'new atheist' position of all or nothing adherence led by R. Dawkins) "There Is a God" Antony Flew, (HarperCollins Pub: New York, 2007) preface p. xvi

5 Ibid. preface p. xvii see R. Dawkins "The God Delusion" (Bantam books:

London 2006) p.140.

6 Paul Davies, “The Mind of God” (Simon & Schuster: London 1992) p.81. see ‘God’s Undertaker’ John Lennox (op cit) p.62.

7 Katherine Tait, (British author/essayist, only daughter of Bertrand Russell) “My Father, Bertrand Russell” (Harcourt Brace Jovanovich: New York 1975) p.189. see “There Is a God” Antony Flew (op cit) (preface by Roy Varghese) preface p. xx.

8 Katherine Tait, (op cit) p.189. (op cit) Antony Flew (Varghese) preface p. xxi.

9 (op cit) p.189. (op cit) preface xxi

10 Bertrand Russell, (British Philosopher- Nobel Laureate) “The Autobiography of Bertrand Russell” (George Allen & Unwin: London 1967) p.146. see “There is a God” Antony Flew, (preface by Roy Varghese) 2007 preface p. xxi.

11 Richard Dawkins, ‘Put Your Money on Evolution’ The New York Times Review of Books April 9, 1989. p.34,35. cited in “God’s Undertaker” (op cit) p.95.

12 Roy Varghese, preface “There Is a God” Antony Flew, (HarperCollins Pub: New York, 2007) preface p. xxiv.

13 Antony Flew, (decorated and world known British philosopher and a former champion for the atheistic worldview before undergoing a profound transformation near the end of his career) “There Is a God: How the world’s most notorious atheist changed his mind” (HarperCollins Pub. New York, 2007) p.75.

14 Ibid. p.88-89.

15 Ibid. p.155.

16 Ibid. p.163. (emphasis added)

17 Paul Davies, “The Fifth Miracle” Allen Lane, (Penguin Press: London 1998) p.61.

18 I.L. Cohen, (engineer, mathematician and a member of the New York

Academy of Sciences who served as an officer of the Archaeological Institute of America) "Darwin Was Wrong – A Study in Probabilities" (New Research Publications: Greenvale, New York. 1984) p.6-8, 209-210, 214-215.

19 Aldous Huxley, (grandson of Thomas Huxley [Darwin's Bulldog] and a brother to Julian Huxley) "Confession of a Professed Atheist" Report June 1966.

20 Thomas Nagel, (a professor of philosophy and law at New York University) "The Last Word" (Oxford University Press, Oxford, England 1997) p.130-131.

21 Alexander Solzhenitsyn, (world famous Russian dissident) spoken in a commencement address to Harvard University students and faculty on June 8, 1978.

22 Kenneth Hsu, (Chinese born geologist and environmental engineer) "Sedimentary Petrology and Biologic Evolution" (Journal of Sedimentary Petrology 56 September 1986) p.730.

23 Henry M. Morris, "The Long War Against God" (Master Books: Green Forest, AR. 4th printing, April 2008) p.24

24 Pierre Paul Grasse, "Evolution of Living Organisms" (Academic Press, New York 1977) p.202 (pathlights.com/evolution/encyclopedia) 2014.

25 Loren Eiseley, (American anthropologist and natural science writer) "The Immense Journey" (Vintage Books, a division of Random House, New York 1957) p.199.

26 Luther D. Sunderland, (American aerospace engineer and avid science writer) "Darwin's Enigma: Fossils and Other Problems" 4th edition (Master Books, 1988) p.9 (allaboutthejourney.org/evolution-and-the-fossil-record) 2014.

27 Louis Agassiz (former longtime Harvard University professor of natural sciences) cited in H. Enoch, "Evolution or Creation" (Evangelical Press, 1986) p.139. (evolutionfacts.com/Evolution-handbook) 2014.

28 Louis Agassiz, on "The Origin of Species" American Journal of Science, 30 (1960) p.154. (pathlights.com/ evolution/encyclopedia) 2014.

29 Nils Heribert-Nilsson, "Synthetic Speciation" 1953, p.31. (evolutionfasts.com/evolution-handbook) 2014.

30 Kenneth Hsu, "Reply" Geology, vol. 15 (1987) p.177. (pathlights.com/ce_encyclopedia/Encyclopedia) 2014.

31 James Perloff, "Tornado in a Junkyard: The Relentless Myth of Darwinism" (Refuge Books: Arlington, Mass. 1999) p.241. (freerepublic.com/focus/f-bloggers)(quotes reveal the credulity of evolutionists) 2014.

32 David M. Raup, (University of Chicago paleontologist, expert on the fossil record) "Geology and Creation" Bulletin of the Field Museum of Natural History, vol.54, Mar. 1983, p.18 (freerepublic.com/focus/f-bloggers)2014.

33 Phillip Johnson, "Objections Sustained: Subversive Essays on Evolution, Law and Culture" (Intervarsity Press: 2000) p.9. see "Mind Siege" by Tim Lahaye, (Word Publishing: Nashville, TN 2000) p.143.

34 Dr. Randy L. Wysong (instructor of human anatomy and physiology) "The Creation-Evolution Controversy" (Wysong Institute, Inquiry Press, 1976) p.418.

35 Dr. Harold Urey (Noble Prize winner for chemistry: UC San Diego) quoted in the "Christian Science Monitor" January 4, 1962 p.4.

36 Michael Denton, "Evolution: A Theory in Crisis" (Adler & Adler Pub: Bethesda, Maryland 1986) p.358. (mountainretreatorg.net/apologetics) 2014.

Chapter Ten

1 Dr. Henry Morris, "The Long War Against God", (Master Books, Green Forest, AR. 2008) p.300.

2 Arthur E. Wilder-Smith, PhD,Dr.Sc.,Dr.es.Sc., "The Natural Sciences Know Nothing of Evolution" (Master Books/ICR, Santee, California:

1981) p.154.

3 Julian Huxley, (professed evolutionary atheist) "At Random, A Television Preview" appearing in Evolution after Darwin 1960, p.41. (evolutionfacts.com/Evolution-handbook) 2014.

4 Collin Patterson - in an address at the American Museum of Natural History, New York City, Nov. 1981. (freerepublic.com/focus/f-bloggers) 2014.

5 Dr. Henry Morris (op cit) 299.

6 Dr. Henry Morris (op cit) 303.

Epilogue

1 Nancy Pearcey, "Total Truth", (Crossway Books (study guide edition), Wheaton, Illinois) 2005

2 Johannes Kepler, AZ Quotes.com/author/quotes. 2014. (use in the public domain)

3 Nelson Glueck, "Rivers in the Desert: A History of the Negev" (Farrar, Straus and Cudahy: New York 1959)

4 Henry H. Halley, "Halley's Bible Handbook", Zondervan Publishing, 1961, p.77

5 Ibid p.80.

6 N. Glueck, "Rivers in the Desert: A History of the Negev" (Farrar, Straus and Cudahy: New York, 1959) p.31.

7 J.O. Kinnaman, found at (geocities.com/Heartland/7234/quotes) accessed October 2009. cited in Grant Jeffrey, "The Signature of God" (Water Brook Press: Colorado Springs, Co. 2010) p.57.

8 Ibid p.57

9 Sir Frederick Kenyon, "The Bible and Archeology" (Harper and Row: New York 1940) p.288-289. cited in Josh McDowell- a Christian evidence study guide 'Research in Christian Evidences' 1979, p.30

10 Peter Stoner, "Science Speaks" (Moody Press, Chicago, Illinois. 1968) p.79.

11 Grant Jeffrey, "The Signature of God" (Water Brook Press: Colorado Springs, Co. 2010) p.70.

12 H. L. Willmington, "Manuscript from Outer Space" (Riverside Book and Bible: Iowa Falls, Iowa. 1974) p.35-37.

Printed in the USA
CPSIA information can be obtained
at www.ICGtesting.com
LVHW090759170624
783368LV00005B/7